MAGNETIC COMPONENTS

MAGNETIC COMPONENTS

Design and Applications

Steve Smith

Richmond, California

VNR VAN NOSTRAND REINHOLD COMPANY

Copyright © 1985 by Van Nostrand Reinhold Company Inc.

Library of Congress Catalog Card Number: 84-5181
ISBN: 0-442-20397-7

Manufactured in the United States of America

Published by Van Nostrand Reinhold Company Inc.
135 West 50th Street
New York, New York 10020

Van Nostrand Reinhold Company Limited
Molly Millars Lane
Wokingham, Berkshire RG11 2PY, England

Van Nostrand Reinhold
480 Latrobe Street
Melbourne, Victoria 3000, Australia

Macmillan of Canada
Division of Gage Publishing Limited
164 Commander Boulevard
Agincourt, Ontario MlS 3C7, Canada

15   14   13   12   11   10   9   8   7   6   5   4   3   2   1

Library of Congress Cataloging in Publication Data

Smith, Steve (Stephen Isaac)
   Magnetic components.

   Includes index.
   1. Electric transformers—Design and construction.
I. Title.
TK2791.S64   1984   621.31′4   84–5181
ISBN 0–442–20397–7

To L. Ron Hubbard, who has done more for mankind than this book ever will.

PREFACE

Magnetic Components Design and Applications is intended primarily for the circuit designer and the power processing systems designer who have found that in order to be more effective they must learn not only to use, but to design their own magnetic components. It will also be useful to the transformer engineer, by showing how to develop high-performance designs quickly and easily by employing optimization criteria.

This book is a design manual, a how-to-build-it manual, and a survey of some common and state-of-the-art practices in magnetic component design and high voltage insulation.

It contains the data necessary to design power transformers on a gradient scale from 60 Hz to several hundred kilohertz, conventional and air-core current transformers, power reactors, saturable transformers and saturable reactors, and air core and conventional pulse transformers. Further, it contains essential information about dielectric materials and fabrication methods, basic heat transfer technology, and electric field gradient control for high voltage applications.

Mathematical methods of optimization are developed, and results are given in a number of areas, particularly in the area of maximizing power density in power transformers and the maximization of stored energy per unit volume for power reactors.

For various reasons, each chapter is written from a different starting level. The chapter on materials and fabrication methods assumes virtually no knowledge of transformer design. Its intent is to introduce the newcomer to the practical side of realizing his design. The chapters on heat transfer and field gradient control also assume no knowledge of the basics. They explain the physical phenomena and techniques from a very basic level, since even practicing engineers have misunderstood concepts in this area. It is important that the fundamental physical phenomena be clearly understood for the art to be well practiced. The chapter on optimization assumes some familiarity with calculus for an understanding of the concepts upon which the derivations are based, but the specific design methods can be used by anyone who can operate a pocket calculator. The chapters on pulse transformers, current transformers, and nonlinear magnetics assume a moderate familiarity with the subject matter.

Since even the most experienced engineer may find that it was a very small and basic bit of information he did not have that kept him from understanding and using material in a subject area, the basics of each subject area are covered in each chapter from the viewpoint of that chapter's subject. This makes the chapters more readable independently.

The order of the chapters is not necessarily from most elementary to most advanced, but rather what I saw as a logical development of the subject. It is recommended that the book be read from front to back without skipping in order to ensure that one has a comfortable grasp of all the basic applications and uses of magnetics.

Some of the equations and derivations in this book are in the English system of units (inches) while others are in the SI system (meters or centimeters). I have made no effort to use only one system of units, for the reason that at this time on this planet our units of measurement are in chaos.

Tape-wound C-core manufacturers dimension their products in inches, as do tape-wound bobbin core manufacturers, but the flux capacity of those bobbin cores is measured in Maxwells (the metric system). The air core pulse transformer design equations were derived from basic physics, and are in the metric system. I have left them thus. There is a particularly handy rise time estimating formula, which uses the coil diameter in inches. Two handy formulas for transformer and inductor design were set up for core dimensions in inches. I have also left them thus. Some heat transfer equations are in English units, some in metric, and some are mixed.

The engineer who prefers to use only the English or only the SI system will find that half his work is done, and he need only convert the other half of the material to his preferred system of units. The engineer who can work comfortably with both systems of units will find both the subject material and the literature of transformer component suppliers ready to use.

STEVE SMITH

ACKNOWLEDGMENTS

I wish to acknowledge the continuing support of Ed Grazda, who talked me into starting this project. My thanks also go to Robert Clintsman, who read the manuscript and made helpful suggestions. My especial thanks to Nancy Schluntz, who edited the rough manuscript, typed it (more than once), did the line drawings, and translated perhaps a thousand pages of my handwriting. I am also indebted to Jon Lambert, who wrote the computer programs and obtained the results of the optimization analysis. Jim Galvin of Lawrence Berkeley Laboratories shared the results of his experimental work with suppressed second harmonic saturable reactor circuits.

We should also acknowledge the lifelong dedication of the thousands of experimenters and pioneers in theoretical and applied mathematics and physics in the last few hundred years, without which none of this would have been possible.

CONTENTS

MAGNETIC COMPONENTS

1

LOW FREQUENCY POWER TRANSFORMERS

1. INTRODUCTION

Electric and magnetic fields are our description of how fixed or moving charges exert forces on other electric charges. We describe these fields by lines of flux, or simply lines. We use the idea of field lines to describe how the influence of our fixed or moving charges is distributed in the surrounding space. Electric field lines start and stop on charged particles, or objects that contain charges. Magnetic field lines do not start or stop. They are all closed curves, and they encircle the path of the moving charges which give rise to them. They tend to concentrate in nearby objects which have certain atomic properties that make them receptive to the presence of these fields. The extent to which magnetic fields tend to concentrate in these materials (ferromagnetic materials) over free space or nonferromagnetic materials is called *permeability*. For electric fields the dielectric constant indicates in a corresponding manner the preference of an electric field for one kind of material over another.

Nothing has either an infinite dielectric constant or an infinite permeability. Some small numbers of the magnetic field lines surrounding the windings in a transformer will surround some or all of the conductors of the primary winding only, rather than reside in the core where they encircle both primary and secondary. Some magnetic field lines will loop outside of the core itself, and constitute the "stray" field of the transformer. These stray field lines are loops like all magnetic field lines, and they will therefore encircle nearby components or circuitry. Because they encircle a conductor, they can induce a current in the conductor, just as they encircled the current flow which gave rise to that field line.

1

The simple thing we call an inductive component, and which we design and construct so readily, is in reality a marvelous consequence of some of the most fundamental forces which make this physical universe behave as it does. The magnetic field is really a relativistic effect, arising out of a remarkable interplay between the electric field and the nature of this space. The interested reader is urged to obtain the three-volume set of the Feynman lectures on physics, and read in particular the second volume.

Since this is a book about magnetic components it might be helpful to define, or at least establish some agreement as to just what magnetic components are.

Magnetic components are those which store or transform energy by utilizing the magnetic fields associated with electric currents.

Electric currents are electrons or charged particles which are moving or caused to move, usually through conductors.

The actual velocity of an electron in a wire is perhaps a tenth of a millimeter per second, but when we push an electron in one end of a wire it repells nearby electrons which in turn repel further electrons, and a different electron pops out the other end of the wire very rapidly. The time it takes for the pushing in of the first electron and the popping out of the first electron is dependent on how fast the *push* propagates down the wire. That push is the electrostatic field of each electron nudging that of the next.

Something else happens here when we do this. We are in effect setting charges in motion, and when something is moving it usually has an energy associated with that motion, called *kinetic energy.* An ordinary piece of mass stores its energy of motion within itself, but charges behave a little differently. You can look at the situation as if the *charge* stores its energy of motion in the *surrounding space,* rather than within itself. The mass of the electron stores kinetic energy in its mass, but the *charge,* which is not mass, behaves differently.

It is this energy of motion of a charge which we call the *magnetic field.* A useful definition of a field is "a physical quantity which takes on different values at different points in space."

Even if we can't see the field we can measure the presence of it. Magnetic fields affect the behavior of charges in such a way as to bring about the behavior characteristics of transformers and inductors.

Physics is not an exact science, and we definitely do not know all the rules. If we set up a very simple, restricted experiment, we find that it behaves in a certain way, and we can say we understand the laws of physics that govern that experiment. The foregoing explanation of the nature of a magnetic field would probably cause Maxwell to roll over in his grave, but for our purposes—understanding the behavior of magnetic components—it is an adequate explanation.

Chapter 1 will introduce the basic considerations of low frequency (50–60 Hz) power transformers up to sereral KVA. It does not give sufficient data to completely design even the simplest transformer. The subject of dielectric materials and fabrication methods is dealt with in the last chapter, while the mathematical tools for choosing the proper core are dealt with in Chapter 2. Chapter 1 is intended as a springboard from which (with the rest of the book) one can evolve sophisticated designs (high voltage, high frequency, optimized form factor, etc.). In order to do this, some common design and construction methods practiced by transformer engineers are presented.

2. IDEAL AND SIMPLE TRANSFORMERS

An ideal transformer is an energy transfer device. Actual transformers have parasitic components which store and/or dissipate small amounts of energy. The action of a transformer is in essence to match a source to a load by changing the voltage-to-current ratio in the power delivered from a source to that required by a load. The impedance of the load is transformed by the square of the turns ratio, while the voltage or current is transformed directly as the turns ratio. The maximum power transfer takes place when the transformed load impedance is equal to the source impedance. This is why some transformers are referred to as "impedance matching" transformers. In truth, they all are. One would usually not speak of power transformers literally in this context, since a power source such as the AC line has an impedance which is usually, to a first approximation, zero.

The transformer itself has series impedances which can limit the available power to a load. The variation in output voltage from no load to full load, expressed as a percentage, is usually referred to as the *regulation* of a power transformer. "Five percent no load to full load regulation" means that 5% of the input voltage is dropped across the series resistances and reactances, the balance being presented to the load.

The parasitic components of a simple transformer are shown in Fig. 1–1. Note that the parasitic components to the right of R_p are due to the presence of the secondary, and from the viewpoint of the source are transformed by the square of turns ratio ($N = N_{sec}/N_{pri}$). From the viewpoint of the load all components as shown would be multiplied by $1/N^2$.

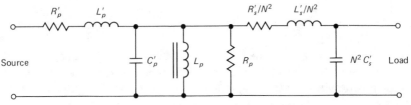

Fig. 1-1. Transformer equivalent circuit.

2.1. Primary Winding

L_p is the inductance of the primary winding at the operating frequency and flux density. Since many core materials are nonlinear, measurement of this parameter should be performed at the design operating conditions. R_p is the equivalent parallel resistance of the core. The power lost in the core is represented as a current flowing through R_p. That current is in phase with the applied voltage. The power due to that current and the applied voltage shows up entirely as heat. The current which flows through L_p is almost 90° out of phase with the applied voltage, and the only thermal loss due to this is the loss from the current flowing through R'_p (the primary winding resistance).

The so-called magnetizing current, the current drawn by a transformer connected to a voltage source, with no load on the transformer secondary winding(s), is actually the vector sum of the currents through R_p and L_p (as well as any reactive currents flowing through shunt capacitances of the windings, usually negligible compared to the other two).

R'_p is the DC resistance of the primary winding. R'_s is the DC resistance of the secondary winding. At 50/60 Hz the current flows essentially in the entire cross section of the conductor, so the DC resistance adequately describes the resistive winding loss parameter of low frequency power transformers. It is important to note that copper (the most commonly used conductor* for designs operating up to ~150–200°C) has a temperature coefficient of resistance of approximately +0.3%/°C. In designs with a relatively high temperature rise, the resistance at the operating temperature must be used when calculating regulation or temperature rise.

---

*In some power transformers where weight is important, aluminum foil is used for the windings. Its density is much less than that of copper, but its resistivity is somewhat higher. In certain situations it can be advantageous, but terminating the foil reliably to tabs or leads is not conducive to the manufacturing of prototypes or small production runs with any degree of facility.

2.1.1. Leakage Inductance. L'_p is the leakage inductance of the primary winding. It represents magnetic field lines encircling primary conductors which are not linked to the secondary winding.

Leakage inductance is a function of geometry alone; it is independent of the magnetic properties of the core. It is a function of the radial build (winding depth or thickness) of the winding, among other things. It represents energy storage potential in the volume of the primary winding and part of the space between primary and secondary. Since primary leakage inductance is a measure of the lines of flux encircling primary conductors only, the energy stored in it will induce a voltage in the primary winding when the primary current is interrupted. That induced voltage can in turn induce a current in the secondary winding of the transformer, but it does so in this indirect manner.

Leakage inductance in a finished transformer is commonly measured by shorting the secondary winding and measuring the primary inductance. This actually gives the total leakage inductance of primary and secondary windings, referred to the primary side. Leakage inductance is not usually a parameter of concern in 60 Hz transformers; however, in large, high-voltage designs or some 400 Hz designs it could be of some concern, as it would affect the load regulation. In saturable reactors it can be a problem even at low frequencies, and should not be ignored. More about this in Chapter 4 (Nonlinear Magnetics). Leakage inductance will also be discussed in more detail in Chapter 3 (Power Reactors) and Chapter 6 (Pulse Transformers).

The considerations involved in the design of high frequency power transformers have much in common with those involved in pulse transformer design, in that leakage inductance and stray capacitance are of great concern. In addition, high frequency power transformers require that attention be given to heat transfer, field gradient control, and the materials of construction. There is, therefore, no one chapter in this book entitled "high frequency transformer design." The reader who wishes to design such components is urged to read the book in its entirety and then apply the data and techniques of the various chapters as appropriate.

2.1.2. Capacitance. C'_p is the equivalent shunt capacitance representing the distributed intrawinding capacitance of the primary winding. In Fig. 1-1 it is shown to the right of the primary leakage inductance L'_p. It is actually distributed across L'_p, but it can usually be approximated as a single lumped element. I have never known it to be a matter of significant concern in 60 Hz low frequency power transformers, and only occasionally in 400 Hz designs.

It may be measured by removing the core from the primary winding, measuring its air core inductance, and then its self-resonant frequency. It may be easier to calculate it from the air core pulse transformer design equations of Chapter 6.

Some apparent components of C'_p and C'_s can be of concern in isolation transformers. The stray capacitance from the ends of windings to the core and thence to other windings or elsewhere can be a matter of great concern when attempting to build a device which will give, say, 100 dB of common mode source noise isolation from a sensitive load. Here one may be concerned about literally a few tenths of picofarads of primary-secondary capacitance, and the windings must be carefully shielded in order to obtain this level of performance. Fig. 1–2 shows some of the stray capacitances in a transformer. Some of the capacitances may be much larger than others, depending on the particular geometry and arrangement of windings. Usually at least half of the capacitances shown are insignificant, but at one time or another the practicing engineer will see and have to deal with each one.

3. SECONDARY WINDING

3.1. Leakage Inductance

L'_s is the leakage inductance of the secondary winding. It is not the total leakage inductance of the transformer as seen from the secondary. That would be the primary leakage inductance reflected through the square of the turns ratio *plus* the secondary leakage inductance.

The action of a transformer is to maintain equal ampere-turns in both

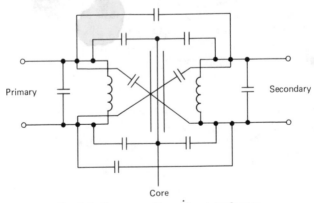

Fig. 1–2. Stray capacitances in a transformer.

primary and secondary windings. If the geometry of primary and secondary windings were exactly equal and symmetrical, then equal ampere-turns would be present within the volumes of both windings, *regardless of the turns ratio of the transformer.* Then one would expect identical magnetic fields to link each winding with an equal portion of the space between them, but not with the other winding. One would then expect the leakage inductance of each winding to be equal within only the square of the turns ratio. In other words, with the transformer in operation equal energies would be stored in both primary and secondary leakage inductances. When a time-rate-of-change of current is forced in one winding of the transformer, the energy stored in the secondary leakage inductance is discharged out of the secondary winding into the load or winding capacitance, just as with the energy in the primary leakage inductance. Depending on the external circuitry, a discharge path for the secondary leakage inductance may or may not be present. The collapsing field will induce a voltage in one winding, which in turn induces a voltage in other windings. The energy definitely will go *somewhere.* Perhaps into charge stray capacitances, perhaps into induced eddy currents or hysteresis loss in the core, perhaps into external components.

This is not usually of great concern in 60 Hz transformers, and only occasionally in 400 Hz transformers. It is definitely of concern in high frequency power transformers and pulse transformers.

A method of calculating leakage inductance can be found in Chapter 6.

3.2. Capacitance

C'_s is the equivalent shunt capacitance representing the distributed intra-winding capacitance of the secondary winding, similar to C'_p.

In high voltage, low current transformers it is possible for C'_s to be high enough that significant excitation current could be drawn by C'_s to be comparable to the load current. But just as primary magnetizing current lags the load current by 90° and adds as a vector, the current drawn by C'_s would lead the load current and also add as a vector. Thus if the current drawn by C'_s were 30% of load current, the apparent current drawn by the transformer would only be about 10% higher.

4. LOSS

It is often part of human nature not to waste something, or to save as much as possible. In our minds we can imagine a device or process which works perfectly, wastes nothing, does exactly what is expected of it, and lasts forever.

The transformers we must design must work not only in our minds, but in this universe as well; and this universe exacts a price from everything in it. As long as the natural laws that govern the structure and function of this universe persist, no physical thing will ever be 100% efficient.

Therefore, one of the first tasks facing the transformer designer is to decide how much *loss* he is willing to have in his transformer.

Some of this loss is reactive, i.e., volt-amperes of energy flow through the transformer but are not converted to heat. Some of the loss is resistive, i.e., manifested as heat. Of the resistive loss, some is dissipated as heat in the core (*core loss*) and some is dissipated as heat in the conductors of the windings.

The designer must then decide how much thermal loss he will budget for the core and how much for the windings, and what limit of reactive loss is acceptable. He then proceeds to design a transformer with losses suitably close to the budget limits. Now we have the moment of truth. The designer looks at his design, considers the physical reality of that transformer, and decides whether his initial loss limits and allocation were appropriate to the physical reality he obtained. Some modification of the original loss budget may become desirable if there is an insufficient agreement between the reality of the designer and that of the transformer.

5. CORES

From time to time, we will be discussing particular design situations, often pulling a particular core material out of thin air for purposes of an example. There are three main factors that influence the choice of core material: the core loss at or over the frequency range of interest, the maximum flux density capability, and the permeability of the core material *at the operating flux density*. There are other factors such as cost, availability of standard or custom shapes, and operating temperature range, but for reasons of brevity we shall take these to be of secondary significance in most discussions. Core manufacturers have such data readily available.

Power transformers operating at 50/60 Hz are able to utilize 0.012 in. (12 mil) thick, tape wound C-cores or 14 mil (20 gauge) stamped laminations, as these materials have quite low core losses at this frequency. They can be comfortably operated at 12,000–15,000 Gauss; less in larger units, depending on available means of heat transfer. More will be said about this in Chapter 8. Twelve mil Z-type material may be operated several thousand Gauss higher with acceptable core loss. The permeability of these materials is quite good for large AC or DC flux densities, but may be unacceptably low for some applications where a very small AC flux density is superimposed on a large DC flux density (for example, power reactors with high frequency ripple).

6. STARTING A DESIGN

As a first design approach to a transformer, one could start by choosing a tentative core type and shape. (A logical method of specifying cores is developed in Chapter 2.) In the case of 50/60 Hz transformers, the material would usually be grain-oriented silicon steel, 0.012–0.014 in. thick, or 0.012 Z-type silicon for higher flux density requirements. After choosing an initial flux density for our core, one would then proceed to calculate the volts per turn. Bear in mind that a final design would consider not nominal operating parameters, but lowest frequency, highest acceptable flux density (usually deriving from core loss and/or exciting current considerations), and minimum core area after mechanical tolerances and stacking factor are taken into account. Tape-wound or laminated steel cores have a thin coating of insulation on the laminated strips or sheets of steel. This takes up space. The space occupied by the core is therefore not 100% iron, but some fraction less. This fraction is called the *stacking factor,* and may be obtained from the catalogs of core material manufacturers.

A useful relationship among core area, turns, frequency, voltage and flux density is

$$\frac{\text{volts}}{\text{turn}} = 2.865 BfA \times 10^{-4}, \tag{1}$$

where B is in kilogauss, f is in Hertz, and A is the *net* core area in square inches. Another way of saying the same thing is

$$\frac{\text{volts}}{\text{turn}} = \frac{\Phi f(\% \text{ saturation}/100)}{45 \times 10^{6}} \tag{2}$$

where Φ is core flux capacity in Maxwells, $[\Phi = B \text{ (Gauss)} A \text{ (cm}^2)]$.

The above expressions are for sine waves, where E is in volts RMS. For square wave excitation the operating flux density is about 11% higher, so V/N must be reduced by that factor, for the same B.

Given a tentative value for volts per turn, one can then proceed to calculate the number of primary and secondary turns. Given the current in each winding, one can pick a tentative wire size and see if the required number of turns of those wire sizes fit within the winding area of the chosen core. If not, a smaller wire size, a higher flux density (to reduce the number of required turns) or more coil winding area is required.

We do not yet have sufficient tools to completely design a transformer or predict the magnitudes of all the parasitic components, but we can use the tools we have. Let us say that we had a core to hand, with 2 sq. in. of cross-sectional area (and a stacking factor of 0.95). Let us further say that we wished to transform the 120 V AC line down to 15 V to provide power to a vacuum tube heater. The number of turns required for the primary winding may be found by calculating volts per turn and dividing that into the primary voltage (120 V). Assume we operate the core at 12,000 Gauss, and that the frequency is 60 Hz. Then,

$$V/N = 2.865 \times 12 \times 60 \times 2.0 \times 0.95 \times 10^{-4}$$
$$V/N = 0.39$$
$$N_{pri} = \frac{120}{0.39} = 306.$$

The number of secondary turns might be

$$N_{sec} = \frac{15}{0.39} = 38,$$

but we will actually get less than 15 V out when the transformer is loaded, since the load current will reduce the effective volts per turn as it drops some voltage across the winding resistances.

Now we need to know the resistance of copper conductors so that we can select conductors which (hopefully) both meet our resistance limit requirements and fit within the available winding area of the core.

7. THE WIRE TABLE

At this point a slight digression is in order, so that the reader may become familiar with the table of wire sizes and its use in transformer design.

There are a number of interesting things about the wire table. Remembering these relationships will allow one to reconstruct it from memory:

1. The diameter of bare #10 is about 100 mils.
2. The diameter of bare #30 is about 10 mils.
3. The diameter of bare #20 is about 31.6 mils. (The square root of 10 multiplied by the diameter of #30.)

Table 1-1. Wire Table

AWG	NOMINAL DIAMETER BARE WIRE DIAMETER	NOMINAL DIAMETER (HEAVY FILM INSULATION)	WEIGHT, LB/1000 FT	RESISTANCE AT 20 °C, OHMS/1000 FT	FREQUENCY AT WHICH RADIUS EQUALS SKIN DEPTH	ALLOWABLE CURRENT (AMPS) AT: 500 CIRCULAR MILS/AMP	750 CIRCULAR MILS/AMP	1000 CIRCULAR MILS/AMP	TURNS PER INCH THEORETICAL	ACTUAL	AREA, CIRCULAR MILS	AWG
10	0.1019	0.1056	31.7	1.00	2608 Hz	21	15	10	9-1/2	8-1/2	10,380	10
11	0.0907	0.0943	25.2	1.26	3293	16	12	8.2	10-1/2	9-1/2	8,230	11
12	0.0808	0.0842	20.0	1.59	4149	13	9.2	6.5	12	10-1/2	6,530	12
13	0.0720	0.0753	15.9	2.00	5226	10	7.0	5.2	13-1/2	12	5,180	13
14	0.0641	0.0673	12.6	2.52	6593	7.0	5.7	4.1	14-1/2	13	4,110	14
15	0.0571	0.0602	10.0	3.18	8308	6.6	4.5	3.3	16	14-1/2	3,260	15
16	0.0508	0.0539	7.94	4.02	10.50 kHz	5.2	3.5	2.6	18-1/2	16-1/2	2,580	16
17	0.0453	0.0483	6.32	5.05	13.20	4.2	2.9	2.1	20-1/2	18-1/2	2,050	17
18	0.0403	0.0432	5.02	6.39	16.68	3.2	2.2	1.6	23	20-1/2	1,620	18
19	0.0359	0.0387	3.99	8.05	21.02	2.6	1.8	1.3	25	22	1,290	19
20	0.0320	0.0346	3.17	10.1	26.45	2.0	1.4	1.0	29	26	1,020	20
21	0.0285	0.0310	2.52	12.8	33.35	1.6	1.1	0.80	32	29	812	21
22	0.0253	0.0277	2.00	16.2	42.32	1.3	0.85	0.64	36	32	640	22
23	0.0226	0.0249	1.60	20.3	53.04	1.0	0.67	0.50	40	36	511	23
24	0.0201	0.0224	1.26	25.7	67.05	0.83	0.52	0.40	44	40	404	24
25	0.0179	0.0201	1.00	32.4	84.54	0.64	0.43	0.30	50	45	320	25
26	0.0159	0.0180	0.794	41.0	107.2 kHz	0.50	0.34	0.25	55	50	253	26
27	0.0142	0.0161	0.634	51.4	134.3	0.40	0.27	0.20	62	56	202	27
28	0.0126	0.0144	0.501	65.3	170.6	0.32	0.21	0.16	69	62	159	28
29	0.0113	0.0130	0.404	81.2	211.2	0.26	0.18	0.13	77	70	128	29
30	0.0100	0.0116	0.317	104	270.9	0.20	0.14	0.10	86	78	100	30
31	0.0089	0.0105	0.252	131	342.0	0.16	0.11	0.079	95	86	79.2	31
32	0.0080	0.0095	0.204	162	423.3	0.13	0.085	0.064	105	95	64.0	32
33	0.0071	0.0085	0.161	206	537.4	0.10	0.067	0.050	117	106	50.4	33
34	0.0063	0.0075	0.127	261	687.5	0.08	0.053	0.04	133	120	39.7	34
35	0.0056	0.0067	0.101	331	863.8	0.062	0.042	0.031	149	135	31.4	35
36	0.0050	0.0060	0.081	415	1.084 MHz	0.050	0.034	0.025	166	150	25.0	36
37	0.0045	0.0055	0.065	512	1.338	0.041	0.027	0.021	182	165	20.2	37
38	0.0040	0.0049	0.052	648	1.693	0.032	0.021	0.016	204	185	16.0	38
39	0.0035	0.0043	0.040	847	2.211	0.025	0.017	0.013	232	210	12.2	39
40	0.0031	0.0038	0.031	1080	2.819	0.019	0.013	0.010	263	238	9.61	40
41	0.0028	0.0034	0.025	1320	3.455	0.015	0.010	0.0075	294	266	7.84	41
42	0.0025	0.0030	0.020	1660	4.334	0.012	0.009	0.0060	333	302	6.25	42

RECOMMENDED LAYER INSULATION (BASED ON STIFFNESS TO GIVE MECHANICAL SUPPORT):

- 0.015" Copaco or Nomex 410
- 0.010" Copaco or Nomex 410
- 0.007 Nomex 410
- 0.005 Nomex 410
- 0.004 Kraft Paper
- 0.003 Kraft Paper
- 0.002 Kraft Paper
- 0.0015 Kraft Paper
- 0.001 Kraft Paper
- 0.0007 Kraft Paper
- 0.010 Nomex 411
- 0.007 Nomex 411
- 0.005 Nomex 411

(hand wound)   (machine wound)

4. The diameter of bare #40 is about 3.16 mils. For each decade of AWG, the area changes by a factor of 10. For each two decades of AWG, the diameter goes down by a factor of 10.
5. The area of #20 is 1000 circular mils.
6. The area of #23 is 500 circular mils. For any wire size, go three sizes up (or down) and the area halves (or doubles).
7. The resistance of #10 is about 1 ohm per 1000 feet, or 1 milliohm per foot.
8. The resistance of #20 is about 10 milliohms per foot. The resistance is inversely proportional to the conductor area.
9. #20 weighs 3.17 lb. per 1000 feet.

If one were to remember only the pattern of the wire table and the fact that #10 has a resistance of 1 milliohm per foot, one would be able to reconstruct most of the wire table from memory.

The typical current densities mentioned here are extremely vague guidelines. Particular regulation and temperature requirements dictate exact current densities.

For smaller 60 Hz transformers, 500–750 cm/A might be an appropriate range to start in. For larger units (where the heat path from the inside of the winding to the outside is much longer, the thermal resistance higher, and hence the internal temperature rise potentially much more than in a smaller unit), perhaps 1000 cm/A or more might be appropriate.

8. CONTINUING THE DESIGN

Once we have found a suitable wire size for the above trial design we may then proceed to calculate core loss and copper loss, and then determine the temperature rise and load regulation of the finished unit. If these final parameters are within the user's limits of acceptance, we have a finished electrical design.

Let us now say that our load requires 12 amperes, and the window of the core (the place where the coil goes) is 1 in. high and 3 in. long. Refer to the wire table, and assume 750 cm/A current density. Choose for the secondary winding #11. The turns ratio is about 8 : 1, so the primary current would be about 1.5 amps. Choose #20. Start with the primary winding, if for no other reason than it will be easier to wind a few turns of heavy wire on top of the lighter gauge. (This is not necessarily true when dealing with extremely fine wires.)

For a 3 in. coil length, assume the actual conductor winding length is 2.5

in. #20 would have about 26 turns/in., or 65 turns/layer. With 306 turns total, 5 layers would be sufficient, with 5 mil interlayer insulation.

Assume 30 mils of something between the primary and secondary windings.

#11 will give about 9.5 turns/in., or about 24 turns/layer. Two layers are sufficient for 30 turns. The total build of the coil is then:

winding form	0.060
5 layers #20 @ 0.0346	0.173
4 layers 0.005 insulation	0.020
interwinding insulation	0.030
2 layers #11 @ 0.0943	0.189
0.015 layer insulation	0.015
outside wrap	0.030
	0.517

One would expect that, even with the wires bulging a bit, such a coil would fit within a 1 in. available height.

In order to calculate the winding resistance we need the length of each winding. We will do this by calculating the mean length of a turn and multiplying by the number of turns of each winding.

In Fig. 1-3, one may see that the inner perimeter of the coil is $2D + 2E$, and the outer perimeter is $2D + 2E$ plus the circumference of a circle of radius r. The mean of those two is $2D + 2E$ plus the circumference of a circle of radius $r/2$, or $2D + 2E + \pi \times$ (radial build of our coil):

$$MLT = 2 \times 2 + 2 \times 1 + \pi \times 0.52 = 7.63 \text{ in.}$$

Assume 8 in., or 2/3 foot. The length of the primary is then about 205 feet and the secondary about 26 feet. From the wire table, the DC resistances are about 2.07 ohms and 0.033 ohms, respectively. The I^2R losses are then about

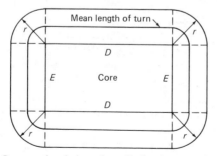

Fig. 1-3. Cross-sectional view of a coil, showing mean length of turn.

4.7 watts and 4.8 watts, respectively, for a total coil loss of 9.5 watts. Without having read the chapter on heat transfer we consider the size of the transformer, estimate that the coil loss looks fairly low, and assume that we won't have any thermal problems.

The voltage drop from the primary current in the primary winding is 1.5 amps × 2.07 ohms = 3.11 volts, or 2.6%. This lowers the effective secondary induced volts per turn by 2.6%. The voltage drop in the secondary winding is 12 × 0.033 = 0.40 volts, or 2.6%. This lowers the secondary voltage under load by another 2.6%. We should add 5.2% to the secondary turns to make up for that, and add 5.2% to the incremental addition to make up for the proportionate loss in the added turns. That is:

$$38 \times 0.052 = 1.976 \text{ turns}$$
$$\text{plus} \quad 1.976 \times 0.052 = 0.103 \text{ turns,}$$

for a total of 2.079 turns additional.

Do not attempt to put fractional turns on a transformer. It is not good practice to drill holes in a core, which is the only way one can realize 0.079 turns (by having a turn link 0.079 of the flux in the core, the hole being drilled off to one side so the turn encloses 0.079 of the cross-sectional area).

We should therefore add 2 turns to the secondary (there is room on the last layer) to obtain 15 volts at full load with a 120 volt input.

9. A CASUAL DISCUSSION

Another method of handling the variables mentioned above and their interaction is to use the design algorithm developed by Col. William T. McLyman in his book, *Transformer and Inductor Design Handbook* (1978). That method is most conveniently applied to cores of standard sizes—those for which the design parameters McLyman developed (the K-factors) have been defined. As will be seen in the next chapter, one can benefit by allowing the core shape itself to be a variable. Manufacturers of C-cores can make virtually any shape, and rarely keep any stock of even "standard" sizes. In many cases it may be more advantageous, especially for 50/60 Hz transformers, to assemble the core out of a stack of punched laminations. This has a number of advantages, not the least of which is that one may be able to obtain delivery on a 50–100 lbs box of laminations in a few days but have to wait six weeks for a C-core.

The reader will find within the chapter on optimization (Chapter 2) a graph of power-handling capacity versus a range of stack heights for a typical lamination size. It will be instructive for the beginning engineer to design a transformer on a suitable stack of standard E-I laminations, and then design the same unit on a C-core with an optimized form factor and compare the two finished designs with respect to volume, weight, and temperature rise for units having the same total losses.

Another convenient beginning point, essentially that which is found or implied in many core manufacturers' catalogs, is to use the tabulated products of core area and window area for C-cores or laminations. The product of core area and window area (in inches) multiplied by 50 is an approximate measure of the power-handling ability of a core at 60 Hz. That total product, *then* multiplied by 7, is an approximate measure of the power-handling ability of a core at 400 Hz. As will be seen in Chapters 2 and 3, this is not a proper figure of merit for transformers or inductors. Hopefully, all core manufacturers will soon begin noting the proper figures of merit in their catalogs.

400 Hz transformers are usually used in military applications. As such, a greater premium is placed on size and weight than for most 50/60 Hz transformers. There is a tendency to push performance harder, to design for higher operating temperatures, or to give greater consideration to heat removal.

As one attempts to carry 60 Hz transformer designs and design guidelines over to 400 Hz, one must bear in mind that not everything carries over in a linear fashion (i.e., at a 1 : 1 ratio). As we will prove in the next chapter, the square of the operating frequency is inversely proportional to a linear dimension to the fifth power; and clearly the surface area is proportional to the square of a linear dimension. Hence, for a constant power loss mapping of a 60 Hz design into a 400 Hz design, the surface area goes inversely as the frequency to the 4/5 power. From 60 to 400 Hz this means a reduction in surface area of about a factor of 4-5 : 1. That would imply a temperature rise over ambient of 4-5 times that of its 60 Hz counterpart. This would be true only for copper losses. Core weight decreases, core materials with different lamination thickness (different watts/pound loss characteristics) are used, and many more variables are introduced.

Fortunately, copper losses need not map over 1 : 1, since the mean length of turn drops to less than half that of its 60 Hz counterpart ($\sim 40\%$), and copper losses drop accordingly. Designing a 400 Hz transformer for about half the copper loss of its 60 Hz counterpart usually gets one into a reasonable temperature rise ballpark. More exact data in this area will be developed later in the book.

One interesting consequence of the attendant size reduction of higher frequency transformers is that the winding resistance of a 400 Hz transformer tends to be lower than that of its 60 Hz counterpart because of the increase in volts per turn and the reduction in mean length of turn. The inrush current experienced by rectifier diodes for capacitor input filters is therefore much higher, and it is occasionally necessary to choose rectifiers for their surge current capability rather than average current rating. This is not necessarily of direct concern to the transformer designer because the thermal mass of copper in the winding is usually sufficient to limit the winding temperature rise to a safe level. But it is a parameter over which he has some control, and the system designer should be aware of the available tradeoffs.

Occasionally one will find requirements for higher frequency transformers operating at power line frequencies of 800–2400 Hz. Several years ago, I became aware of a system which required 125 kV DC at 1 amp from a three-phase, 5000 V, 5 kHz power source. Fortunately, most low frequency power designs are at 60 to 400 Hz. The designer can usually obtain higher frequency designs by an extension of the 400 Hz design considerations.

Skin effect in 400 Hz transformers is not usually a problem. If the designer is in doubt, he can check the wire table in this chapter.

Leakage inductance is not usually significant in lower voltage, lower power 400 Hz designs. In high voltage, high power designs it can be a significant factor in the load regulation. A good example is a three-phase transformer with a wye secondary, delivering 33 kV at 12 kW. Each secondary had 3200 turns on a form about 2.5 in. by 4.5 in. The secondary leakage inductance was calculated to be on the order of one Henry per coil! At 400 Hz, the impedance of one Henry is about 2500 ohms; at 360 mA that gives a voltage drop of 900 volts per leg, or 1800 volts for the wye. This is 5.5% of the output, and even though the inductive drop is in quadrature with the resistive drop, this may need to be taken into consideration by both the transformer designer and the system designer. Series leakage inductance in square wave high frequency power transformers can be treated as an impedance which drops a direct proportion of the input voltage, for regulation purposes. Its energy storage effects cause power loss in high frequency square wave inverters by momentarily blocking the power flow from the DC source to a load that accepts a power flow most of the cycle. This can be a major cause of power dissipation in the switches of many current-fed inverters.

Leakage inductance is roughly a linear function of the coil length, build, or mean length of turn, but goes as the square of the number of turns. Therefore the best way to reduce leakage inductance is to reduce the number of turns by increasing the core area. This will increase the mean length of turn linearly; therefore, to a first order the leakage inductance will go down linearly as the core area goes up. The length of the winding can also be increased, since this will reduce the leakage inductance approximately linearly. One can go only so far in this direction, for as we do this the number of volts per layer increases and demands thicker layer insulation. (It should not be considered unusual to see layer insulation on the order of one to two wire diameters in some high voltage designs.) When a large amount of layer insulation is required, it can increase the radial build of the coil such that one is no longer gaining but begins to lose, and an increase in the winding length buys nothing. As the number of turns is reduced and the core area increased proportionately, the primary magnetizing current goes up, and one will soon reach a point where that magnetizing current is unacceptably high, and then one may reduce leakage inductance further only by decreasing the radial winding build and/or increasing the coil length. In current-fed inverters which must operate over a wide load range, excessive magnetizing current can be a problem as the transformer regenerates its magnetizing current back to the source for half of each cycle. If the average load current (including core loss) is not sufficient to give a net positive current into the inverter, then the current feed choke is forced by the inverter to accept an instantaneous current reversal, which of course it will not. The result is a large voltage spike. The simplest solution is to keep the inductance high enough that the magnetizing current is less than the average current into the inverter. A minimum load may be needed, or additional circuit components to handle the regenerative flow.

Leakage inductance can be very significant in 400 Hz saturable reactors. This will be discussed in Chapter 4.

Those geometries which tend to reduce leakage inductance (relatively few layers, relatively long) also raise the shunt interwinding capacitance. One will rarely get in trouble because of excessive interwinding capacitance in high voltage, low frequency transformers, but it can happen. A three-phase, 400 Hz design, delivering 27 kV at 20 mA DC from a wye secondary ended up being built on a core with about 2¼ sq. in. of core area and 5 sq. in. of winding area per coil. The coil length was 5½ in. and the radial build of the coil about ¾ in. This is a very poor set of proportions for a high voltage transformer. The coil is too long for the winding build, and the ratio of iron to copper area is somewhat low because the designer in a fit of extreme conservatism used a design flux density of 8 kilogauss. The secondary windings were each self-resonant at about 5 kHz, and this resonance was pumped fairly well by the 2400 Hz component of the rectification process, giving about 25% ripple from a circuit that was supposed to have about 5%.

The designer could have recognized that the proportions looked instinctively wrong, or he could have simply calculated everything and discovered the resonance. If the designer does not have a well-developed instinct for transformer design parameters, he should calculate everything; he will rapidly develop a sense of what "looks good" and what needs to be checked.

Interleaving of windings is sometimes used in resonant charging reactors and some high frequency power transformers.

The main purpose of an interleaved winding structure is to reduce the leakage inductance between windings. The main drawbacks are two. First, it increases the coil complexity and hence the manufacturing cost. Second, it increases the capacitance between windings in an often surprising manner. Depending on the voltage developed between windings or winding portions during circuit operation, the energy stored ($\frac{1}{2}$ CV^2) in a particular stray capacitor may be rather high, and the transformer will have the behavior of a larger capacitor than had been intended. Discussion of leakage inductance and related parameters may be found in Rippel and McLyman (1982).

A rough rule of thumb is that the leakage inductance is reduced as the square of the number of interleaves.

Interleaving of windings should only be used where the required leakage inductance or the space available does not allow the component to be designed in the normal manner, with one winding on top of another. As we will see in the chapter on pulse transformers, one can make the leakage inductance of a winding structure arbitrarily low by increasing the winding length and reducing the radial build.

Interleaving may be used to great advantage when the induced voltages of the windings to be interleaved are in phase or very small. In that case, there is little or no energy stored in the interwinding capacitance. One example of this is current balance transformers or high frequency power combining transformers. During normal circuit operation the induced voltage is relatively small. One-to-one non-inverting pulse transformers are another example. The reader must beware of using this indiscriminately, however. High voltage isolation transformers will store considerable energy in the interwinding capacitance, and here interleaving is usually contraindicated. This will be discussed in more detail in the chapter on field gradient control (Chapter 7).

The leakage reactance per se of an inductor is that associated with the radial build of the coil, and represents flux not coupled to the core. It may show up as a high frequency resonance (with the feed-through or interlayer capacitance) or it may cause coupling of high frequency noise to improperly dressed nearby wiring. Aside from making the inductor longer and thinner, one can reduce the stray field by using a core structure which envelops more of the coil, and placing the gap in the core leg which is inside the coil.

Pot cores offer the most effective self-shielding structure.* Next most effective are the Type III constructed units (also known as coil-type windings—a winding on each leg of a U–U or U–I core) and the Type II units (also known as shell type—one coil on an E–core). Least effective is the Type I construction (also known as core-type—one coil on one leg of a U-core). The effectiveness of a Type III design in reducing stray fields (or reducing pickup from external fields) should not be underestimated. All single-coil magnetic structures have an external field shape which is that of a dipole. Type III structures have a quadrupole field which falls off with distance much faster than a dipole field. A toroid is a special case of a Type III structure.

The last few pages have given a rather casual discussion of some of the considerations that affect magnetic component design. It is hoped at this point that the novice designer has an awareness of the notion that there are more than two or three such considerations. This does not mean the subject is complex; complexity is only due to the subject not being fully seen or describable. The following chapters will take those various considerations and develop them in sufficient detail that the designer can deal with them systematically.

---

* See Chapter 2, Figs. 2–2, 2–3, 2–4 for pictures of Type I, II, and III structures.

2
OPTIMIZATION

1. INTRODUCTION

The word *optimize* has been used so frequently in the last few years to mean such a variety of conditions that its popular meaning appears to be "something the author has developed." The accurate meaning is, "to achieve the best or most satisfactory balance among several factors."

Optimum form factor is not necessarily the optimum amount of mechanical redesign necessary to make it fit into the space available, and optimum efficiency may or may not be optimum cost. In most cases we will be seeking an optimum which is a maximum or a minimum of something, with something else taken as a variable and other things allowed to float independently or held constant as the situation indicates. This requires no more extensive mathematical background than understanding differential calculus and being able to differentiate simple functions to follow the derivation, and only a command of simple algebra to be able to produce results.

We shall be using mathematical tools to evaluate magnetic components, and we define *optimize* as, "to seek a maximum or minimum for some parameter or weighted combination of parameters." Sometimes the functions we will deal with have a single maximum or minimum, so the decision-making process will be straightforward. We must recognize that the factors to which we wish to optimize our design must be included in the initial formulation. If the least expensive design for a given temperature rise is the object, then a mathematical statement of the cost must be formulated as a function of core sizes and shape, number of turns of different wire sizes, etc.

Our basic approach will be to derive a general expression for the parameter to be optimized (maximized or minimized). We use known design relationships to reduce the number of variables so that the relationship can be expressed in terms of one parameter. We then take the first derivative of that expression with respect to our parameter and set it equal to zero. This will locate a point of zero slope on our expression, which is a maximum or minimum. This is the basic process used in all applications of this concept in this book. In more complex situations we will derive that which is to be optimized and then present the results of computer processing.

A concept familiar to mathematicians and physicists, but new to most electronics engineers, is introduced in order to analyze magnetic components of varying characteristics and sizes. It is important that the reader understand it, since it is the keystone of the optimization analysis.

Triangles are said to be similar if they have equal angles, even if some are larger or smaller than others. Let us say we wish to discuss the general class of all similar triangles with one right angle. We take the larger triangles and scale them down to the size of a standard triangle of unit height. We take the smaller triangles and scale them up to the size of our standard triangle. The standard triangle of unit height is said to be the *normalized* size, all our different triangles have been *normalized,* and the process is called *normalization.*

We keep track of the *scaling factor* by which we multiplied each of our original triangles in order to normalize them, and when we are done studying our different triangles we wish to return them to their original sizes. We multiply the dimensions of each by the reciprocal of its original scaling factor. This is called *denormalization.* It is a returning to the original, real-world size. The reciprocal of the scaling factor is called the *denormalization factor.*

This is an extremely powerful mathematical tool. It allows us to easily obtain a much deeper understanding of how to go about designing a magnetic component, and why some designs are better than others.

The decision as to what to optimize, what to allow to be a variable, and what to hold fixed, must not be lightly made. In some cases the choice is clear, such as maximizing the power-handling capability of a single-phase or three-phase power transformer. By looking at watts per cubic inch of volume at constant copper losses and allowing the relative shape to vary, we can generate a family of normalized designs. From this the more efficient shapes can be selected and then scaled to whatever power level is required. We will repeat this procedure, looking at combined core and copper losses, and derive another family of normalized designs.

In the case of a power reactor, we will address efficiency of the power reactor as an energy storage element. This is done by seeking a criterion which maximizes the number of Joules of stored energy per cubic inch of volume, with respect to the rate of energy loss due to the current flowing through the winding resistance. The most efficient shapes for this purpose can be found by allowing the form factor to vary. Interestingly enough, the geometric criterion thus derived is identical to that for transformers, wherein only copper losses were considered. This will be derived in Chapter 3.

We have not sought to optimize anything for minimum weight, although it could be done. That is left as an exercise for the interested reader. Our first development of the subject will deal only with copper losses. Our second development will cover combined core and copper losses.

It is probably implicit in any effort to apply rigorous, precise mathematical tools to something crafted by man, that there will be assumptions about relative significance and approximations made in the name of expediency. A totally precise generalized design would take into account every scrap of paper and gram of impregnant, would consider only discrete wire sizes rather than a continuum, and so forth. To avoid boring the reader to death and then presenting him with a single equation three pages long containing a meaning known only to God, we simplify and make assumptions in order to obtain a generally useful tool.

For special design considerations such as high voltage, the reader can with little difficulty modify the derivations along the lines indicated, to suit his purpose.

One general assumption made in most of the work in this chapter is that the cross-sectional area of the copper in the winding area is roughly 40% of the maximum available winding area. This is based on the concept that enough space and insulation to give the required degree of mechanical support to a coil usually also gives sufficient electrical insulation for the most common range of voltages (up to a few hundred volts in small units, or a few thousand volts in larger units). In small units the copper occupies somewhat less than 40% of the available area, but the mounting surface tends often to be larger in proportion to the unit than in large transformers. Therefore the internal temperature rise is somewhat less due to the increased heat transfer area. In some bobbin-wound designs with rectangular wire, or some foil-wound designs, the copper fill factor is somewhat higher.

2. DERIVING THE FIGURE OF MERIT FOR WINDING LOSS

Consider a transformer with identical primary and secondary windings. Let the mean length of turn be U (the average of the primary and secondary mean length of turn in this development). The length (l_w) of a winding is NU, where N is the number of turns in that winding. The resistance of that winding is:

$$R = \frac{\varrho l_w}{A_w} , \qquad (2.1)$$

where ϱ is the volume resistivity of the conductor material and A_w is the area of one turn of the conductor. For copper, ϱ is 7×10^{-7} ohms per square inch-inch.

Hence,

$$R = \frac{\varrho N U}{A_w} . \qquad (2.2)$$

Let each winding occupy 20% of the available winding area of the core, whose dimensions shall be F (height) and G (length). Then

$$A_w = \frac{0.2FG}{N} \qquad (2.3)$$

and

$$R = \frac{\varrho N^2 U}{0.2FG} \qquad (2.4)$$

where R is either the primary resistance or the secondary resistance referred to the primary.

Let the input power to the transformer be $P_{\text{watts}} = V_{\text{volts}} I_{\text{amperes}}$. Note that we do not need P or I in these equations.

The required number of turns is

$$N = \frac{V}{2.865 BfDES \times 10^{-4}} , \qquad (2.5)$$

where B is the flux density in kilogauss, f is frequency in Hertz, D and E are the dimensions of the core cross section (that which the turns are wound around) in inches, and S is the space factor of the core (less than unity for all tape-wound cores, unity for ferrites).

Square Eq. (2.5) and substitute for N^2 in Eq. (2.4),

$$R = \frac{\varrho V^2 U}{(0.2FG)\,(2.865^2 B^2 f^2 D^2 E^2 S^2 \times 10^{-8})} \cdot \qquad (2.6)$$

Rearranging to separate constants and given conditions from geometric parameters,

$$R = \left\{ \frac{\varrho V^2}{1.64 \times 10^{-8} B^2 f^2 S^2} \right\} \left\{ \frac{U}{D^2 E^2 FG} \right\}. \qquad (2.7)$$

All terms in the left set of brackets may be considered set as initial conditions.

3. USING THE FIGURE OF MERIT

Choose a set of normalized core dimensions from one of the tables in this book. Those sets of dimensions have been selected to satisfy the criterion that, within each given overall form factor, each set of core dimensions gives the lowest value of $U/D^2 E^2 FG$. Each set gives the lowest resistance design in that form factor.

That's not quite the end of the road.

In the real world the engineer would specify a total copper loss for the transformer, either from regulation or temperature rise or from efficiency considerations. Any of these give a specific numerical maximum for the winding resistance. In light of this, Eq. (2.7) can be rearranged to reflect our understanding:

$$\frac{D^2 E^2 FG}{U} = \frac{\varrho V^2}{1.64 \times 10^{-8} R B^2 f^2 S^2} \cdot \qquad (2.8)$$

Since all terms on the right are given, the right-hand side of Eq. (2.8) has a specific numerical value. This numerical value is associated with, and is a measure of, the real-world transformer we wish to build. *That* is the denormalized design. It is in fact power-handling capability at constant efficiency.

The values of $D^2 E^2 FG/U$ and $D^2 E^2 FG/UPQ$, which are given in various tables within this chapter, are normalized. The former is a figure of merit for transformers. The latter is a figure of merit per unit volume (PQ representing the volume, as we will see later), showing that some overall form factors have better volumetric efficiency than others. They should be considered to have arbitrary dimensions. Normalized designs are used to compare one with another. We select one and scale it up or down to fit our specific requirement. Here is how that is done:

1. Pick a normalized design with what appears to be a suitable overall form factor (shape) PQ and a suitably high value of D^2E^2FG/UPQ.
2. For that design, note the value of D^2E^2FG/U.
3. Take the numerical value of the right-hand side of Eq. (2.8) and divide it by the value noted in step 2. You have divided a denormalized value of D^2E^2FG/U by a normalized value of that same parameter. Call that quotient K^5.
4. Take the fifth root of the number obtained in step 3 (i.e., raise it to the 0.2 power). The resulting number K is the *denormalization constant*. (I recommend five significant figures to minimize accumulation of round-off errors.)
5. For the normalized design chosen in step 1, take the given values of D, E, F, and G and multiply each by K to obtain new dimensions KD, KE, KF, KG. Those are the dimensions (in inches) of the real-world core upon which the transformer of Eq. (2.8) may be constructed.

Those are the basics. The material which follows develops those ideas in the light of specific circumstances.

4. SPECIFIC GEOMETRIES

There are three basic geometries of a single-phase transformer (not including the toroid, which is most closely related to Type III). The types are shown in Figs. 2–1 through 2–4. Type I is sometimes called a *core type,* Type II a *shell type,* and Type III a *coil type.* The designations for core dimensions given are those commonly applied to tape-wound C-cores (see Fig. 2–1). It may be noted that Type II (Fig. 2–3) corresponds also to a transformer constructed from E-E or E-I stamped laminations. Observe that Type I designs (Fig. 2–2) are the simplest to construct with C-cores, while Type III designs (Fig. 2–4) offer roughly 50% higher power-handling capability at constant efficiency per unit volume. It has been known that the latter type is better than the former, but until this analysis was done no one knew how good they really were. The shell designs (Type II) are intermediate in volumetric efficiency, but offer reduced core weight since the magnetic circuit length is less than for single-core designs.

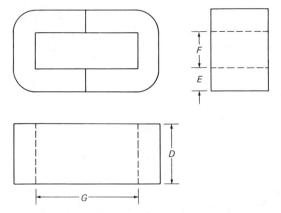

Fig. 2–1. Basic core dimension nomenclature. The cross-sectional area of magnetic material is *DE;* the coil winding area (sometimes called the window) is *FG.*

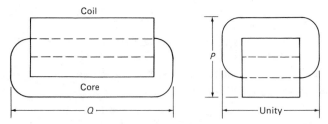

Fig. 2–2. Normalized overall dimensions, Type I. $P = 2E + 2F$, $Q = 2E + G$, $1 = D + 2F$, core area $= DE$, and coil area $= FG$.

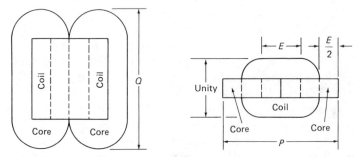

Fig. 2–3. Normalized overall dimensions, Type II. $P = 2E + 2F$, $Q = E + G$, $1 = D + 2F$, core area $= DE$, and coil area $= FG$.

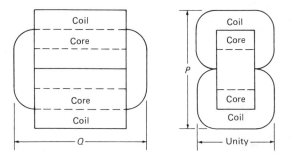

Fig. 2-4. Normalized overall dimensions, Type III. $P = 2E + 2F$, $Q = 2E + G$, $1 = D + F$, core area $= DE$, and coil area $= FG$.

5. OPTIMUM CORE DIMENSIONS

We have seen that the power-handling capability at constant efficiency is represented by D^2E^2FG/U. With the volume as PQ, the maximum value of D^2E^2FG/U can be calculated for each of the many sets of (P, Q). Hence optimum values of D, E, F, and G can be found for each set (P, Q).

Finding the particular values of D, E, F, and G which give the highest value of D^2E^2FG/U for 500 different form factors in each of 4 cases is a task best left to a computer. It turns out that an operating system with six significant figures can deliver only four significant figures in results, because of the accumulation of round-off errors. The computer was told to locate the maximum value of D^2E^2FG/U to within 1 part in 10,000. It came fairly close. We have checked some of the optima on a pocket calculator (more tedious but much more accurate, with eight significant figures). The values given in the tables appear to be within a 0.1% or better of the true optimum values, which is at least 20 times more accurate than needed for transformer design. The width of the peak at maximum is such that a variation of D on the order of 3% (with E, F, and G varying correspondingly to hold the overall form factor constant) produces a variation on the order of 0.3–3% in D^2E^2FG/U.

Since the power-handling capability sensitivity factors are small, we can round core dimensions up or down to convenient nearby fractional dimensions with little change in the actual power-handling capability. The technique is therefore a useful one in the real world of tolerances and standard fractional dimensions.

We may compare these relative optima by forming the quotient D^2E^2FG/UPQ, power-handling capability per unit volume at constant efficiency. A plot of the values of this factor over a range of proportions (P, Q) is presented in the Tables 2–1 to 2–3 for each of the three previously outlined

Table 2-1. Case I, Choke Data.

Q/P	1.0	1.2	1.5	1.8	2.2	2.7	3.3	3.9	4.7
1.0	106	75	26	02					
1.2	143	118	61	15					
1.5	186	177	129	64	07				
1.8	216	221	193	134	45				
2.2	244	263	258	221	134	28			
2.7	268	299	316	302	243	130	15		
3.3	288	329	363	369	340	258	122	14	
3.9	301	349	396	417	409	356	248	114	02

Form Factor : $F = (1-D)/2$; $E = (P - 1 + D)/2$; $G = Q - P + 1 - D$; Function = (Power/Vol.) $\times 10^5$)

Table 2-2. Case II, Choke Data.

Q/P	1.0	1.2	1.5	1.8	2.2	2.7	3.3	3.9	4.7
1.0	231	244	228	179	87	07			
1.2	255	279	284	256	180	63			
1.5	279	315	342	338	296	198	59		
1.8	295	340	381	395	377	311	188	55	
2.2	310	362	417	446	452	418	333	216	50
2.7	322	381	447	490	516	509	460	378	235
3.3	332	395	471	524	567	583	562	511	408
3.9	339	406	488	548	602	634	633	603	531

Form Factor: $F = (1-D)/2$; $E = (P - 1 + D)/2$; $G = Q - (P + 1 - D)/2$; Function = (Power/Vol.) $\times 10^5$

types. Note the ridge of relative maxima. There are obviously some preferred form factors.

Exact optimization core dimensions for a range of values (P,Q) are given in Appendix A.

Figure 2-5 shows D^2E^2FG/UPQ for various E-I laminations over a range of stack heights. The interested reader might compare the plotted values of pwr/vol with those given in the Appendix for Case II. In order to do this it will be necessary to determine the normalized dimensions of each lamination

Table 2-3. Case III, Choke Data.

Q/P	1.0	1.2	1.5	1.8	2.2	2.7	3.3	3.9	4.7
1.0	184	172	132	83	31	02			
1.2	226	226	191	134	62	10			
1.5	274	293	278	225	132	40	01		
1.8	309	346	355	317	222	96	12		
2.2	343	396	435	426	351	208	60	03	
2.7	372	440	507	530	494	372	183	45	
3.3	395	476	568	619	625	550	376	180	20
3.9	411	502	610	683	721	690	560	369	120

Form Factor: $F = 1-D$; $E = (P - 2 + 2D)/2$; $G = Q - P + 2 - 2D$; Function = (Power/Vol.) $\times 10^5$

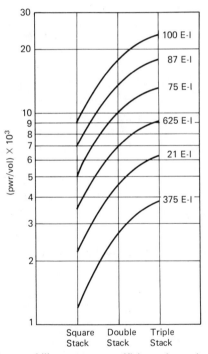

Fig. 2-5. Power-handling capability at constant efficiency for various lamination sizes and stack heights.

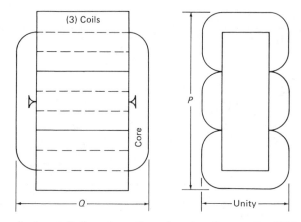

Fig. 2–6. Normalized overall dimensions, three-phase transformer. $P = 3E + 3F$, $Q = 2E + G$, $1 = D + F$.

stack size. This provides not only an exercise in normalizing a set of transformer dimensions, but an incentive to obtain the literature of some lamination suppliers.

The three-phase transformer is now considered. Note that in most manufacturers' catalogs the leg thickness is designated as $2E$. Since it is the total cross-sectional area per leg we have considered previously, for consistency we shall also call the leg thickness E. Be sure to translate this when talking to manufacturers. They wind one C-core on top of another to produce a tape-wound E-core, and they call each C-core tape buildup E. (See Fig. 2–6.)

The power-handling capability per unit volume at constant efficiency is given in Table 2–4, with exact core dimensions for various form factors shown as Case IV in Appendix A.

6. HIGH VOLTAGE CONSIDERATIONS

For a high voltage transformer we have somewhat more difficulty in starting with a normalized design. Different application requirements have different operating voltage levels, and those require real-world physical spacings which cannot be scaled at different power levels. Similarly, the number of turns per layer leads to some number of volts per layer, which requires a certain thickness of layer insulation. Layer width is therefore not an arbitrarily scalable parameter. The particular rectifier and shield configuration dictates

Table 2-4. Case IV, Choke Data.

Q/P	1.0	1.2	1.5	1.8	2.2	2.7	3.3	3.9	4.7
1.0	204	231	245	230	183				
1.2	228	268	301	301	261	179			
1.5	254	308	366	391	376	300			
1.8	271	336	412	461	475	423	296		
2.2	288	261	457	528	578	568	468		
2.7	301	383	495	587	670	707	659	535	
3.3	312	401	526	635	746	825	834	762	574
3.9	320	413	548	669	800	910	962	936	806

Form Factor: $F = 1-D$; $E = (P - 3 + 3 D)/3$; $G = Q - 2(P - 3 + 3D)/3$; Function = (Power/Vol.) × 10^5

certain insulation thicknesses between windings, shields, and core, and these also cannot be arbitrarily scaled.

A reasonably decent design can be done in some cases by selecting a core based on the previously outlined power-handling criteria and arbitrarily increasing the F and G dimensions to allow for insulation. This is most useful where a relatively low voltage winding floats at a moderately high potential, and where the insulation area in the core window is less than the copper area. For high voltage, low capacity isolation transformers, the result tends to be an approximately square core area, an approximately square core window, and a secondary winding of roughly square cross section, sometimes wound on a circular form or in a toroidal shell and spaced symmetrically from core and primary.

In the case of a high voltage transformer which develops higher voltages (above perhaps 10 kV), the form factor of the transformer begins to be influenced very strongly not only by the insulation around the secondary winding, but also by the required form factor of the secondary winding itself. The secondary wire size has been determined by a successive approximation/ trial and error initial design effort. At so many volts per turn a layer of some length develops so many volts. The layer insulation must not only provide mechanical support for the wire, but must also have sufficient thickness that when impregnated it will withstand the electrical stress of the layer-to-layer potential. In conventional coils, wound alternately left to right and right to left, the maximum stress on the layer insulation is due to the voltage developed by two layers of windings. In some cases, either to reduce the layer-to-layer voltage stress (by one-half) or the intrawinding capacitance (by one-quarter) each layer is wound from left to right, the layer finish insulated and crossed over to the left, and the next layer wound again from left to right.

The maximum length of a layer is governed in most cases by the thickness of layer insulation necessary to hold off the voltage developed by the number of turns in that length. When the volume of layer insulation begins to exceed the volume of the conductor, the point of diminishing returns is close for most 60/400 Hz transformers and some higher frequency situations.

The most common procedure we have used for high voltage plate transformer designs is this:

Select a core with $E = F$, $D = (E$ to $2E)$, and G to be determined later. Juggle dimensions as necessary to fit into customer-specified overall dimensions.

If no overall dimensions are given, select a core by assuming a copper space fill factor of 20% instead of the 40% used for transformer derivation earlier in this chapter. Start with P somewhere in the range of 1–2 and Q on the ridge of maxima. Use the D and E dimensions of that core.

Convince the customer to use a full-wave doubler circuit, as it eases transformer insulation requirements.

Determine secondary RMS current from Schade's curves in Appendix C,* and pick a wire size of about 400 circular mils per ampere for 400 Hz designs or 800 circular mils per ampere for 60 Hz designs. Determine volts per turn from core area and flux density, using 12–14 kilogauss for grain-oriented silicon steel, and 14–16 kilogauss for Z-type material. Pick a layer insulation somewhere between half the wire diameter and twice the wire diameter. Use enough turns per layer to give an assumed 200 volts per mil layer insulation dielectric stress. That gives the winding length for a chosen wire size. Assume 50 volts per mil of stress in the margins. That gives the G dimension of the core. Add up the required insulation thicknesses and builds of the primary and secondary windings, and see if it fits into the chosen F.

Then calculate the approximate intrawinding secondary capacitance, using the air core transformer formulas for pulse transformers as developed in Chapter 6. The exception is that for the factor S (the layer insulation thickness of a one-turn-per-layer design), use the layer insulation thickness of your trial design *divided by the number of turns per layer,* since one layer

---

*Shade's curves are used in rectifier circuit design. Any designer who deals with rectification of sine wave power and capacitor input filters should be intimately familiar with these curves. They are an invaluable design tool.

of insulation is associated with many turns per layer. Then calculate the leakage inductance of the transformer. The self-resonant frequency of the transformer must be well above the 12th harmonic of the input frequency for single-phase designs, or the 36th harmonic of the input frequency for three-phase designs.

Check the winding resistance to determine whether energy losses are acceptable and rework the design as appropriate.

Check the core loss to ensure it is within acceptable limits for overall losses. Since the thermal conductivity of epoxy-impregnated paper is relatively poor, about 50–65% of the copper loss should be placed in the primary, where it can be conducted to the core. This gives a chance of roughly equal temperature rises in both primary and secondary. If at all possible, mold the entire unit in a high thermal conductivity epoxy resin to aid in heat flow from both the high voltage coil and the core to the mounting surface. Keep in mind that the thermal conductivity of the core is greater out to the edges of the laminations than across the laminations. If possible, make the winding form somewhat wider in the E dimension so that the high thermal conductivity potting can surround both sides of the core in its E dimension and aid in heat flow from both the core and the primary.

For high frequency, high voltage designs, proceed along similar lines. Here the exception is that the self-resonant frequency must usually end up high enough that the rise time of the transformer will be a fairly small fraction of the period of the input waveform (no more than 10%). For high power, high frequency designs the rise time will usually become limited by the L/R time constant of the leakage inductance and the load resistance.

In some high power, high frequency designs it may be necessary to use a larger wire size than would at first seem to be called for based on current-carrying considerations. The design must be forced to enough layers that the distributed intrawinding capacitance is low enough, while at the same time the length of the coil is kept long so that leakage inductance will be low. This tends to occur in high volts-per-turn designs with very high turns ratios (in excess of 100 : 1), where leakage inductance must be minimized. In those cases secondary copper losses become almost negligible. The dominant dissipative loss is core loss, with primary copper loss the second major dissipative factor. Core loss can usually be fixed based on core surface area and heat transfer means available, thereby giving flux density, since a particular core, frequency, and core loss imply a particular flux density.

In practice, I start a high frequency, high power, high voltage design by picking an allowable core temperature rise based on available cooling means.

From that I work out a trial flux density and volts-per-turn. I then design the secondary winding and the insulation around it, and attempt to place a primary winding with acceptably low resistive losses in the remaining space. If the design does not fit, there are several options.

If the ferrite core has an appreciable portion of its surface smooth and flat, as in pot cores, U cores, or E cores, aluminum shapes can be bent to those surfaces to aid in heat conduction or convective heat transfer. A higher flux density is thereby allowed in the core, and fewer turns.

If the coil does not fit in the core window, and an appreciable fraction of the window area is taken up with the margin insulation, there are two solutions. One is to look for a core with a greater G dimension so that the winding length can be increased without reducing the margins. The other solution is to keep the original core size and increase the winding length of the design by reducing the margins. This will raise the electrical stress in the margins.

The more highly stressed the design, the more care must be taken to ensure not only adequate dielectric strength, but also a more uniform electric field in the stressed region. The electric field between coil and core can be shaped by putting a round conductor next to the sharp edge of the core facing the coil. That, and a number of layers of Nomex-Kapton-Nomex (or Kraft-Mylar-Kraft, or Kraft-Kapton-Kraft) laminate inside the window going up one edge of the coil, along its length, and down the other edge (the F dimension, the G dimension, and again the F dimension) with a width about 1.5–2 times the D dimension, will allow the margin stress to be raised to about 100–200 volts per mil with good epoxy impregnation.

If the secondary resistance loss (including skin effect) is still small, the wire size can be reduced. If necessary the layers can be wound in the same direction to halve the layer insulation stress.

High frequency, high voltage power transformers done on toroidal cores do not yield the most efficient designs. The major heat source will be the core, and the thermal conductivity of all the insulation and windings on top of it is poor. The entire voltage developed by the turns around 360° of the core is developed across the layer of interwinding insulation, and it is difficult to control insulation thickness while ensuring good impregnation on a toroid. Further, the many turns on a high voltage toroid tend to be piled somewhat helter-skelter on top of each other, so that turns far apart electrically will find themselves in physical contact. The winding machine can occasionally scrape or scratch a piece of wire, reducing the film thickness of insulation on the wire.

In designs of sufficiently high voltage or power that these factors would be significant, toroids are less than desirable.

At higher frequencies, ferrites or tape-wound C-cores (one-half or 1 mil alloy thickness) are the preferred materials. At lower frequencies tape-wound C-cores of 1–4 mils do best. One might be tempted to use thin-gauge, tape-wound bobbin cores at frequencies of 100 kHz or more for high voltage, but it really becomes academic since leakage inductance and intrawinding capacitance go up so fast with turns ratios that pushing the frequency higher and higher does not help. For turns ratios on the order of 100 : 1 and power levels of a few hundred watts to a few tens of kilowatts, the best operating frequency ranges downwards from 200 kHz toward 2 kHz. (One would here define "best" as an equality of dissatisfaction between the physical size, the high parasitic losses due to the frequency not being lower, the core loss, and the excessively large filter components needed due to the frequency not being higher.)

Given a fairly specific design situation, some parameter could of course be optimized with respect to some other, using differential calculus as we have done here and in Chapter 3. One fairly general item is the optimization of layer insulation thickness for maximum high frequency response. This very useful tool for high frequency designs is developed in Chapter 3.

7. A DESIGN EXAMPLE

Let us now work out an example of a power transformer design. Let us say that out input is 208 volts, 400 Hz, and we choose a flux density of 12 kilogauss. The power to be transferred is 1 kw, the turns ratio is 1 : 1, and we want 2% copper losses. For about 5 amps, with 1% of the copper loss in the primary, our primary loss would be 10 watts, for a DC resistance of 0.4 ohms. Using Eq. (2.8), and assuming a stacking factor for our 4 mil core of 0.90,

$$\frac{D^2 E^2 FG}{U} = \frac{7 \times 10^{-7} \times 208^2}{1.64 \times 10^8 \times 0.4 \times 12^2 \times 400^2 \times 0.81}$$

$$= 0.2474.$$

Assuming a Case I configuration, look at Table 2–1 and select (somewhat arbitrarily, but following the ridge of maxima) a form factor of $P = 1.5$, $Q = 2.7$. For this normalized core, $D^2 E^2 FG/U$ (labelled "Pwr" in the table) is 0.01281. That core has dimensions of $D = 0.735$, $E = 0.617$, $F = 0.132$, and $G = 1.465$. Since the power capability of our normalized core is less than that required, we will obviously have to scale it up. Our denormalization coefficient will be $(0.2474/0.01281)^{1/5} = 1.8079$. Our normalized core dimensions become $D = 1.3288$ (1 ⅜ in.), $E = 1.1155$ (1 ⅛ in.), $F = 0.2386$ (¼ in.), $G = 2.6485$ (2 ⅝ in.), and our overall dimensions about $1.8 \times 2.7 \times 4.9$ in.

From Eq. (2.5),

$$N = \frac{208}{2.865 \times 12 \times 400 \times 1\frac{3}{8} \times 1\frac{1}{8} \times 0.9 \times 10^4}$$

$$= 109 \text{ turns.}$$

From Eq. (2.3),

$$A_w = \frac{0.2 \times \frac{1}{4} \times 2\frac{5}{8}}{109} = 1{,}200 \text{ square mils, or } 1{,}534 \text{ circular mils.}$$

Pick #18. At 21 turns per inch and a winding length of 2 ¼ in., we expect 47 turns per layer, or 2.32 layers for each winding.

This is distinctly inconvenient, as it means that two-thirds of a layer in each winding is wasted space. Further, we see that if we have a total of six layers of #18, we're going to come out right about 0.25 in. build, without any room left over for the interwinding insulation or the winding form. A quick estimate of the winding resistance tells us to expect a length of about 50 feet and a resistance of 0.32 ohms per winding. So we are a little on the conservative side and can afford to drop down to #19 AWG. At 24 turns per inch we expect 54 turns per layer or 108 turns for two layers. Excellent. Change the number of primary turns to 108 from 109 (a hair higher flux density) and run with it.

Assume a winding form of 0.030 epoxy fiberglass and layer insulation of 0.007 in. Nomex 410. Assume interwinding insulation of two wraps 212 Nomex-Kapton-Nomex laminate and an identical outside wrap. Our total build would then be:

winding form	0.030
2 #19	0.076
layer insulation	0.007
interwinding ins.	0.010
2 #19	0.076
layer insulation	0.007
outside wrap	0.010
	0.216

This will fit in a 0.25 in. high space. The mean length of turn U (average of both windings) is 5¾ in. The total length of each winding is about 52 ft. At 8.05 milliohms per foot we have 0.417 ohms, just about what we wanted.

We can take a look at core loss here, since this method does not predict it analytically and leads to a class of designs with a relatively large core compared to the amount of copper. In this case our core volume is 12.77 cu. in. Assuming a specific gravity of 0.276 pounds per cubic inch and a stacking factor of 0.9, we expect a core weight of 3.17 pounds. If our core loss is around 10–15 watts per pound, our core loss for this transformer would be in the range of 30–50 watts.

Our total copper loss for the transformer is only about 20 watts, so we are on the wrong side of the maximum efficiency criterion where copper losses should be about equal to core losses. Since core loss goes (very roughly) as the 2.5 power of the flux density, we would expect that a reduction of flux density to 0.76 of its present value would reduce our core loss roughly by half. Close discussion with several C-core manufacturers (Arnold, Magnetic Metals, Magnetics Inc., National Magnetics) would tell you who has a good batch of 4-mil silicon steel that gives better core loss, and whether someone's Z-type silicon might have less core loss—or one could go to a 2-mil core annealed for minimum core loss. (In special cases some manufacturers can anneal tape-wound cores for minimum loss. Be sure to discuss this with the core manufacturer in critical applications.)

With a little work on this matter we might be able to get the core loss down to the 20 watt neighborhood.

If we were to specify a higher allowable resistance (say, 0.6 ohm) for a total copper loss of 30 watts, our denormalized core would be somewhat smaller than this one. The core weight would drop, and with it the core loss.

Since our core has a significant radial build, the mean magnetic circuit length (and hence the core weight for the same DE and FG products) could be reduced by going to shell-type construction (type II).

At lower frequencies where core loss comes out somewhat less than the design copper loss, core and copper losses can be balanced out somewhat by going to a type III design (coil type), and reducing the mean length of turn of the windings.

One would conclude from an examination of the tables and the core loss data for 12 mil, 4 mil, and 2 mil silicon steels at various frequencies that Types I and II designs are more favored at lower frequencies (60 Hz). Type I or III designs will be more favored for maximum efficiency designs with equal core and copper loss from 400 Hz up to where parasitics (leakage inductance and shunt capacitance) begin to dictate coil geometry; at which point one usually ends up with a Type I or II geometry based on a single coil for simplicity and lower capacitance at high step-up ratios, or a Type III geometry for minimum leakage or saturation inductance at high step-down ratios.

Looking back at our design, we might compare the core and copper aspects of this class of designs. The coil volume (UFG) is 15.09 cu. in. About 40% of that, or 6 cu. in., is copper. Our total copper losses are about 20 watts; the copper weight is about 0.4 pounds, for 50 watts per pound of copper losses. On the other hand, our copper losses are about 1.3 watts per cubic inch of total coil volume, and our core losses are roughly 3–4 watts per cubic inch (at 400 Hz), or about 10–15 watts per pound.

You can see that there is a significant difference in the loss density when comparing core losses and copper losses. One might expect that something or other would become optimized if the transformer generated its thermal losses uniformly throughout its volume. This idea is a thermodynamic concept, discussed in greater detail in the chapter on pulse transformers.

8. AN OPTIMIZATION ALGORITHM FOR MINIMUM VOLUME WHICH INCLUDES BOTH CORE AND COPPER LOSS

It can be shown* that the winding loss in a linear magnetic component may be predicted and a specific core size selected to deliver that performance. Further, the volumetric efficiency of a magnetic component with respect to winding resistive losses has been shown to be a function of the relative overall proportions of the unit. Computer analysis has given exact core dimensions such that the volume is minimized for a given level of losses, for each of a wide range of form factors.

These data, while expedient for much of magnetic component design, take no account of core loss, neither predicting it nor attempting to hold any optimum relationship (for minimum overall loss) between core and copper loss. Calculations can be made which show that the efficiency of a transformer is a maximum when core and copper losses are equal. This may or may not be valid. One can argue, based on thermodynamic considerations, that the transformer will be in its lowest energy state (and therefore, presumably, at maximum efficiency) when the rate of thermal energy generation per unit volume is equal in both core and coil. Whether only the copper volume of the

*In the preceding portion of this chapter and portions of Chapter 3, following.

coil or the entire coil volume is to be considered could also be discussed. Further, these two lines of reasoning do not necessarily give the same result.

Instead of attempting to pass on the validity of these considerations, we take a more objective approach and set up a general expression for the total loss of a transformer (the sum of core and copper loss). The advantage of such a general approach is that we do not force our preconceived ideas into the analysis, but simply state the problem in a totally objective manner. The application of mathematical tools to our problem will then give us the answer which is a logical consequence of the physical relationships of our problem as we have stated it.

What we will do, specifically, is this:

1. We will set up a general expression for total loss.
2. We will add the condition that, for this total loss, we want the flux density to be as high as possible. This second condition means that we want to utilize our core material capability to the greatest extent possible. These results will be used to simplify our general expression.
3. We will then solve our general expression for the denormalization constant, which will allow us to scale a normalized core of desired relative proportions up or down to give the actual core upon which the transformer is to be constructed.
4. We will observe that our expression for the denormalization constant, in terms of circuit parameters and core dimensions, has not presupposed any particular proportions. We then use the same technique for varying core dimensions to minimize the denormalization constant as we did when considering winding loss only. We will then obtain relative proportions for cores which give minimum total loss per unit volume, and we will be able to see which relative proportions are to be preferred.

Nowhere in this procedure will we assume any fixed ratio between core and winding losses. We let that ratio fall out as it may, and are then free to speculate on what it means.

8.1. Copper Loss

We begin by recalling a general expression relating the core dimensions, winding resistance and denormalization constant of a transformer:

$$R = \frac{0.24 \times 10^8 \varrho V^2 U}{(D^2E^2FG)(B^2f^2\sigma K^5)}. \tag{2.9}$$

The power loss of such a resistance is I^2R, as the coil loss is

$$P_{coil} = I^2R = \frac{0.24 \times 10^8 I^2 \varrho V^2 U}{(D^2E^2FG)\,(B^2f^2\sigma K^5)}$$ (2.10)

8.2. Core Loss

The core losses will have to be taken off the manufacturer's graphs unless analytic descriptions of their material properties have been furnished. It can be put in the form

$$\text{Loss/unit volume} = C_1 B^{2.5},$$ (2.11)

where C_1 is a frequency-dependent constant which will vary with frequency and flux density, but can be determined with adequate accuracy. The 2.5 exponent of B is an approximation taken for the purposes of this analysis. It, too, will vary for one material or another, and depends somewhat on f, and B itself. These can be determined from the core loss data published by core manufacturers or by exact measurement of material properties in the neighborhood of the desired operating point.

The actual volume of our denormalized core is DEU_m, where U_m is the mean length of the magnetic circuit. If we use normalized core dimensions and a denormalization constant, our actual core volume is

$$\text{core volume} = K^3 DEU_m$$ (2.12)

and our core loss is

$$P_{core} = C_1 K^3 DEU_m B^{2.5}.$$ (2.13)

8.3. Total Loss

The total loss is $P_{coil} + P_{core} = P_{total}$, or

$$P_{total} = \frac{0.24 \times 10^8 I^2 \sigma V^2 U}{(D^2E^2FG)\,(B^2f^2\sigma K^5)} + C_1 K^3 DEU_m B^{2.5},$$ (2.14)

where P_{total} is understood to be a total allowable power loss for the transformer *which the designer specifies.* It will therefore be treated as a constant.

8.4. Application of the Optimization Criterion

Equation (2.14) is not directly soluble for K in the general case as it is an eighth degree polynomial, and such things apparently do not have general solutions. One could obtain a graphical solution for specific numerical circumstances, but that would not be very useful and we have many tools remaining.

Let us say that we wish B to assume the maximum value possible, consistent with the total loss limitation. We will differentiate the expression for P_{total} with respect to B, set it equal to zero, and find the requisite relationship between B and the other parameters.

Rearrange Eq. (2.14), multiplying both sides by B^2K^5, and call P_{total} simply P:

$$0 = -B^2K^5 + \frac{0.24 \times 10^8 I^2 \varrho V^2 U}{(D^2 E^2 FG)\,(f^2 \sigma P)} + \frac{DEU_m C_1}{P} K^8 B^{4.5}. \qquad (2.15)$$

Differentiate with respect to B, and set equal to zero:

$$0 = -2BK^5 + \frac{4.5 DEU_m C_1}{P} K^8 B^{3.5} \qquad (2.16)$$

$$B^{2.5} = \frac{2PK^{-3}}{4.5 DEU_m C_1} \qquad (2.17)$$

$$B^2 = K^{-2.4} \left(\frac{2P}{4.5\, DEU_m C_1} \right)^{0.8} \qquad (2.18)$$

For the moment, let

$$C_2 = \frac{0.24 \times 10^8 I^2 \varrho V^2 U}{(D^2 E^2 FG)(f^2 \sigma)} \qquad (2.19)$$

Equation (2.14) may then be written as

$$P = \frac{C_2}{B^2 K^5} + K^3 DEU_m B^{2.5} C_1. \tag{2.20}$$

Substitute Eq. (2.17) and (2.18) in Eq. (2.20), then solve for K:

$$P = C_2 \left(\frac{2P}{4.5 DEU_m C_1} \right)^{-0.8} K^{-2.6} + \frac{4P}{9} \tag{2.21}$$

$$K^{2.6} = \frac{9C_2}{5P} \left(\frac{2P}{4.5 DEU_m C_1} \right)^{0.8} \tag{2.22}$$

$$K = \left(\frac{9C_2}{5P} \right)^{1/2.6} \left(\frac{2P}{4.5 DEU_m C_1} \right)^{0.8/2.6} \tag{2.23}$$

8.5. Discussion of Results

Equation (2.23), the expression for the denormalization constant, is the result of step 3 as discussed in the introduction to this section. Note that it is independent of the flux density term B. Knowing the relationship of flux density to core loss, once the designer has specified a desired total loss he does not need to choose B; he has already done that implicitly in the loss specification.

Depending on the total loss level required, the mathematics may ask for a sufficiently high level of B that the selected core material is required to operate at a flux density above its capability, i.e., in saturation. In this case the designer must select a higher maximum flux density material, or choose a somewhat higher K in order to force a lower B.

Equation (2.23) gives the denormalization constant of a core such that the maximum flux density is used, a given total loss obtained, and we have not assumed the results of the minimum copper loss transformer form factor analysis. Neither have we assumed any fixed apportionment of loss between coil and core. It is instructive, however, to look closely at Eq. (2.21). P is the total power dissipated, and the second term on the right is the core loss term after the criterion that B be maximized was invoked. Note that this core loss term is $\frac{4}{9}$ of the total power, roughly 44.5%. It would appear that when the exponent of B in the core loss expression is 2.5, the core loss should be a bit less than 45% of the total loss for maximum efficiency.

CORE LOSS VERSUS COPPER LOSS FOR ARBITRARY CORE LOSS FLUX DENSITY EXPONENT

$$P_{core} = K^3 DEU_m B^X C_1 \qquad (1)$$

$$P_{total} = \frac{0.24 \times 10^8 I^2 \varrho V^2 U}{D^2 E^2 FGB^2 f^2_\sigma K^5} + K^3 DEU_m B^X C_1 \qquad (2)$$

Multiply by $B^2 K^5$, divide by P:

$$B^2 K^5 = \frac{0.24 \times 10^8 I^2 \varrho V^2 U}{D^2 E^2 FGf^2_\sigma P} + \frac{DEU_m C_1}{P} K^8 B^{X+2} \qquad (3)$$

Differentiate with respect to B:

$$2BK^5 = \frac{(X+2)DEU_m C_1 K^8}{P} B^{X+1}$$

$$2K^5 = \frac{(X+2)DEU_m C_1 K^8 B^X}{P} \qquad (4)$$

$$\frac{2P}{(X+2)DEU_m C_1 K^3} = B^X \qquad (5)$$

Recalling Eq. (2),

$$P = \frac{C_2}{B^2 K^5} + K^3 DEU_m C_1 B^X \qquad (6)$$

Substitute Eq. (5):

$$P = \frac{C_2}{B^2 K^5} + \frac{2K^3 DEU_m C_1 P}{(X+2)DEU_m C_1 K^3} \qquad (7)$$

$$P_{total} = \underbrace{\frac{C_2}{B^2 K^5}}_{\text{Copper loss fraction}} + \underbrace{\frac{2P_{total}}{X+2}}_{\text{Core loss fraction}} \qquad (8)$$

Optimum core loss is $2/(X+2)$ of the total loss:

$X = 3$,   core loss $= 40\%$
$X = 2.5$,   core loss $= \%$ = approx. 44%
$X = 2$,   core loss $= 50\%$
$X = 1.5$,   core loss $= 2/3.5 =$ approx. 57%
$X = 1$,   core loss $= \frac{2}{3} =$ approx. 67%

8.6. Finding the Optimum Core Dimensions for Various Overall Form Factors

The fourth step in our analysis is to vary the relative core proportions to minimize K in Eq. (2.23).

Substitute the original expression for C_2:

$$K = \left(\frac{9 \times 2.4 \times 10^8 I^2 \varrho V^2 U}{5 f^2 \sigma P D^2 E^2 FG} \right)^{1/2.6} \left(\frac{2P}{4.5 DEU_m C_1} \right)^{0.8/2.6} \quad (2.24)$$

$$K = \left(\frac{9 \times 2.4 \times 10^8 I^2 \varrho V^2}{5 f^2 \sigma P} \right)^{1/2.6} \left(\frac{U}{D^2 E^2 FG} \right)^{1/2.6} \left(\frac{2P}{4.5 C_1} \right)^{0.8/2.6}$$
$$\left(\frac{1}{DEU_m} \right)^{0.8/2.6}. \quad (2.25)$$

Since we are going to take the derivative of K with respect to D and express the remaining core dimensions (E, F, and G) in terms of D, the first and third bracketed quantities in Eq. (2.25) are independent of core dimensions and may be treated for the moment as a constant. Accordingly, let

$$C_3 = \left(\frac{9 \times 2.4 \times 10^8 I^2 \varrho V^2}{5 f^2 \sigma P} \right)^{1/2.6} \left(\frac{2P}{4.5 C_1} \right)^{0.8/2.6} \quad (2.26)$$

Equation (2.25) then becomes

$$K = C_3 \left(\frac{U}{D^2 E^2 FG} \right)^{1/2.6} \left(\frac{1}{DEU_m} \right)^{0.8/2.6}. \quad (2.27)$$

When the derivative of K with respect to D is set equal to zero, C_3 will clearly drop out. Our expression then becomes

$$0 = \frac{d}{dD} \left\{ \left(\frac{U}{D^2 E^2 FG} \right)^{1/2.6} \left(\frac{1}{DEU_m} \right)^{0.8/2.6} \right\}. \quad (2.28)$$

At this point the drudgery of solving Eq. (2.28) about six hundred times for each of four basic geometric configurations is best left to a computer. One would obtain the D, E, F, and G dimensions of a core such that the value of K is minimized for a design which delivers a specified total loss. Our last remaining question is whether some overall form factors are better than others. Eq. (2.25) gave K as a function of circuit parameters and geometry. For each (P,Q) form factor we have obtained a set of (D, E, F, G) values. For each such set we have the computer calculate the geometric portion of Eq. (2.25):

$$\left(\frac{U}{D^2 E^2 FG}\right)^{1/2.6} \left(\frac{1}{DEU_m}\right)^{0.8/2.6}$$

We then prepare a P-Q array of the value of this parameter per unit volume for each geometric configuration, and look over the array for the region where this parameter has a relative minimum. The values of (P,Q) corresponding to such minimum will be the preferred form factors for transformers designed in accordance with the foregoing concepts. For aesthetic reasons, the Tables 2–5 through 2–8 give the *reciprocal* of the above function. The reader's attention is commended to the ridge of maxima for each of the four cases, and to the relative volumetric efficiency.

As will be discussed by those who read McLyman (1982), the flux density exponent of core loss varies from near 1 to almost 3 for available materials. We see (in the boxed discussion at the end of Section 8.5) that, as that exponent varies, the optimum apportionment of loss between coil and core also varies. If we are concerned with maximum volumetric efficiency, it is apparent that designs with relatively more core volume are penalized more if their material has a higher loss exponent (all other things being equal), forcing a lower flux density for proper loss balance. That implies more turns and thereby lower volumetric efficiency. One might expect, therefore, that the

Table 2-5. Case I, Function $1/[PQ\ f(D)]$.

Q/P	0.56	0.68	0.82	1.00	1.20	1.50	1.80	2.20	2.70
1.00	729	709	663	574	448	243	67		
1.20	721	715	688	627	532	352	166		
1.50	702	706	696	662	602	476	315	100	
1.80	681	692	690	670	632	546	429	234	19
2.20	655	670	675	666	643	588	512	378	163
2.70	627	644	653	652	639	604	554	470	328
3.30	597	617	629	633	627	604	570	513	425
3.90	573	593	607	614	612	596	572	529	465

Table 2-6. Case II, Function 1/[PQ $f(D)$].

Q/P	0.56	0.68	0.82	1.00	1.20	1.50	1.80	2.20	2.70
1.00	835	852	856	841	806	722	608	403	112
1.20	802	824	833	828	807	751	673	537	304
1.50	761	785	800	803	793	758	709	624	488
1.80	727	753	769	777	773	751	715	655	563
2.20	690	716	735	746	746	733	709	666	601
2.70	653	679	698	712	716	709	692	661	614
3.30	618	643	663	677	684	681	670	647	611
3.90	589	614	634	649	657	657	649	631	602

Table 2-7. Case III, Function 1/[PQ $f(D)$].

Q/P	0.56	0.68	0.82	1.00	1.20	1.50	1.80
1.00	797	796	775	723	645	513	380
1.20	784	794	787	755	695	580	452
1.50	760	779	785	772	737	652	543
1.80	736	760	773	772	752	694	609
2.20	706	734	753	761	754	719	661
2.70	674	704	726	741	742	724	688
3.30	642	673	697	716	723	717	694
3.90	616	646	672	692	703	703	690

location of the ridge of optima [on a (P,Q) plot] would be different for different core loss exponents, and the core and coil dimensions within those (P,Q) envelopes would also be different. This first complete analysis of optimum transformer geometry for low-frequency power transformers covers only the case of a core loss exponent of 2.5. Later work will examine how geometry is affected by core loss exponents between 1 and 3.

This procedure gives an electrical design for a linear transformer or AC ballast reactor which meets a given total loss specification, and provides the minimum volume for such a design.

One sees for transformers something which I personally was surprised to

Table 2-8. Case IV, Function 1/[*PQ f(D)*].

Q/P	0.56	0.68	0.82	1.00	1.20	1.50	1.80
1.00	573	596	611	614	603	563	505
1.20	554	579	598	610	608	584	541
1.50	528	555	578	595	603	595	570
1.80	506	534	558	579	591	593	579
2.20	481	509	534	557	573	582	578
2.70	456	484	509	533	551	565	568
3.30	432	459	484	509	528	545	552
3.90	412	439	464	488	508	527	536

discover from the computer plots of volumetric efficiency (for that is what I believe the reciprocal of the transformer function per unit volume to be). The volumetric efficiency actually increases as one moves toward smaller (P,Q) values. There is over the (P,Q) range an absolute maximum, which certainly was not the case when resistive winding losses only were considered. Just as we had no (optimum) solutions for large P and small Q, we also have no solutions for very small P and Q. The volumetric efficiency rises slowly up to an absolute (or global) maximum (as opposed to the relative or local maximum of the inductor plots) and then drops off more or less abruptly. The plots were cut off at $(P,Q) = (0.33, 0.33)$ because of space limitations. The absolute maxima are on the plots for Cases III and IV.

Extended computer runs showed that the apparent maxima at the edges of the (P,Q) plots for Cases I and II were essentially the global maxima. One need not be too concerned about requiring all designs to be at the outer fringes of theoretical peak efficiency, however. Note that from (Case I, $P = 0.33$, $Q = 0.56$, Case II) to $P = 2$, $Q = 3.3$, the volumetric efficiency drops about only 18%, and the latter design may well be vastly easier to fabricate, let alone to realize. The ubiquitous ridge of local maxima is present, and it is suggested that where possible the designer attempt to place his designs in its vicinity.

3
POWER REACTORS

1. INTRODUCTION

A capacitor stores energy, functioning as a charge reservoir—a charge is introduced into it, and is stored there. The voltage of the reservoir changes little.

A series inductor is the analog of a shunt capacitor. Just as a capacitor stores energy as charge, so an inductor stores energy in its magnetic field as a result of the current flowing through it. The stored energy is $\frac{1}{2} LI^2$, where L is in Henries, I in Amperes, and the energy in Joules.

Just as a capacitor tries to hold a constant potential using its stored energy, or energy storage capability, so an inductor tries to hold a constant current using *its* stored energy or energy storage capability.

All power supply ripple filtering chokes function this way: A pulsating voltage source from a rectified sinusoidal or "square" wave AC source attempts to pass a pulsating current into it. The stored energy smooths out the pulsating current, and a more nearly constant current is the result. The stored energy resists the efforts of the fluctuating voltage to vary the current.

There is a complementary use of which the designer should be aware. In this application, the idea is to have the net energy stored in the reservoir be small compared to the reservoir's capacity. Recall that the voltage an inductor tolerates across itself for a time and the resultant change in current is $E\Delta T = L\Delta I$, where E is in volts, L in Henries, ΔT is the period of time involved in seconds, and ΔI is the change in current in amperes. If L is sufficiently large, then for some E and ΔT, ΔI will be small. This application is called a *current balancing transformer*. A similar application is called a *common mode choke*. We will look at their similarities and their differences.

2. BALANCING TRANSFORMERS AND COMMON MODE CHOKES

Note that currents in the choke (see Fig. 3-1), due to the intended energy delivered from the power supply to the load, are equal and opposite and therefore cancel. (This is assuming exactly equal numbers of turns and exactly balanced leakage inductance.) If the high frequency impedance of the inductor is much more than that of the source that couples the common mode noise into our system (so called because the noise tends to be common to all lines from source to load), then we will have attenuation of the noise current flowing through the load. That attenuation depends on the ratio of those impedances. This is the case where the "arbitrary ground return impedance" and the high frequency (HF) noise voltage source impedance are small.

The ganged switches (Fig. 3-2) could be a pair of vacuum tubes or semiconductors, chosen because the demands of the load were more than one could handle. R_1 and R_2 are the internal resistances of each; R_1 or R_2 is usually much smaller than R_{load}. Another common embodiment of this concept is where a single switch serves two loads (Fig. 3-3). In most applications, R_1 and R_2 differ by only a small amount.

In Fig. 3-2, the action of L is to keep the *net* current constant, which means close to zero, since the load currents are opposed. A difference between R_1 and R_2 would cause different currents to flow through each switch. This is not often desirable. The different voltages developed across R_1 and R_2 place different voltages across the two windings of the choke—opposite but not quite equal. The choke tends to resist a change in the initially zero

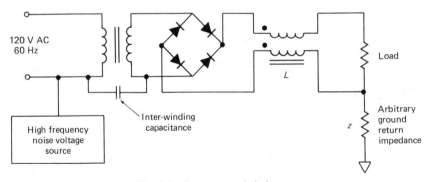

Fig. 3-1. Common mode inductor.

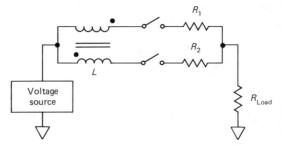

Fig. 3-2. Balancing transformer, current combining.

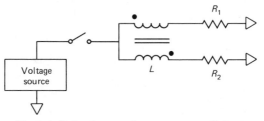

Fig. 3-3. Balancing transformer, current splitting.

current. The net current in the choke, which is the switch difference current, increases at a rate governed by

$$\frac{\Delta I}{\Delta T} = \frac{I_{\text{load}}(R_1 - R_2)}{L} . \qquad (3.1)$$

In Fig. 3-3 the action of L is to attempt to maintain equal currents in each load by attempting to maintain a zero current difference. The current from the switch is thus split into two equal portions and that tends to compensate for differences in two loads. Again, the difference current ΔI is governed by $\Delta I/\Delta T = (I_{\text{load}}(R_1 - R_2)/L$.

Balancing transformers work equally well in AC or pulsed DC applications, provided the inductor is properly designed to accommodate the flux density resulting from the actual AC and DC components. By a simple transformation of Fig. 3-2, we see that a balancing transformer can combine the currents from two AC sources if they are nearly synchronized. Small voltage and phase imbalances are readily accommodated by the configuration of Fig. 3-4. Any number of AC voltage sources may be combined by an appropriate iteration of this technique.

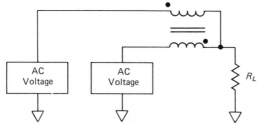

Fig. 3-4. Power combining.

3. AC BALLAST REACTORS

Power reactors are also used as impedances to be inserted in AC power lines such that current limiting or control may be obtained with minimal power dissipated as heat. Many power reactors used to filter DC currents in power supplies have a relatively small AC flux component and a relatively large DC flux component. The usual procedure is to design the reactor as if the DC flux component were the only contribution to the peak flux density in the core. After completing the design, the AC flux component is calculated to insure that it was indeed negligible, and that there was sufficient margin between the maximum flux density of the core material and the operating DC flux density to accommodate the AC flux superimposed on the DC flux level.

In AC ballast reactors there is no DC component. The approach used here is to design the AC ballast reactor as if it were the primary of a power transformer. The wire size is chosen to handle the maximum current flowing through the reactor, such that an acceptable level of thermal loss is maintained. Then, insert such gap as may be required to obtain the desired inductance, using Eq. (3.3b), which is given later in this chapter.

4. INDUCTOR DESIGN CONSIDERATIONS

4.1. Air Core Inductors

The inductance of an air core inductor is given by

$$L = \frac{\mu_0 \, \pi r^2 N^2}{r + b} \text{ Henries,} \qquad (3.2)$$

all dimensions in centimeters, where μ_0 is the permeability of free space, $4\pi \times 10^{-9}$ Hy/cm, r is the radius of the coil, b the length, and N the number of turns. *r* should be seen as the *mean* radius (the radius of the inside of the coil plus half the radial build of the winding).

Conditions under which air core inductors are useful include: where fairly small inductance is required; the frequencies involved are quite high (pulse width is short); the physical presence of a core is inconvenient or undesirable; space may not be at a premium; large thermal losses may be allowed and the presence of a core would hinder heat removal from the conductors; no suitable core materials exist; or some interlocking combination of the above. The designer will recognize such circumstances either by instinct or after a few attempts to design the inductor on a core. If there is some uncertainty, I would recommend designing an air core inductor first to obtain a physical reality on this version.

4.2. Ferromagnetic Core Inductors

For inductors with cores, a useful relationship is

$$L = \frac{3.2N^2A \times 10^{-8}}{l_g + l_c/\mu_\Delta} \text{ Henries,} \qquad (3.3a)$$

which sometimes may be simplified to

$$L = \frac{3.2N^2A \times 10^{-8}}{l_g} \qquad (3.3b)$$

where N is the number of turns, A is the *net* core area in *square inches* (the gross core area times a "stacking factor" or volume fill factor, available from manufacturer's catalog data), l_g is the gap in the magnetic circuit in inches, l_c is the length of the magnetic circuit in inches, and μ_Δ is the incremental permeability (defined as shown in Fig. 3–5).

We have here the hysteresis loop of some arbitrary material. With B in units of Gauss and H in units of Oersteds, the permeability of air (vacuum) is taken as unity. Note that if our operating DC flux density is at point I, any change in B requires a smaller change in H. At point II, however, the same change in B would require a larger change in H. The ratio of B to H is the material permeability μ, but you notice that it may not be a constant for any given material over the full operating range. At a given point, the ratio of an increment of B to the corresponding increment of H is the incremental permeability μ_Δ. You must know the permeability of your core material *at the operating point* to an accuracy appropriate to your calculations.

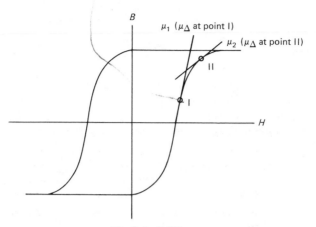

Fig. 3-5. *B-H* loop.

The choice of core material for an inductor must also be based on the maximum flux density at which the material permeability is still high, and on the frequency of the AC signal which the inductor is handling. The frequency will have an influence, sometimes a strong influence, on the permeability at that proposed maximum flux density.

The core loss associated with the AC signal, due to the AC flux density arising out of the AC voltage appearing across the inductor, must not be ignored because that core loss is, in effect, a resistor in parallel with the inductor and it allows an AC current to bypass the inductor. In extreme cases it may reduce the effective inductance to almost nothing.

Note that in Fig. 3-7 it may turn out that l_g is 0.1 in., l_c is 6 in., and all we need to know is that μ_Δ is more than 600 to ensure an accuracy of 10%. On the other hand, l_g may be 0.01, and if μ_Δ is only 600, then the material properties and core dimensions cannot be ignored. In most of the following derivations in this chapter, the l_c/μ_Δ term will be dropped for the sake of simplicity. In many design situations it is not insignificant, and the designer must remember to calculate the actual inductance of a proposed design using Eq. (3.3a).

Another necessary and useful relationship for DC inductors is

$$B = \frac{.6NI}{l_g} \qquad (3.4)$$

where B is in units of Gauss, N is the number of turns, I is the DC current in amperes, and l_g is the gap in inches. for a given core material, a desired B_{DC}

can be set. The designer would pick the material which would support the highest flux density and still have high permeability and low core loss at the ripple frequency. The current is presumed known, so a relationship between N and l_g can be formed. This can be substituted in Eq. (3.3b) and an inductance found for some given core area.

Clearly, a core must be chosen, and that makes the situation somewhat ambiguous. Two options are available. In the first edition of *Transformers and Electronic Circuits* (Lee, 1955), an interesting nomograph was given. A version of it is reconstructed here as Fig. 3-6 (for B_{DC} = 12,000 Gauss) and Fig. 3-7 (for B_{DC} = 2,000 Gauss). In neither case is any account made for the material permeability or the AC flux density which would have to be calculated separately and added to B_{DC}, to ensure that the reactor would not saturate under actual operating conditions. Using this nomograph, one can choose any two of the four parameters—turns, gap, current, and Henries per square inch of core area—and obtain the remaining ones. For example: Pick some core or core area dimensions as a trial. Knowing the desired inductance, form the quotient *Henries per square inch*. Find the light line sloping from upper left to lower right which corresponds to this number. Knowing the desired operating DC current, find the heavy line sloping from upper left to lower right. Find the intersection of those two lines. Then read horizontally to the left to obtain turns, and read vertically down to obtain the gap.

The second option is to apply the tools of optimization theory to the power reactor, just as we did to single- and three-phase transformers.

5. THE MAXIMAL EFFICIENCY POWER REACTOR

A figure of merit for power reactors is the energy stored, $\frac{1}{2} LI^2$. Another is the power lost, I^2R.

We can construct a very interesting figure of merit by expressing the "effi-

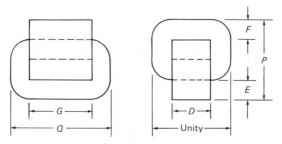

Fig. 3-6. The maximal-efficiency power reactor.

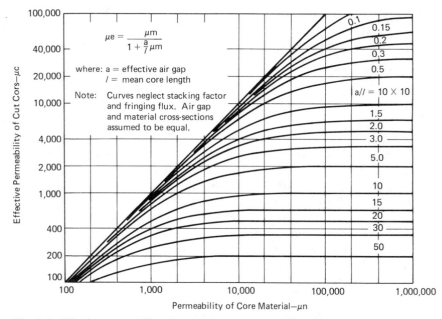

Fig. 3-7. Effective permeability of cut core versus permeability of core material. (SOURCE: "Silectron Cores," Arnold Engineering Bulletin SC-107A, Marengo, Illinois; Figure 23, p. 50.)

ciency" of the power reactor as energy stored per unit of energy lost per unit time. That is just:

$$\frac{\frac{1}{2} L I^2}{I^2 R} . \tag{3.5}$$

It is true that the current cancels in this expression. Since we introduce it again in some of the following equations, let's just carry it along and cancel later.

It is likely that some overall form factors are more efficient than others, in that the expression (3.5) might be higher for some than others. Further, with several design variables (turns, core area, flux density, gap, etc.) it would be expected that several different designs might be obtained within a given form factor, for which (3.5) might be different in each case.

We shall develop an expression for (3.5) in terms of electrical constants and dimensional parameters, and study the manner in which the relative proportions of the finished unit influence the magnitude of the maximum value of (3.5).

We shall use the geometry of a Type I power transformer and those terms (see Fig. 3-6).

We know that $L = (3.2N^2A \times 10^{-8})/l_g$ and $l_g = (0.6NI)/B$ from Eqs. (3.3b) and (3.4); hence $L = (5.33NDEB \times 10^{-8})/I$, and

$$\tfrac{1}{2}\, LI^2 = 2.67NIBDE \times 10^{-8}. \tag{3.6}$$

The mean length of the turn of the coil is U, and the total length of the winding is NU. The resistivity of copper is $\varrho = 10.7$ ohms/circular mil-foot, or 0.7 ohms/square mil-inch, or 7×10^{-7} ohms/square inch-inch. $R = \varrho l_w/A_w$ where A_w is the area of one turn, i.e., the winding area divided by the number of turns and multiplied by the winding space utilization factor. Call the space utilization factor σ. We suggest the designer try a space utilization factor of 0.4 initially. If some other space utilization factor is appropriate for a particular class of designs, the engineer can modify this equation and those following with the appropriate factor. So

$$R = \frac{\varrho N^2 U}{\sigma FG}$$

or $R = 7 \times 10^{-7}N^2U/\sigma FG$, and I^2R is just

$$I^2R = \frac{7 \times 10^{-7}N^2I^2U}{\sigma FG}. \tag{3.7}$$

What we want is zero rate of change of $\tfrac{1}{2}\, LI^2/I^2R$ with respect to geometry. Since E, F, and G can be expressed in terms of the fourth core dimension D and the external form factor constants P and Q, we take

$$\frac{d}{dD} \left[\frac{\tfrac{1}{2}\, LI^2}{I^2R} \right] = 0. \tag{3.8}$$

The differentiation rule for quotients is

$$d\frac{A}{B} = \frac{B\, dA - A\, dB}{B^2}.$$

For $U = 2D + 2E + \pi F$,

$$\frac{d}{dD} (\tfrac{1}{2} LI^2) = 2.67NIB \times 10^{-8}(E + {}^{D}\!/_2)$$

$$\frac{d}{dD} (I^2R) = 14 \times 10^{-7} N^2 I^2 \frac{FG(3 - {}^{P}\!/_2) + (2D + 2E + \pi F)(\tfrac{1}{2}G + F)}{F^2 G^2}$$

and $\dfrac{B\, dA - A\, dB}{B^2} = 0$ when $B\, dA - A\, dB = 0$, or

$A\, dB = B\, dA$, so

$$\frac{14 \times 10^{-7} N^2 I^2 (2D + 2E + \pi F)(2.67NIB \times 10^{-8})(E + {}^{D}\!/_2)}{FG} =$$

$$\frac{2.67NIB \times 10^{-8} DE \times 14 \times 10^{-7} N^2 I^2 [FG(3 - {}^{P}\!/_2) + (2D + 2E + \pi F)(\tfrac{1}{2}G + F)]}{F^2 G^2}$$

$$(2D + 2E + \pi F)(E + {}^{D}\!/_2) = \frac{DE}{FG} [(FG(3 - {}^{P}\!/_2)$$

$$+ (2D + 2E + \pi F)(\tfrac{1}{2} G + F)], \qquad (3.9)$$

which is a function of geometry alone, proving in principle the existence of a geometric criterion. [Eq. (3.9) *does* have solutions.] Since $F = (1 - D)/2$, $E = (P - 1 + D)/2$, and $G = Q - P + 1 - D$, we locate values of D (and the corresponding values of E, F, and G) that satisfy Eq. (3.9). This is most easily done by computer.

To denormalize from dimensions of $1 : P : Q$ inches to dimensions of $K : KP : KQ$ (the desired real-world design), the energy ($\tfrac{1}{2} LI^2$) is multiplied by K^2 and the power loss (I^2R) is multiplied by $1/K$. To find the desired K for an actual (denormalized) design of some given L and R, (given the optimum D, E, F, and G, which will be presented shortly) we take Eq. (3.6) and denormalize:

$$\tfrac{1}{2} LI^2 = K^2 \times 2.67NIB \times 10^{-8} DE \qquad (3.10)$$

where now only K and N are unknown (B having been picked for a specific core material, all othe parameters now set), take Eq. (3.7) and denormalize:

$$I^2R = \frac{7 \times 10^{-7}N^2I^2U}{\sigma KFG}$$ (3.11)

where again, only K and N are unknown. We now combine Eqs. (3.10) and (3.11), and solve for K and N. From Eq. (3.10),

$$N = \frac{LI}{5.33 \, DEBK^2 \times 10^{-8}},$$ (3.10a)

and from Eq. (3.11),

$$K = \frac{7 \times 10^{-7}N^2\,U}{\sigma RFG}.$$ (3.11a)

Inserting the expression for N into Eq. (3.11a),

$$K = \frac{7 \times 10^{-7}L^2I^2U}{28.44 \times 10^{-16}RFGD^2E^2B^2K^4}$$

$$K^5 = \left[\frac{10^9L^2I^2}{4.06\,\sigma RB^2}\right]\left[\frac{U}{D^2E^2FG}\right].$$

Note that the *first* term on the right consists only of the data given to the engineer by the application—the inductance, current, and resistance—and the flux density which he must pick based on the choice of a core material of suitable properties for the application.

Note that the *second* term is a pure geometric criterion for any core. The smaller it is, the smaller the design will be and therefore the better the design.

Hence we have

$$K = 47.67\left[\frac{L^2I^2}{\sigma RB^2}\right]^{1/5}\left[\frac{U}{D^2E^2FG}\right]^{1/5},$$ (3.12)

The value of D^2E^2FG/U for various overall form factors has been tabulated in Appendix A, along with the dimensions of cores which fit those form factors and maximize this geometric parameter. *This is power handling or energy storage ability at constant efficiency.* It is interesting to note that this criterion is identical to that developed for transformers, which are energy transfer devices rather than energy storage devices.

N may be obtained by substituting the normalized value of DE in Eq. (3.10a).

Note that the terms DEK^2 in the denominator of Eq. (3.10a) are the *normalized* terms D and E from the tables of Appendix A. If you use D and E as denormalized (i.e., the actual dimensions that the real core will have for the desired design), recognize that you are using KD and KE, which contain K^2.

The expression $l_g = 0.6NI/B$ [Eq. (3.4)] now gives the gap required, and the area of the conductor is just

$$A_W = K^2 \frac{FG}{2N} \qquad (3.13)$$

square inches.

6. DESIGN EXAMPLES

To construct the design, choose the wire size nearest that specified by Eq. (3.13), select a core with dimensions reasonably close to the optimum specified, calculate turns per layer and number of layers, and determine the total winding build (including layer insulation if a bobbin is not used), and check for fit. The wire size may need to be adjusted up or down depending on the space fill factor of the particular situation. The next step is to calculate the inductance of the actual design to ensure that no numerical errors have been made, using Eq. (3.3a).

In summary, to construct an inductor design we would:

1. Pick a (trial) set of *proportions* for the finished unit, assuming Type 1, 2, or 3 construction as experience, performance or construction simplicity may dictate.
2. From the P,Q chosen in the previous step, look up the value of D^2E^2FG/U.
3. Substitute that value in Eq. (3.12). Multiply the normalized core dimensions (listed in Appendix A) by K to obtain the denormalized core dimensions.
4. From Eq. (3.10a), calculate turns.
5. From Eq. (3.13), calculate conductor cross-sectional area.
6. Work out turns per layer or strip width. Determine interlayer insulation thickness required for mechanical support or voltage capability. Check winding build; rework if necessary.
7. Calculate winding resistance; rework if necessary.

8. Calculate AC flux density. Rework if necessary with lower DC flux density.
9. Evaluate heat flow problems (if any) from thermal losses in coil (and core, if AC core loss is significant).

For example, let's say we need a 100 mH choke for a DC filter application. The current is 750 mA DC, and the allowable DC resistance is 2 Ω. We choose a 4 mil C-core; and for winding convenience we wish a single coil. We choose proportions of $1 : 1.8 : 3.3$, for which $D^2E^2FG/U = 2.191 \times 10^{-2}$. Assume a σ of 0.5.

For that core $D = 0.723$, $E = 0.761$, $F = 0.138$, $G = 1.777$. We can operate grain oriented silicon steel at 12,000 Gauss easily in this application, so, using Eq. (3.12),

$$K = 47.67 \left[\frac{0.01 \times 0.5625}{0.5144 \times 10^6} \right]^{1/5} \left[\frac{1}{2.191 \times 10^{-2}} \right]^{1/5}$$

$$= 0.848.$$

Our denormalized core dimensions are therefore $D = 0.613$, $E = 0.645$, $F = 0.117$, $G = 1.507$. Choose close fractional dimensions for the convenience of the C-core manufacturer and to eliminate unnecessary tooling costs:

$$D = \tfrac{5}{8} \text{ inch,} \qquad E = \tfrac{5}{8} \text{ inch,} \qquad F = \tfrac{1}{8} \text{ inch,} \qquad G = 1\tfrac{1}{2} \text{ inch.}$$

From Eq. (3.10a), inserting the *net* core area ($D \times E \times 0.9$ stacking factor),

$$N = \frac{10^{-1} \times 0.75}{5.33 \times 0.352 \times 1.2 \times 10^4 \times 10^{-8}}$$

$$= 334 \text{ turns}$$

$$A_W = \frac{FG}{2N} = 280 \text{ square mils.}$$

Let's try #26 AWG wire. At 54 turns/inch and a 1¼ inch winding length, we would have 68 turns/layer, and five layers. Use 0.003 Nomex 410 for layer insulation, so we have 0.021 per layer × 5 layers = 0.105 in. winding build. This clearly fits, but ignores some practical matters like the thickness of the winding form (0.020 minimum if we use epoxy/fiberglass) and the fact that the coil has some "spring" to it. Unless we impregnate it with epoxy and cure it in a C-clamp, we'll never get a coil with 0.125 build to fit a 0.125 window height. Even at that, we haven't put any outside wrap on the coil. *Designing for an 80% fill factor on coil height gives us some room to accommodate dimensional tolerances on both the coil components and the core,* as well as the inevitable "bowing" of wire wound over a rectangular coil form.

Let's check the winding resistance first and make sure we have only one thing to fix. Mean length of turn = 2.9 in., total length = 81 ft, #26 has 0.041 Ω/ft, so $R = 3.31\ \Omega$. Too high.

This is because we calculated that we needed 280 square mils of cross-sectional area for our wire, but picked #26 with 253 circular mils, so our conductor had 200 square mils instead of 280. Let us look more closely at that: 200 square mils $\times$ 334 turns = 66,800 square mils; coil winding area = 187,500 square mils, so we were only using about one-third of the space, and the equations assumed a space utilization factor of 0.5. That is just about how much our resistance was off by. The point here is that in many layer-wound coils using magnet wire, the space utilization factor for copper may be in the 25–40% range, not 50%. This is the nature of the interface between theory and reality.

One option is to raise the F dimension of the core to allow a larger wire size. We estimate #24 would be required. At 54 turns per layer, 6.19 layers would be needed. This is mildly annoying. We have to squeeze the turns per layer up. We can increase the winding length and reduce the (⅛ in.) margins, or we can lengthen the core to allow a greater winding length and keep the ⅛ in. margins, or we can try to wind more turns in the specified 1¼ in. winding length than the wire table recommends. We could, of course, simply leave it at 6.19 layers and put the few remaining turns on a seventh layer, but that would waste the remainder of the space on that layer.

There is another option. We can go back to our initial choice of flux density, and decide that 12,000 Gauss was conservative. We can get by and meet the specification with perhaps as high as 14,000 Gauss. The incremental permeability is still considerably higher than what we anticipate the effective permeability of the core will be when gapped (calculate l_g from $l_g = 0.6NI/B$, then divide the mean length of the core by the gap. That is μ_e if μ_Δ is still much higher—perhaps 50 times higher; see Fig. (3-7). Along those lines, we could also specify our C-core to be 4 mil Z-type material. That is a square loop, grain-oriented material, and looks quite good several thousand Gauss higher than "round loop" silicon steel.

Let's see what we can do with this approach. Keeping the same core size for the moment, from Eq. (3.10a),

$$N = \frac{10^{-1} \times 0.75}{5.33 \times 0.352 \times 1.4 \times 10^4 \times 10^{-8}}$$

$$= 285 \text{ turns at } 14,000 \text{ Gauss.}$$

Let us try this on for size. With #24, we had 54 turns per layer. For 285 turns we need 5.28 layers. If we increase the G dimension of the core to $1\frac{1}{16}$ we can get 57 turns/layer and still keep our $\frac{1}{8}$ in. margins. Now we have five layers. With #24 and 0.003 Nomex 410 for layer insulation and wire support, we have a total thickness of 0.026 in./layer, and a winding build of 0.130 in. A winding form would raise that to 0.150 in., and the outside wrap would make it 0.160 in. If 0.160 in. were 85% of the available height, then we would want the F dimension to be $\frac{3}{16}$.

Our DC resistance for this design is now 1.77 Ω. Our overall dimensions are now very close to those specified as optimum proportions. The finished unit has grown $\frac{1}{16}$ in. in length and $\frac{1}{8}$ in. in width and height.

Let's take another example. Let's say we want a 20 μH, 15 amp choke with 0.010 Ω DC resistance. Since this will be used in a 500 kHz switching regulator, we select a ferrite core (Magnetics, Inc. type F material for highest DC flux density) and a DC flux density of 3,000 Gauss. Assume $\sigma = 0.5$.

Let us try core proportions of $P = 1.2$, $Q = 1.8$, and a type I geometry. For that core $D^2D^2FG/U = 0.004774$ (from the Appendix), and $D = 0.742$, $E = 0.471$, $F = 0.129$, $G = 0.858$.

Substituting in Eq. (3.12),

$$K = 47.67 \left[\frac{4 \times 10^{-10} \times 225}{0.5 \times 10^{-2} \times 9 \times 10^6} \right]^{1/5} \left[\frac{1}{0.004774} \right]^{1/5}$$

$$= 47.67 \left[\frac{9 \times 10^{-8}}{4.5 \times 10^4} \right]^{1/5} \times 2.9122$$

$$= 0.6347.$$

Our denormalized core dimensions are therefore $D = 0.471$ in., $E = 0.299$ in., $F = 0.082$ in., $G = 0.545$ in. From Eq. (3.10a),

$$N = \frac{2 \times 10^{-5} \times 15}{5.33 \times 0.471 \times 0.299 \times 3000 \times 10^{-8}} = 13 \text{ turns.}$$

Area of one turn $A_w = FG/2N = 1718$ square mils is required. Use 0.0040 copper laminated to 0.001 Kapton, a spiral strip coil; copper width = 0.450, winding build = 0.065, mean length of turn ≈ 1.79 in., DC resistance = 0.0045 Ω/foot; 1.9 ft = 0.0086 Ω, slightly less than 0.010 Ω because in this case our copper area was slightly better than 50%.

Admittedly, it is not often that one can specify such a custom shape in ferrite and be able to afford it. Still, it is now possible to look through the manufacturer's catalog for various ferrite cores which come close to the designed size. The search for the right size core to base the design upon has been vastly simplified.

Another unusual thing done in this example was to specify a copper-Kapton laminate as the winding material. Round wire could have been used, but the foil is much more convenient for fabrication purposes in this case. The material can be obtained from a company such as Scheldahl. Cut it to about 0.520 wide, with the foil etched back to a width of 0.450, centered on the Kapton film.

As a final check, calculate the AC flux density for this choke. The RMS voltage is 30 V at 500 kHz, so [using Eq. (1.1)]

$$\frac{30}{13} = 2.865 \times B \times 5 \times 10^5 \times 0.14 \times 10^{-4}.$$

$B = 110$ Gauss, which is small compared to 3,000, so we can indeed use 3,000 Gauss DC.

The thermal loss in the winding is ~ 2.25 watts, and the final packaging and environmental requirements happen to be compatible with this level of heat and the coil surface area.

Let's look at another example. We need an 80 mH choke to handle 8 amps DC, and our DC resistance is allowed to be 0.2 Ω. Further, we have only a 2⅜ in. wide space available for this choke, but it can be as tall or long as we wish. Let us try this on a Type I configuration and see where we are. Pick $P = 2.2$, $Q = 5.6$, and see what we get: for this form factor $D^2E^2FG/U = 0.06477$, $D = 0.720$, $E = 0.960$, $F = 0.140$, $G = 3.680$. Use 12,000 Gauss. Assume $\sigma = 0.5$.

$$K = 47.67 \left[\frac{0.08^2 \times 8^2}{0.2 \times 0.5 \times 144 \times 10^6} \right]^{1/5} \left[\frac{1}{0.06477} \right]^{1/5}$$

$$= 2.55.$$

We see from this that our normalized dimensions would become 2.55 in. $\times$ 5.61 in. $\times$ 14.28 in., which wouldn't quite fit in our 2⅜ in. space. We elect to use a Type III geometry for this unit, to take advantage of its higher volumetric efficiency. For Case III, $P = 2.2, Q = 5.6, D^2E^2FG/U = 0.1092$, $D = 0.772, E \doteq 0.872, F = 0.228, G = 3.896$. Then, $K = 54.75 \times 0.02695 \times 1.5572 = 2.2977$, which would fit in 2⅜ in. Our length is 12.9 in., and our height is 5.06 in. Our denormalized core dimensions are $D = 1.774$, $E = 2.004, F = 0.524, G = 8.860$.

Note that the overall volume of the type III version is 143.6 cubic inches, where for the same electrical functions implemented in a type I design the overall volume is 204.3 cubic inches. Where space is at a premium the two-coil single-core geometry is clearly superior.

If we wished, we could specify 4 mil Z-type silicon for the core and operate our reactor at about 14,000–15,000 Gauss. If cost were no object, we might even go to Supermendur, operate around 19,000 Gauss and reduce the volume even further.

At these current levels the larger wire size will give a better space fill factor than the previous example. Finishing the design, and selecting wire size, layer insulation, gap, etc., is left to the reader as an exercise. The reader is also urged to examine Table 2-2 and determine whether other proportions than $P = 2.2$, $Q = 5.6$ might be more advantageous for this Type III design.

It must be emphasized that there is no substitute for common sense. (Common sense may be defined as looking objectively and observing what really is, instead of assuming "what everyone knows" or assuming that because something has always been done a certain way that is the only way or the best way to do it.)

Say we are given proportions of $0.68 \times 0.68 \times 1$ in which to put a transformer or choke. From Table 2-1, we notice that the power handling capability per unit volume for $P = 0.68$, $Q = 0.68$ is about 75×10^{-5}. But this assumes a certain *orientation* of the core and coil within that volume! $0.68 \times 0.68 \times 1$ has the same proportions as $1 \times 1 \times 1.47$. Looking up $P = 1$, $Q = 1.5$ in the same table, we find the power handling capability per unit volume to be 186×10^{-5}, more than double!

7. A HIGH FREQUENCY CONSIDERATION

We can see another interesting property of inductors by looking at some of the high frequency parasitics. Assuming the construction is fairly uniform, our inductor at higher frequency would look like Fig. 3-8. In the figure, C' is the equivalent lumped shunt feed-through capacitance and L' is the equivalent lumped leakage inductance of the winding. If we wished to max-

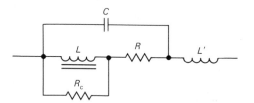

Fig. 3-8. High frequency parasitics of an inductor.

imize the parasitic self-resonant frequency of our inductor we would proceed as follows:

The leakage inductance of a cylindrical solenoid of radius r (cm), winding build Δ (cm), length b (cm) is

$$L' = \tfrac{2}{3} \pi \mu_0 \, \frac{r}{b} \, \Delta N^2 \text{ Hy,} \qquad (3.14)$$

where $\mu_0 = 4\pi \times 10^{-9}$ Hy/cm. The equivalent shunt capacitance is

$$C' = \frac{\epsilon_r \epsilon_0 2 \pi r b}{NS} \text{ Farads,} \qquad (3.15)$$

where ϵ_r is the dielectric constant of the layer insulation, $\epsilon_0 = 8.85 \times 10^{-14}$ F/cm (the permitivity of free space), S is the thickness of insulation between turns in cm, and Δ (the winding build) in cm is just $\Delta = N(t + s)$, where t is the conductor thickness.

The self-resonant frequency is proportional to the $L'C'$ product.

The characteristic impedance is $Z_0 = \sqrt{L/C}$, so $L' = Z_0^2 C'$. Z_0 may also be seen to be the load impedance seen by the winding. Taking S as our variable, the $L'C'$ product is smallest where

$$\frac{dL'}{dS} = -Z_0^2 \frac{dC'}{dS}$$

$$= \frac{d}{dS} \left[\tfrac{2}{3} \pi \mu_0 \, \frac{r}{b} \, N^3 (t + s) \right] = \tfrac{2}{3} \pi \mu_0 \, \frac{r}{b} \, N^3,$$

$$-Z_0^2 \frac{dC'}{dS} = -Z_0^2 \frac{d}{dS} \left(\frac{\epsilon_r \epsilon_0 2 \pi r b}{NS} \right) = \frac{Z_0^2 \epsilon_r \epsilon_0 2 \pi r b}{NS^3} \qquad .$$

So,

$$\frac{Z_0^2 \epsilon_r \epsilon_0 2\pi r b}{NS^2} = \frac{2}{3} \pi \mu_0 \frac{r}{b} N^3 \qquad (3.16)$$

$$\frac{3\epsilon_r \epsilon_0 b^2 Z_0^2}{\mu_0 N^4} = S^2$$

$$S_{\text{optimum}} = \frac{bZ_0}{N^2} \left(\frac{3\epsilon_r \epsilon_0}{\mu_0} \right)^{1/2} \qquad (3.17)$$

There is a lower self-resonant frequency involving the shunt capacitance and the inductance of the inductor itself at that frequency. In tape-wound cores, eddy current losses tend to short out the upper resonance. In ferrite core inductors this resonance may dominate the resonance discussed above.

Note that for this concept to be meaningful, we have to have some understanding of the impedance seen by the winding. The higher the load impedance, the higher the allowable leakage inductance and the less desirable the shunt capacitance. A flyback transformer may find itself disconnected from the driver at intervals, as would pulse width modulated inverter transformers. An RC snubber would be used to damp the oscillations, and here Z_0 would be related to R (preferably somewhat higher). This should facilitate the design of dissipative snubbers, the tradeoff between transformer parasitics and snubber power loss becomes apparent and can be dealt with in a predictable manner. This saves much "cut and try" time and permits rapid evolution of potential designs.

In some circuit design situations the apparent self-resonant frequency of an inductor can be extended by placing a much smaller inductor, usually on a ferrite core, in series with the main inductor. This also has the effect of filtering out the higher frequency components that could cause eddy current "losses" and attendant high frequency feed-through in the larger choke, allowing the use (sometimes) of less expensive core material in the larger choke. These eddy current losses (and thermal core loss as such) comprise the shunt resistance of Fig. 3–1 shown as R_c. It corresponds to the parallel resistance representing core loss in Chapter 1, Fig. 1–1.

8. SWINGING CHOKES

In many applications it is desired that only a certain minimum inductance be maintained in an inductor, and that this inductance minimum gets less as the current through the inductor increases. An example is input filter chokes in which only a certain critical inductance need be maintained for the choke to look like a choke-input filter over a range of operating currents. There are a number of methods of obtaining such performance. Core materials with a relatively large coercive force, such as grain-oriented silicon steel, behave acceptably as swinging chokes over a limited range. For higher frequency applications, molypermalloy powder cores are available from a number of manufacturers. High flux density powder cores are available from Magnetics, Inc.

Molypermalloy powder (MPP) cores are very useful in a variety of DC filter applications. They are toroids, with effective permeabilities from a few tens to a few hundreds. They have in effect a distributed gap, eliminating the proximity effect of a gapped inductor heating up turns near the gap, and they can tolerate DC coercive forces in the range of 10–100 Oersteds. Simple DC filter reactors may be fabricated with a minimum of design work. On the other hand, they are toroids, and as such their parasitic capacitance can not easily be controlled. Toroids almost invariably have higher feed-through capacitance than layer-wound structures on E or U cores. In high frequency converters, the shunt capacitance in the energy storage reactor must be charged and discharged each cycle, and external snubbers* are often necessary to damp the ringing associated with inductor parasitic capacitance. In most modern converters and inverters, chopping frequencies are in the range of 30–300 kHz, and the self-resonant frequencies of the associated inductors should be on the order of 1–30 mHz or higher. High frequency power processors will have abnormally high switching losses unless great care is taken to minimize the feed-through capacitance of power reactors. If one asks "How much capacitance is too much?" the answer is simply to set an allowable power loss limit, and then calculate $\frac{1}{2} CV^2 f$.

In toroidal inductors of a single layer, the dominant shunt capacitance is not turn-to-turn, but from the beginning turns to the core and thence to the ending turns.

The permalloy powder cores have a very thin insulating coating and the capacitance can be relatively high. If this type of core is to be used the designer would do well to add a layer of some dielectric tape over the core to increase the spacing from winding to core. If it is compatible with other processing requirements, Teflon, FEP, or polypropylene tapes would be preferred, as they have dielectric constants of about 2 With negligible AC core loss, non-adhesive-coated fiberglass cloth tape could be used, as the air

---

* Snubber: That which snubs or damps an overshoot or oscillation.

spaces between the fiberglass strands would give an effective dielectric constant of perhaps 1.5. Increasing the average turn-to-core spacing from a few mils to a few tens of mils can do wonders for the feed-through capacitance of such designs. For lowest feed-through capacitance, layer-wound coils of magnet wire or foil would be preferred, where the winding geometry can be closely controlled and the turns nearer the core provide shielding for those farther away from the core. The dominant feed-through capacitive component then becomes the layer-to-layer capacitance, and not only is thicker layer insulation used to reduce the capacitance, but higher core flux density is used to reduce the number of turns (and the copper loss!), thus leaving more of the core winding area available for insulation. It should not be considered unusual in high frequency magnetic component design, where capacitance is a critical problem, for the copper space factor to drop down to the 0.1–0.2 range.

For applications in which the range of inductance variation with current approximates linearity over a current range of more than about three to one, or in which molypermalloy powder toroids are not suitable, there are a number of other options, all of which amount to constructing a series set of reactors which saturate at different current levels. One can obtain an inductance which drops linearly (with a slope of –1) with increasing current, or slopes of –½ or –2 or whatever may also be obtained. Instead of actually winding separate reactors and putting them in series, we can use the same winding and let it encompass magnetic circuits with varying gaps. In this manner we obtain a structure where, at high currents, the portion of the core with little gap will saturate (giving a very low permeability for its portion of the core area), while the remaining portion of the core area (which has a greater gap) may not be saturated and provides the desired level of inductance.

A design procedure would be to design a choke for the lowest inductance needed at the highest DC current. For that number of turns, find the additional core area and gap for the higher inductance required at the lower current. Increase the width of the core to give the additional area and gap that new portion appropriately less. One continues in steps, depending on how smooth one wishes the inductance-versus-current characteristic to be. In the actual fabrication of the design, one could use C-cores with identical E, F, and G dimensions and different D dimensions, placing them side by side in the common coil. The inductor can be placed on a stack of E-I laminations with butt gaps or even a wedge-shaped piece of gap material in the gap region. Fig. 3–9 shows a stack of E-I laminations with gaps of several different thicknesses; Fig. 3–10 shows a stack of E-I laminations with a wedge in the gap.

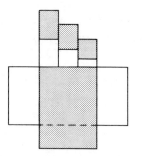

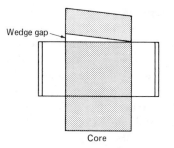

Fig. 3-9. Core with stepwise graded gap.   Fig. 3-10. Core with linearly graded gap.

In higher frequency applications where ferrites might be desirable because of their low core loss, some number of cores could be assembled in parallel with a common coil wound around all of them. Each core would have a different thickness of gap material (shown in Fig. 3-9 or 3-10). Another method is to take a standard ferrite core and grind the gap region to produce a nonuniform air space across the width of the gap. In the case of a pot core, a contour could be ground on the face of the center leg. In the case of ferrite U-cores, for convenience of assembly the outer edges of the cores would be left in contact or separated by the minimum gap, and the middle portion contour ground as shown in Fig. 3-11.

Calculation of performance of such nonlinearly gapped structures is relatively tedious. It consists of integrating the contribution of inductance for each increment of gap height across the width of the core face. In the case of a core such as a stack of laminations with two or three separate thicknesses of gap, it is a simple matter to calculate the inductance of each of the inductors and add them.

In the case of a stack of laminations with a wedge-shaped piece of gap material, one must once again perform an integration in order to calculate

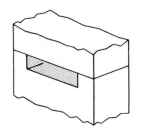

Fig. 3-11. Ferrite core with step-ground face.

the actual inductance. As current increases, progressively more of the core saturates and may be assumed to have permeability of one. The remaining unsaturated portions of the core, together with the total number of turns on the reactor, then gives the inductance remaining at that current level. This is oversimplified. The saturated permeability is more likely to be on the order of 5–20, and the flux from the larger portion of the gap will fringe out into the narrow gap regions where the material is saturated. The simplest method of obtaining a finished design for a swinging choke is to calculate what will do the job, test it, and modify the design accordingly. As one gains experience, the first design will come acceptably close to the design requirements more frequently.

Methods of machining ferrites to obtain nonuniform gaps are discussed in Chapter 9.

In inductors with relatively large gaps (comparable to the distance from the core to the first turn) the fringing flux from the gap has a component perpendicular to the winding conductor and therefore induces surface eddy currents, especially in foil-wound coils. These may be greatly reduced by bevelling the gap. The fringing flux can never be totally eliminated. It can be greatly reduced by using a bevel angle of 45° and a bevel depth of about 2–4 times the length of the gap. This is only an approximate rule of thumb—exact calculations are very difficult. Spacing the winding away from the core obviously helps, but just as obviously eats up valuable winding space.

Since the fringing flux occupies volume outside of the gap between core faces, it reduces the effective gap length by increasing the effective core area at the gap, and increases the inductance over that calculated from our formulas (which ignore fringing). One could choose a bevel angle, say 45° to start, and then increase the bevel depth until the measured inductance was equal to the calculated inductance. Then at least half of the fringing flux would be pulled into the gap. Increasing the bevel depth to obtain equal increments of decreasing inductance would give roughly equal increments of reduction in the fringing flux, and corresponding reductions in the eddy current losses due to this proximity effect. In critical situations one may have to initially design the inductance somewhat high to allow for the core face bevelling necessary to obtain very low levels of eddy current damping and losses. Many separate smaller (distributed) gaps are also used.

9. RESONANT CHARGING REACTORS

The design of resonant charging reactors (for energy storage capacitors, pulse-forming networks, and the like) is an application of inductors that deserves special consideration. The reactor contains both an AC flux compo-

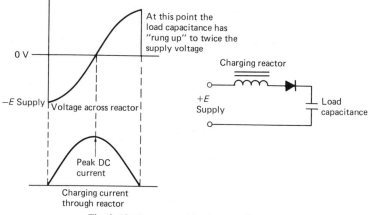

Fig. 3-12. Resonant charging waveforms.

nent due to the impressed voltage, and a DC flux component due to the DC current flowing through it. The designer must budget a portion of the core's flux capacity to handle each flux component. The division of flux between AC and DC components in this application is not obvious.

With reference to Fig. 3-12, assuming the capacitor initially discharged and $+E$ voltage applied instantaneously, we see that the voltage across the reactor reverses during the charging cycle, while the current is a unidirectional half-sine pulse.

The resonant charging time is one-half the resonant period of the charging reactor and the load capacitor, chosen by the designer to give a required charging time for the given load capacitance.

The charging period is given by $T = \pi\sqrt{LC}$. The angular frequency ω is of course

$$\omega = 1/\sqrt{LC}. \qquad (3.18)$$

The required inductor value is given by

$$L = \frac{T^2}{4\pi^2 C}, \qquad (3.19)$$

where T is the charging period. The charging current is given by $I = E/Z_0 \sin \omega t$, where $Z_0 = \sqrt{L/C}$, $\omega = 1\sqrt{LC}$.

The peak current is given by

$$I = \frac{E}{\omega L} e^{-\pi R/4L}, \qquad (3.20)$$

where the exponential term is seen to be a damping term reducing the peak current slightly, due to the resistance of the choke and the diode.

Assuming R is small, we will neglect the exponential term for the sake of simplicity.

$$I = \frac{E}{\omega L} \tag{3.20a}$$

Simplifying and substituting for ω,

$$I = E\sqrt{C/L}. \tag{3.21}$$

This form allows the designer to conveniently calculate I for circuit design purposes. For purposes of the derivation, we will use Eq. (3.20a) since we seek an expression independent of L or C explicitly.

The DC flux through the reactor reaches a peak at half the charging period, while the AC flux (the integral of reactor voltage over time) reaches a maximum at half the charging period and then decreases. During the first half of the charging period the AC and DC flux add, and the maximum flux density in the core is reached at half the charging period. The peak flux density due to the AC component may be found by integrating the voltage over half the charging period:

$$\int E \, dT = \int_0^{\pi/2} E_s \cos \omega T \, dT = E_s \sin \omega T \Big|_0^{\pi/2} = E_s, \tag{3.22}$$

which shows that the cosine waveform over $\pi/2$ second, E volts peak, has a volt-second product equal to E volts over one second. For a *charging time* of T seconds, the ET product is

$$\text{¼} ET. \tag{3.23}$$

From $10^8 ET = N\Phi$ (T now in *seconds*) and Eq. (3.19), the AC flux is $10^8 ET = \pi NAB_{AC}$.

Where E is the DC supply voltage, T is the full charging period, N is the number of turns on the reactor, B is the peak AC flux density in Gauss, and A is the net core area in *inches,* we have

$$4.94 \times 10^6 ET = NA B_{AC}, \tag{3.24a}$$

$$B_{AC} = \frac{4.94 \times 10^6 ET}{NA}. \tag{3.24b}$$

The DC flux density is given by $B_{DC} = 0.6\,NI/l_g$, where I is the peak current from Eq. (3.20). Substituting,

$$B_{DC} = \frac{0.6NE}{\omega L l_g} .$$

T, being one-half the resonant period (the full charging period), is just π/ω. We have

$$B_{DC} = \frac{0.6NTE}{\pi L l_g} . \tag{3.25}$$

We have a relationship connecting L and l_g:

$$L = \frac{3.2\,N^2 A}{l_g} \times 10^{-8}$$

$$L l_g = 3.2\,N^2 A \times 10^{-8}. \tag{3.26}$$

Substituting for the $L l_g$ product in Eq. (3.25),

$$B_{DC} = \frac{5.97ET \times 10^6}{NA} . \tag{3.27}$$

The AC and DC flux densities may be combined, using Eqs. (3.20b) and (3.27):

$$B_{DC} + B_{AC} = \frac{5.97ET \times 10^6}{NA} + \frac{4.94 \times 10^6 ET}{NA}$$

$$= B_{max} = \frac{ET}{NA} (5.97 \times 10^6 + 4.94 \times 10^6)$$

$$B_{max} = 1.091 \times 10^7 \frac{ET}{NA} , \tag{3.28}$$

which shows that the DC flux density contributes about 55% and the AC flux density about 45%. It is now apparent that the designer does not, *can not* partition so much of the flux for AC and so much for DC components. This type of circuit forces the AC and DC flux components in the proportions shown in Eq. (3.28).

The design of charging reactors is now greatly simplified. Let us say that we have a supply voltage of 11,500 volts, a charging time of 714 microseconds, and we wish a maximum flux density of 15,000 Gauss. Then

$$15,000 = \frac{1.091 \times 10^7 \times 11,500 \times 714 \times 10^{-6}}{NA}$$

$$NA = 5983.$$

Whatever core we pick, we must have that product of turns and core area.

As far as an exact design goes, we still do not have an exact core picked out, we have not mentioned a (user-specified) winding resistance, and surely there must be something here to optimize.

Rearranging Eq. (3.28),

$$DE = A = \frac{1.091 \times 10^7 ET}{NB_{max}}. \tag{3.29}$$

For some specified DC resistance R,

$$R = \frac{\varrho l_w}{A_w} = \frac{\varrho NU}{A_w}.$$

Since $A_w = \sigma FG/N$, where σ is the winding space utilization factor, 0.2–0.5 for single winding reactors (or about half that for dual winding charging reactors used in energy scavenging circuits),

$$R = \frac{\varrho N^2 U}{\sigma FG}. \tag{3.30}$$

Rearranging,

$$\frac{R\sigma FG}{\varrho U} = N^2. \tag{3.31}$$

Substituting for N in Eq. (3.28),

$$N^2 = \frac{1.19 \times 10^{14} E^2 T^2}{B^2 D^2 E^2} = \frac{R\sigma FG}{\varrho U}. \tag{3.32}$$

Rearranging,

$$\frac{D^2E^2FG}{U} = \frac{1.19 \times 10^{14}E^2T^2\varrho}{B^2R\sigma} \ . \tag{3.33}$$

For this application we discover that our geometric figure of merit is identical to that for inductors and transformers. In order to obtain a realizable design one would take our circuit-defined parameters, select a B_{max}, and plug them into Eq. (3.12) to obtain the required value of D^2E^2FG/U. Bear in mind that B_{max} can be somewhat on the high side, as the AC core loss is that due to 45% of B_{max}, albeit at the high end of the loop closer to saturation. Then, determine the construction type (I, II, or III) and consult the computer runs for a value of D^2E^2FG/U for a suitable form factor. That D^2E^2FG/U value is for a *normalized* core, just as before for transformers and chokes. One then forms the ratio of D^2E^2FG/U for the normalized core and D^2E^2FG/U for the actual requirement.

The ratio obtained is K^5, the denormalization constant. K is then obtained, and each of the dimensions of the normalized core is multiplied by K. That gives the dimensions of the denormalized core for the actual application.

We now have a core area, and using Eq. (3.28) we obtain turns. The inductance follows from Eq. (3.19), and the gap follows from those data and Eq. (3.26).

We have seen in this chapter some very interesting conclusions. There is an optimization criterion for inductors which have a flux density derived from DC operating conditions. When both AC and DC operating conditions are combined, as in resonant charging reactors, we have the same optimization criterion. For transformers where winding loss only is considered, we have the same optimization criterion. All linear magnetic components fall into one of these three classes (even pulse transformers!). It would therefore appear that the power-handling ability at constant resistive loss can be improved for any such component by choosing core dimensions which maximize the D^2E^2FG/U figure of merit within the volume given for that component. At operating voltages where dielectric stress is significant, or at frequencies where capacitance, leakage inductance or core properties (permeability, core loss, interlaminar voltage stress) must be considered, it will become necessary to deviate from this criterion. It is nonetheless an extremely useful concept.

4
NONLINEAR MAGNETICS

1. INTRODUCTION

You may recall that in a conventional transformer some relationship between turns, voltage, core area, and frequency determined the flux density in the core. That flux density was either set as high as possible or chosen to give some desired level of core loss.

We can no longer deal with these concepts on such a simplistic level. As far as the transformer is concerned, the applied voltage across some number of turns reduces to an awareness of volts per turn. This causes a magnetic field (flux) which increases at a rate depending on the magnitude of the voltage per turn around the core. Another way of saying this is that a space rate of change of an electric field can give rise to a time rate of change of a magnetic field. This observation forms part of the basis of Maxwell's equations, the most profound advance of physics in the nineteenth century. A time rate of change of a magnetic field can also cause a space rate of change of an electric field, which is why one winding can induce a voltage in another.

The lower the frequency, the more time one-half cycle will take, and the more time there is for the flux to increase.

The flux is the total flux inside the area of the coil. The smaller that area, the greater the flux *density*. It will appear that a current is required to drive the core through a flux excursion. In fact, it is the volt-second product developed across the winding *by* that current which causes a change in flux, and this relates back directly to Maxwell's equations. The current required to cause this is simply a consequence of the properties of the empty space or the material filling that space.

We see now that we can look on a coil of wire wound around a core as a *volt-second integrator*. This is true whether we have one or more windings, and regardless of the circuit function. The product of voltage impressed and elapsed time (or the integral of voltage over time, more exactly) is stored in the core as the flux density at the end of that period of elapsed time. Depending on the application, or the design approach taken to that application, it may be more useful to approach the design in terms of voltage, time, and total integrated flux. The end result will be the same. The difference is one of viewpoint taken on a problem for the purpose of solving that problem.

No magnetic material has an infinite capacity for information storage, and this implies that sooner or later our core material will run out of flux capacity. When this happens, the material is said to saturate. Saturation is seen as a gradual onset of the condition, or sometimes as a relatively abrupt transition. At or beyond this point, the continued presence of a voltage across the winding gives no further increase in the flux *stored* in the core. In point of fact, the flux in the core does increase, but only at 10^{-3}-10^{-6} as rapidly as before, as the core in saturation has a permeability which is approaching one. When the impressed voltage is removed, the flux density in the core falls back to a residual value. This is shown in Fig. 4-1 for both square-loop and round-loop materials. For fairly decent square loop materials $B_{residual}/B_{max}$ would be above 0.90, approaching 1.0. For round loop materials $B_{residual}/B_{max}$ would be perhaps on the order of 0.7.

For the material of Fig. 4–1(a), this is similar to introducing a gap into the magnetic circuit. Note that the material is still "square loop material" and the loop width is still the same, but the loop has been "sheared over," giving a much lower $B_{residual}/B_{max}$ while retaining the loop width and linearity of the square loop material.

To see the effect of core saturation on a transformer we can look at the equivalent circuit of a transformer. (See Fig. 4–2). Before the core saturates, the impedance of L_p is usually much higher than R_L or any of the series impedances, so very little of the current from the source is diverted. Most is delivered to the load.

When the core saturates, the impedance of L_p becomes much less than R_L, preferably even much less than the series impedances. Under these conditions most of the current from the source is diverted by $L_{p\,sat}$ and very little flows thorough the load.

2. BASIC APPLICATIONS

Figure 4–2 applies to a transformer (or inductor) in which the winding is across the lines carrying current to and from the load. When saturable magnetic devices are used to control the flow of power from a source to a

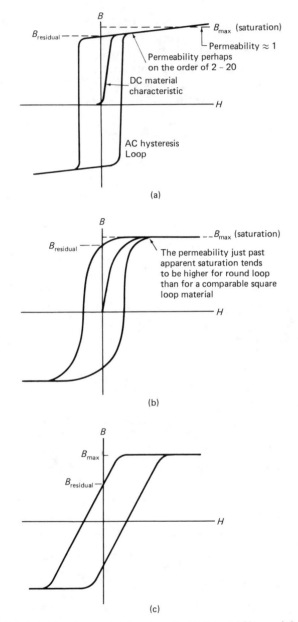

Fig. 4-1. (a) Square hysteresis loop. (b) Round hysteresis loop. (c) Hysteresis loop of a square material with a significant air gap.

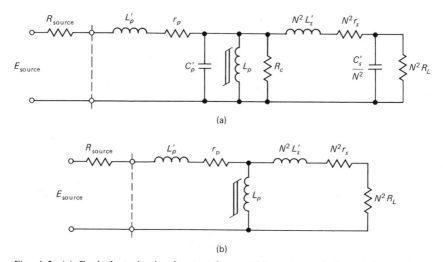

(a)

(b)

Fig. 4-2. (a) Equivalent circuit of a transformer. L_p' = primary leakage inductance; C_p' = primary equivalent shunt capacitance; r_p = primary winding resistance; L_p = primary inductance (note the slightly different symbol for the core; the slanting lines on each end of the core symbol indicate a core in which saturation has something to do with circuit function; R_c = the equivalent parallel resistance of the core loss phenomenon; $N^2 = N_{pri}/N_{sec}$; $N^2 L_s'$ = the seconding winding leakage inductance as seen from the primary side; $N^2 r_s'$, C_s'/N^2 are similar to $N^2 L_s'$; $N^2 R_L$ = the secondary load resistance as seen from the primary side. (b) simplified equivalent circuit.

load, it is customary (for efficiency purposes) to place the saturable core in series with source and load. In this case we have the equivalent circuit of Fig. 4-3.

In both suppressed second harmonic and proportional phase control circuits, the power flows from the reactors to the load. In a suppressed second harmonic circuit, one core is always unsaturated and the other saturated (as

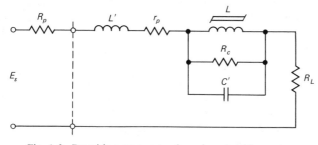

Fig. 4-3. Parasitic components of a series saturable reactor.

long as the circuit is in its linear region), but the output phase is shifted in single-phase systems by the reactors driving current back into the source for a portion of each cycle. High frequency inverters driving such regulators must be capable of handling bidirectional power flow.

In proportional phase control circuits, one core is saturated; the other core is unsaturated at first and then saturates. Then power flows from source to load for the remainder of that half-cycle.

In the suppressed second harmonic circuits, the reactors take turns being out of saturation, acting as DC-to-AC current transformers, and constant AC current control elements.

In proportional phase control circuits, the reactors act as switches. In both cases the saturated inductance L' appears in series with source and load. The effects of L' can be far more deadly than that of R_p, with its simple resistive loss and consequent temperature rise. L' is not only the leakage inductance of the winding due to its nonzero radial build over the bare core dimensions; it is the inductance of the coil without a core (the "air core" inductance) multiplied by a number ranging from 1 (with luck) to perhaps 10 (the saturated permeability of the core). At load or control currents sufficient to give perhaps 100 times the loop width in Oersteds (manufacturer's data B-H loop plots), almost any material will have a saturated permeability approaching one. However, at a mere ten times overdrive, the nickel alloys begin to come close to 1; some ferrites may also, but 4 mil and 12 mil Z-type silicon steel may have effective permeability on the order of 5. This can be exceedingly inconvenient. The saturation inductance drops the supply voltage at high load currents and limits the maximum power transfer. The solution to having a design which is close, but has too much saturation inductance, is to go to the squarest possible core material affordable. Then raise the core area and reduce the number of turns as much as possible. Since the inductance goes down as the square of the turns, one need only hold the product of turns and core area constant. The overall effect will be that the saturation inductance goes down roughly linearly with decreasing turns. In many cases it also helps to make the minimum and maximum magnetic circuit lengths close, so the entire core saturates more abruptly.

Carried to extremes, this can lead to ridiculous geometries. One must maintain a sense of proportion.

As we reduce the number of turns we also reduce the unsaturated inductance of the saturable reactor. That controls the amount of power which leaks from the reactors in their off state and appears across the load when the control circuitry says "deliver zero power to the load." The usual procedure to trim in an existing design with excessive saturation inductance is to reduce the number of turns to get the desired saturation inductance. (On C-core designs the air core inductance can be halved by putting half the turns on each leg of the C-core, since the coils become decoupled as the permeability drops.) Then check the off-state inductance. On toroids a particular material or permeability may have to be specified, or the strip width raised while reducing the inside diameter and strip build. On silicon steel C-cores one would raise the off-state inductance by very carefully lapping the core faces on a sheet of #400 silicon carbide paper on a surface plate or a sheet of heavy plate glass, then clean the faces with solvent. Nickel alloy C-cores are much softer, more difficult to lap, and may require their faces to be etched. Such processes are best left to the manufacturer. Bond with an epoxy adhesive which is free of fillers or pigments, clamping tightly or passing a current through the coil until cured. A smooth radius on the edges of the core faces rather than a square edge or 45° chamfer will increase the bond strength significantly. With these techniques one can achieve a control range of, for example, 10 μH to 600 mH for a 10 kW, 400 Hz control reactor using 4 mil Z-type silicon steel. At an on/off ratio of 60,000 : 1 this would be doing very well indeed.

3. "MAGNETIC AMPLIFIERS"

Saturable inductive components are capable of performing many tasks no other component can, with great simplicity and reliability. A few examples will be discussed. The first (the main control element in a power processing system) was the most popular some years ago, but has unjustly fallen out of favor because of a lack of current awareness and the advent of high speed, high power switching transistors. Semiconductor power control circuits are usually smaller, lighter, more complex, more fragile, and less reliable. It is occasionally useful to have magnetics as an option for trading off these parameters.

The first basic type of "magnetic amplifier" or saturable reactor control circuit gives an output waveform theoretically identical to that obtained by a triac or SCR used in a proportional phase control circuit. Referring to Fig. 4-4, the "firing angle" of the control elements is governed by the control current. In the most common application of this configuration, both reference and control ampere-turns are much larger than the ampere-turns (Oersteds) required to actually swing the cores over their dynamic control range. This gives *gain* in the ratio of the actual control or reference current to the control range, since only a very small fraction of the control current at threshold (where control and reference currents differ by the control range) is needed to control.

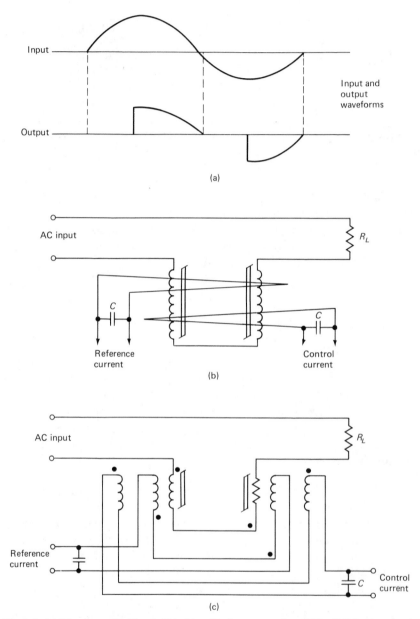

Fig. 4-4. (a) Waveforms for Fig. 4-4(b); (b) stylized representation of Fig. 4-4(c); (c) magnetic amplifier circuit; (d) *BH* loop and H_L curves for square permalloy (SOURCE: "Magnetic Metals Tape Wound Core Design Manual and Catalog," Bulletin C1 (4/15/68), pp. 32, 36).

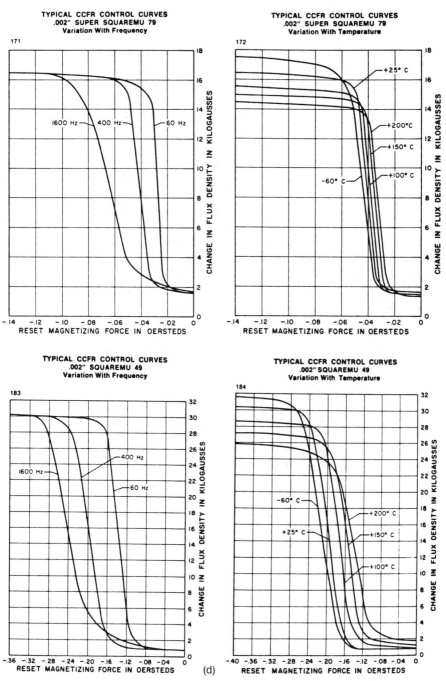

Fig. 4-4. (continued)

Figure 4–4(b) is the customary stylized representation of this configuration: two toroids are wound with their AC windings, then placed one on top of the other and the DC windings wound over both. It is electronically equivalent to Fig. 4–4(c), with the separate windings and phasings shown for clarity. The advantage of a DC winding over both cores is that in the case of Fig. 4–4(c), each DC winding on each core develops a very high induced voltage. The voltages cancel because of the phasing, but each winding has a rather high voltage across it, giving insulation problems and necessitating some attention to winding capacitance. With a DC control winding going through both cores each turn, the induced AC voltages cancel on a turn-by-turn basis, eliminating the above problems. This geometry also gives less leakage inductance between AC and DC windings, which is not so important for the phase control configuration. It is more important for the suppressed second harmonic configurations. (Apropos of which, note the capacitors on the DC windings. They define a low impedance source, so the AC second harmonic currents are free to circulate within the DC control windings.) In the suppressed second harmonic configuration, the DC control windings are fed from a high impedance current source (a series reactor, usually) and second harmonic currents essentially cannot circulate. A second harmonic voltage (for single-phase circuits), or a sixth harmonic voltage (for three-phase circuits) therefore appears across the reactors. Since this configuration functions as a DC-to-AC current transformer, the output voltage waveforms to the load are, surprisingly enough, rectangular waveforms. The rise and fall times are related to the switching speeds of the saturable reactor cores, and in high frequency applications *are critically dependent* on the shunt capacitance in the DC control winding and the isolation reactor. Fig. 4–5 shows the elementary configuration of a single phase control circuit.

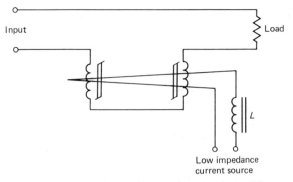

Fig. 4–5. Suppressed second harmonic magnetic amplifier.

The impedance of the inductor of Fig. 4-5 or the capacitors of Fig. 4-4 are of considerable significance. If the inductor of Fig. 4-5 has an impedance of less than twice the maximum load impedance, reflected through the square of the turns ratio of the reactor, the circuit of Fig. 4-5 loses its current transformation characteristic and tends to become a proportional phase control circuit. A similar transition in the circuit of Fig. 4-4 takes place if the source impedance of the DC control windings is allowed to become too high due to insufficient capacitance.

4. THE THREE-PHASE SUPPRESSED SECOND HARMONIC SATURABLE REACTOR POWER CONTROLLER

The three-phase applications of Fig. 4-5 deserve special consideration. First, let us consider a Δ-Δ system, with a single three-phase transformer and the saturable reactors in series with each leg. (See Fig. 4-6.)

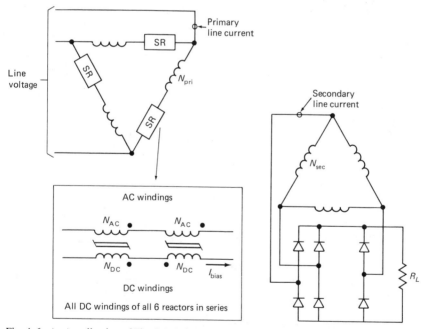

Fig. 4-6. Δ - Δ realization of Fig. 4-5. Primary current $= I_{bias} N_{DC}/N_{AC}$, full 180° conduction. Secondary leg current $= I_{pri}N_{pri}/N_{sec}$, full 180° conduction. Output DC current $= I_{pri}N_{pri}/N_{SEC} \times 2$. Output DC voltage (neglecting IR drops and diode drops) $= E_{pri(pk)}N_{sec}/N_{pri}$. Diode current $= I_{0\,DC}$, 120° conduction.

Ripple voltage across the choke in series with the bias current supply feeding the bias windings is triangular, at six times line frequency and approximately $E_{AC(line)}N_{DC}N_{AC}$.

Now let us consider the situation identical to Fig. 4–7 except that the secondary is wye instead of Δ. This is a Δ-Y system (Fig. 4–8 and 4–9). Many configurations are possible. For the rectifier transformer a single three-phase transformer may be used, with Y or Δ secondaries, or both. In the event that both Y and Δ secondaries are used they should be separately rectified and the individual DC outputs connected in series. Two single three-phase rectifier transformers may be used, one with Y and one with Δ secondaries, the primaries in parallel and the secondaries handled as above. Three single-phase transformers may be used, making three single-phase systems.

The reactors can be inside the primary Δ, or in the line with either a Δ or Y primary. They all work *except* three single-phase transformers with secondaries connected Y. Looking from secondary neutral-to-line, that gives three in-phase signals at three times line frequency, so there is nothing line-to-line to rectify. In all cases where reactors are in the line, line current is 120° conduction with 60° of zero current. With reactors inside a primary Δ, transformer primary conduction is 180°, giving 120° conduction in the line.

Not everything about these configurations has been reduced to a cut-and-dried set of formulas. (That doesn't prevent us from using the data to deliver

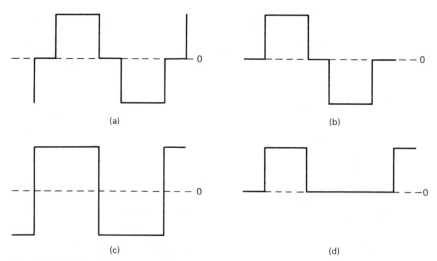

(a)   (b)

(c)   (d)

Fig. 4–7. Waveforms associated with system of Fig. 4–6. (a) Primary line current; (b) transformer primary and secondary winding voltage; (c) transformer primary and secondary current; (d) diode current.

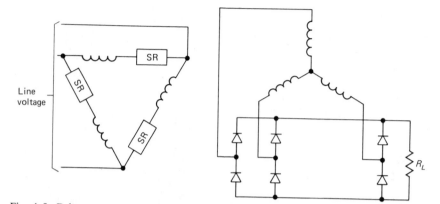

Fig. 4-8. Delta-wye system. Primary current = $I_{bias} N_{DC}/N_{AC}$; full 180° conduction; Secondary current = $(I_{pri} N_{pri}/N_{sec}) \times \frac{2}{3}$, for 60°, $(I_{pri} N_{pri}/N_{sec}) \times \frac{4}{3}$ for 60°; $(I_{pri} N_{pri}/N_{sec}) \times \frac{2}{3}$ for 60°; Output DC current = $(I_{pri} N_{pri}/N_{sec})$, $\times \frac{4}{3}$; RMS secondary winding current = $1.225 I_{out\ DC}$; RMS diode current = $0.612 I_{out\ DC}$; Output DC voltage = $(E_{pri(pk)} N_{sec}/N_{pri}) \times 1.5$.

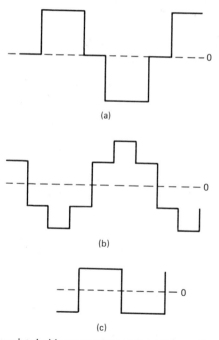

Fig. 4-9. Waveforms associated with system of Fig. 4-8. (a) Primary line current and secondary line voltage (average two bridge AC inputs); (b) Primary winding voltage, secondary line current, and secondary line voltage; (c) Primary winding current.

a product!) Consider the case of reactors in the line, a three-phase, four-wire Y source, and a three-phase transformer with a Y primary. If the neutrals are tied together, a square wave current at three times the line frequency flows from neutral to neutral. Saturable reactor conduction goes to 180° —without changing secondary currents or load currents! This was a situation the original investigator did not pursue, and which I have not had time to. All of my applications to date have been 400 Hz and I find Δ primaries more attractive, so I have not needed to pursue the question.

When one three-phase core is used for the rectifier transformer with saturable reactors inside the primary Δ, the primary currents are the same for Y or Δ secondary, but the primary voltages are different (Fig. 4-10). The wave form with a Y secondary has a better fit inside the line voltage sine wave envelope, allowing a closer approach to the load voltage obtained with full line voltage on the transformer. One obtains ~ 72% with a Δ secondary, and ~ 83% of peak with a Y secondary. Three transformers, making three single-phase systems with outputs combined at the DC buss, gives 180° conduction (and separate bias supplies, giving three times the bias current), but 180° primary voltage (square wave) which fits a sine wave rather poorly, so one obtains only ~ 60% of peak line voltage as DC output.

The basic working relationships are these: Assume all windings, both transformer and saturable reactor, are 1 : 1, with a Δ primary and saturable reactors inside the Δ. With a three-phase bridge on a Δ secondary, load cur-

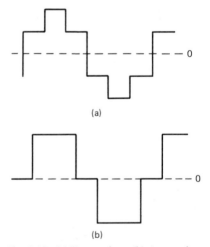

(a)

(b)

Fig. 4-10. (a) Y-secondary; (b) Δ-secondary.

rent is twice bias current and load voltage is peak primary voltage. Using a Y secondary, three-phase bridge load current is ⅔ the bias current and load voltage is 1.5 times peak primary voltage. For both Δ and Y secondaries and equal output voltages, Δ leg turns should be 1.5 times Y leg-to-neutral turns (note that $\sqrt{3}$ has no place here!). There are no off-phase relationships going on, so that with Y and Δ secondaries together you still get 6 times line frequency ripple, not 12 times.

Bias ripple voltage (which appears across the choke) for 1 : 1 reactors is about one-half line voltage at six times line frequency.

Regulation of current (as load varies from minimum to maximum to shorted) is a function of the quality of the iron in the choke. Ordinarily grain-oriented silicon steel can give 25% or worse regulation, whereas with Z-type silicon steel regulation may approach or exceed 1%. So, high-μ square loop material is preferred, especially for open-loop current source applications.

One interesting point here—since both saturable reactors in each pair share the voltage equally, one would think that each reactor need only be designed for one-half the AC line voltage. If this is done and the load is abruptly shorted, a fault current flows for about one-fourth to one-half cycle. Since semiconductors blow before most fuses or circuit breakers, this is an undesirable condition. If the windings on the reactors are designed for three-fourths of the line voltage each, this condition seems to vanish and there appears to be no fault current when the load is shorted in a three-phase, sine-wave excited system. In square-wave excited systems, I would expect a fault current for a portion of the cycle unless the reactors were designed for full voltage. In any case, the reactors rebalance themselves after one cycle.

5. OTHER APPLICATIONS

Another application of saturable magnetics which comes to mind is coupling a trigger pulse into a high energy discharge circuit. The specific application we will discuss is a series injection trigger transformer for a flash lamp.

Most flash lamp triggering systems (Fig. 4–11) use series injection (instead of parallel) because the triggering energy is more reliably injected *into the flash lamp,* and the trigger transformer itself can perform a pulse-shaping function for the flash lamp pulse current. Also, shunt triggering of flash lamps is usually accomplished with a wire on the flash lamp exterior and the attendant capacitive coupling. It has been found that this causes crazing of the flash lamp glass envelope and premature lamp failure.

At t_1, the trigger circuitry generates a high voltage pulse (typically

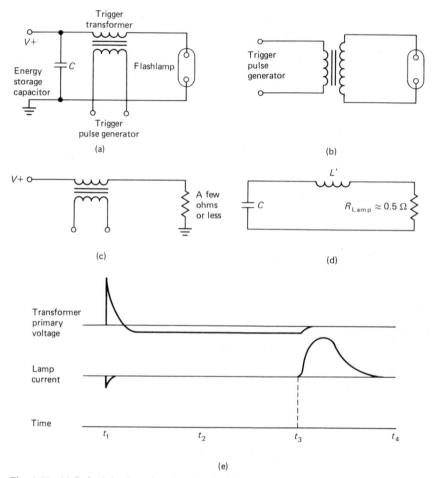

Fig. 4-11. (a) Series injection triggering of a flashlamp; (b) equivalent circuit for trigger pulse $t = t_1$; (c) equivalent circuit after trigger pulse, $t = t_2$; (d) equivalent circuit after transformer saturation, $t \geq t_3$; (e) simplified timing diagram.

10–30kV on the secondary) such that the pulse polarity on the lamp is the same as the supply polarity. The lamp then breaks down, and a voltage opposite to the trigger pulse polarity and equal to the supply voltage appears across the secondary. This resets the core to $-B_{max}$, taking placing during t_2. See Fig. 4-12. The flash lamp does not deionize during this brief period, since during this time a current (determined by the secondary inductance and the volt-second support of the core) flows. At the end of t_2 the core is at

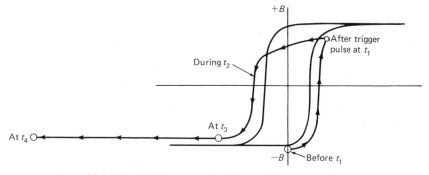

Fig. 4-12. *B-H* loop excursion of core during t_1-t_4 cycle.

$-B_{max}$, saturates, and a current [determined by the capacitor, the lamp impedance (~ 0.5 Ω), and the air-core inductance of the secondary winding] flows during $t_3 - t_4$. The permeability of the core goes to essentially unity as it is normally driven to usually more than 1000 Oersteds. At the end of the discharge cycle the core is properly reset for another trigger pulse.

The high energy discharge pulse from the secondary is isolated from the primary because the core is saturated (and presumably the primary is not physically too close to the secondary).

The basic idea is that one can couple a signal from one circuit to another, and the action of the core saturating decouples the two circuits so some other action may take place in one circuit without then being coupled to the other.

Another use is in pulse sharpening circuits, where one has a relatively slow rising pulse and is willing to accept some delay and lose some of the pulse width in return for sharpening the leading edge. Since the core is a volt-second integrator, it can offer a high impedance for some number of volt-seconds and a low impedance thereafter. If the integrated area of the (slow) rise time of the pulse is set roughly equal to the volt-second support of the saturable reactor, then one has the beginnings of a viable design. The sharpened rise time depends on the switching speed of the core, which in turn depends on the core diameter (the smaller, the faster, all other things equal); the ID/OD ratio (the closer to unity the faster, all other things equal); the drive level (load current times turns gives a multiple of that required to saturate the core; the higher the multiple, the faster); the alloy tape thickness, switching speed being on the order of the skin depth for unlimited available overdrive; and the material properties (some core materials such as the high nickel alloys switch faster than others). Since in the unsaturated state the reactor has a less-than-infinite inductance, some current will flow through it before it saturates. This will give a pedestal before the leading edge of the output pulse, as shown in Fig. 4–13. In most applications, this is not objectionable. In high current SCR modulators it is *desirable,* as priming the SCR with a low current before the main load current flows through it allows the SCR to turn on more fully, giving greatly reduced switching losses in most types of SCRs.

In some circuits a square wave input is desired. From that, a pulse-width modulated output signal is generated, either for use as a variable duty cycle control signal or for power regulation purposes with a saturable reactor as the pass element. The basic circuit would be that of Fig. 4–4 or one of its many variations, such as the self-saturating configuration of Fig. 4–14. The reset current in that example is supplied by a feedback amplifier and the supply voltage is held off by the volt-second support of the core (controlled by the DC control circuit) until the core saturates. Current is then passed to the

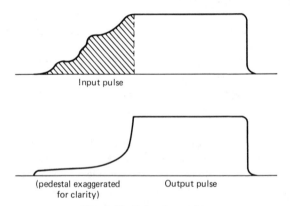

Input pulse

(pedestal exaggerated   Output pulse
for clarity)

Fig. 4–13. Pulse sharpening.

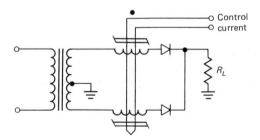

Fig. 4–14. A magnetic amplifier power control circuit.

load. In higher frequency circuitry this would be most common. The core material would be likely to be 1 mil low nickel or high nickel alloy, or possibly one-half mil or even one-fourth mil if frequency and switching speed requirements dictated it.

As a particular case of our first example, let us say that we specifically wished to limit the volt-second product or charge coupled from one circuit to another, or that we wished to integrate the areas of a number of pulses in a pulse train to form a duty-cycle limiter. It might be that a saturable reactor or transformer could perform that function more expediently than some other method. It depends on the specific application, but should be kept in mind as an often viable option.

6. MATERIAL PROPERTIES

The max/min ratio of an unsaturated/saturated core gives a max/min inductance ratio which is a vital parameter in determining the performance of the finished product. The properties of the core material have a great deal to do with the maximum inductance and the attainable ratio. Fig. 4–15 shows a few interesting hysteresis loops which show some typical phenomena, and are characteristic of most materials. Certain proportions are exaggerated for clarity.

The value of μ at some point between $-B_{max}$ and $+B_{max}$ is called the *incremental permeability* (denoted by μ_Δ, which is the slope of the B-H curve at that point). Recall the discussion of that point in Chapter 3. More rounded loop materials or gapped materials would have lower permeabilities and

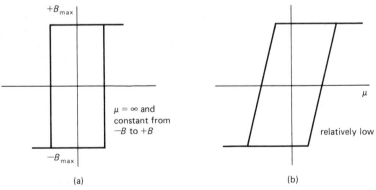

Fig. 4–15. Hysteresis plots. (a) Ideal square loop material; (b) effect of air gap on (a) material; (c) typical round loop materials; (d) some square loop materials.

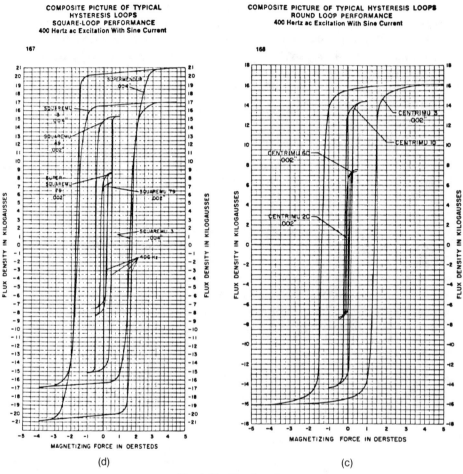

(d)   (c)

Fig. 4–15. (Continued)

SOURCE: Magnetic Metals Company, Tape Wound Core Design Manual and Catalog, Bulletin C1 (4/15/68), p. 24.

hence a lower max/min ratio (assuming the core was driven in all cases to $\mu \approx 1$) than cores with less or no gap, or some square loop materials. The permeability comparison curves in Fig. 4–15 (c,d) show the behavior of some typical materials. The magnetizing current waveforms of pulse transformers using the core materials of Fig. 4–15(a–d) points out some interesting differences between them.

For the core material of Fig. 4–15(a) the permeability is infinite, hence the inductance is infinite and the pulse magnetization current (given by: $ET/L = I$) is zero.

For the core material of Fig. 4–15(b) the permeability μ_Δ is constant and finite until saturation. Hence the inductance is constant. The pulse magnetization current rises linearly with time to saturation, and then rises abruptly to a level determined by external circuit impedance levels (Fig. 4–16).

For the core materials of Fig. 4–15(c) the permeability near $-B$ or $+B$ is much lower than around mid-range, hence the inductance will be lower in those areas and the pulse magnetization current higher than the average linear value (Fig. 4–17).

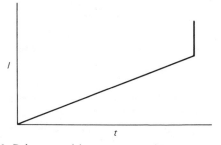

Fig. 4–16. Pulse magnetizing current waveform of Fig. 4.15(b).

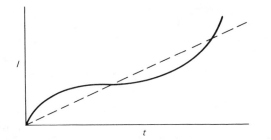

Fig. 4–17. Pulse magnetizing current waveform of Fig. 4.15(c).

The core materials of Fig. 4–15(d) are representative of some square loop 50 nickel and most square loop 80 nickel alloys in that the incremental permeability rises on the average linearly with B. This gives an inductance which increases roughly linearly with time, and hence a roughly rectangular magnetization current pulse. The observed range of wave forms for this class of materials is shown in Fig. 4–18.

With some square loop high nickel alloys the peak-to-valley ratio of Fig. 4–18(b) may be somewhat more than 2 : 1.

When one is concerned about the droop on a pulse support capacitor because of the pulse current drawn by such a load, the more rectangular current pulse wave form will give a more linear (average) droop on the pulse support capacitor. It is much easier to cancel out the pulse droop with other simple circuitry when the perturbations are linear instead of wildly nonlinear.

Not all materials are specified such that one can in so many words predict *how* round or *how* square one "round" or "square" loop material is compared to another, but there are several useful published parameters. First, look and see whether the manufacturer says anything like "round loop material" or "square loop material" or just says "our material." Next, look for a B_r/B_m ratio, i.e., the ratio of the residual flux density (when the core has been driven to saturation and then the drive reduced to zero) to the maximum flux density. This is a measure of how flat the top of the loop is. Square loop materials should be expected to be above 0.90; round loop materials are considerably less than that.

Next, look for a μ versus B set of curves, such as that in Fig. 4–19 (reproduced courtesy of Magnetic Metals). Note that some of the materials show a much wider excursion of incremental permeability than others. Comparing μ at 1000 Gauss with μ_{max}, one would expect that Super Square 80,

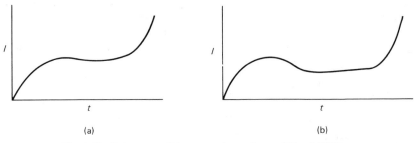

(a)   (b)

Fig. 4–18. Pulse magnetizing current waveform of Fig. 4.15(d).

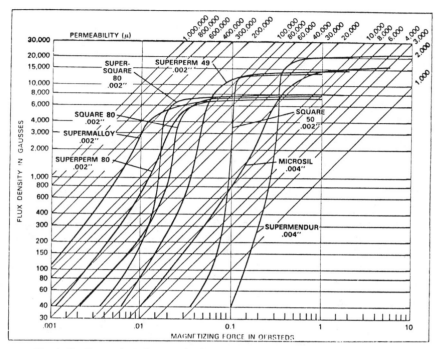

Fig. 4-19. Composite DC magnetization curves (from Magnetic Metals data book).

Square 80, Supermalloy, Square 50, and Supermendur would have pulse magnetizing current wave forms somewhat more rectangular than the others. In fact, under pulse conditions the magnetizing current waveform will correspond to the hysteresis loop of Fig. 4-15(d). The actual material from a particular manufacturer will vary somewhat from lot to lot, and different manufacturers' "equivalent" materials will also vary in this respect. If this characteristic is critical each manufacturer will have to be qualified on a lot-by-lot basis. If the core is a small toroid it may be necessary to go to a tape-wound bobbin core made by a specialist.

In any case, do go to several manufacturers and discuss your needs. They know what they can do with their materials and annealing cycle variations, and most can accommodate special requirements.

Referring back to Fig. 4-3, we see that we can now draw equivalent circuits for the saturated and unsaturated states of a saturable reactor in a series circuit. Consider the feed-through capacitance as being *distributed* across L'. Then refer to Fig. 4-20.

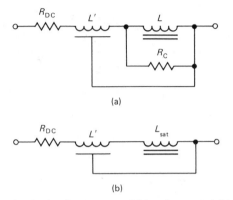

(a)

(b)

Fig. 4-20. Equivalent circuits for the unsaturated (a) and saturated (b) states of a saturable reactor in a series circuit. (a) R_{DC} is the winding DC resistance, C is the equivalent shunt capacitance of the winding, L is the inductance, and R_C is the core loss. (b) Here L' is the winding leakage inductance, and L_{sat} is the saturated inductance of the coil due to the presence of the saturated core. R_{DC} is the winding resistance.

7. USEFUL RELATIONSHIPS

There are some useful design equations we shall review. $100ET = N\Phi$ gives the flux and turns required to support a rectangular pulse of E volts for T microseconds. Note that Φ is the *full* flux capacity from $-B_{max}$ to $+B_{max}$. This relationship is also useful for square wave AC excitation, where ET is one-half cycle of the impressed AC waveform. In the event that the impressed waveform is a sine wave, multiply the RMS value by 1.11 to obtain the average value, which is the amplitude of a rectangular waveform of equal volt-seconds. (RMS stands for *root mean square,* which is the effective *heating* value of the waveform. The RMS current is the square root of the product of the peak current and the average current.)

The flux capacity of a core in Maxwells, herein called Φ, is the product of the change in flux density in Gauss and the *net* core area in square centimeters. A core which saturates at 8000 Gauss may swing from -8000 to $+8000$ Gauss. If its area were one square centimeter, it would have a flux capacity of 16,000 Maxwells.

The coercive force in Oersteds resulting from a certain number of ampere-turns on a core is given by $0.4\,\pi NI = Hl$, where l is the mean magnetic path length of the core in cm, I in amperes, and N is in turns. A table of milliampere-turns required to give one Oersted of drive on various core sizes follows (Table 4-1), along with graphs of switching speed versus drive level

Table 4-1. Milliampere-turns for ¼, ½, and 1 Oersted, Based on Bobbin Groove Diameter.

ID, INCHES	¼ OERSTED	½ OERSTED	1 OERSTED
250	397	794	1588
313	497	994	1988
375	595	1191	2381
438	695	1391	2781
500	794	1588	3175
563	894	1788	3575
625	992	1984	3969
687	1091	2181	4362
750	1191	2381	4763
875	1387	2778	5556
1000	1588	3175	6350
1125	1786	3572	7144
1250	1984	3969	7938

NOTE: These numbers are theoretical, for a core with unity OD/ID ratio. For a real core with finite radial build, l is higher for the last wrap of alloy than the first, and hence H is higher. Taking the ID dimension in the table for the OD of the core we would come closer to reality. Recognize also that the core does not saturate all at once; the inside goes first and saturation progresses radially outward to the outside wrap of the core. If switching speed is critical, a core with ID/OD ratio as close to unity as practical should be used.

for various alloys and tape thicknesses (Fig. 4-21). These are part of a design algorithm for high performance bobbin cores, furnished courtesy of Infinetics, Inc.

8. SUITABLE CORE MATERIAL SUGGESTIONS

Suitable core materials for various frequency ranges are listed somewhat in order of decreasing loss, increasing switching speed, and increasing cost:

60 Hz: 12 mil grain-oriented silicon steel
12 mil Z-type silicon steel
12 mil 50 nickel

400 Hz: 4 mil Z type silicon steel
4 mil 50 nickel
4 mil 80 nickel

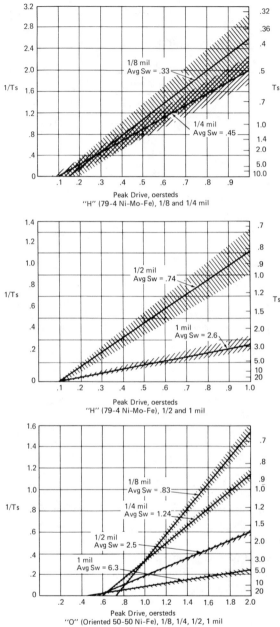

SWITCHING TIME CURVES

Curves showing the effect of Pulse Excitation drive level on the switch time (Ts) in microseconds. The average value of the slope or Switching Coefficient (Sw) in oersted-microseconds is also noted.

Materials covered:
H (79-4 Ni-Mo-Fe) and (Oriented 50-50 Ni-Fe) in gages from 1/8 mil to 1 mil

While drawn as a straight line, the actual value of 1/Ts tends to blend asymptotically with the abscissa. This occurs below 0.25 oersted for H material, and 0.8 oersted for 0 material.

Particularly in the case of H material, the slope Sw can be altered to some degree and its intercept point with the abscissa shifted markedly by selection of alloy heat and processing. Intermediate values of Sw can be obtained by custom rolling of the alloy to non-standard gages.

$$Sw = \frac{H1 - H2}{\dfrac{1}{Ts1} - \dfrac{1}{Ts2}}$$

Sw = oersted-microseconds
H = oersteds
Ts = microseconds

Fig. 4–21. Switching time curves. [SOURCE: Infinetics, Inc. Technical Data Form 87-1, Magnetic Core and Nickel alloy tape-wound toroids. (1601 Jessup Street, Wilmington, Delaware. 1965)]

100

(In some cases the lower loss of 1 or 2 mil materials may make them preferable.)

10–20 kHz: 1 mil or ½ mil 50 nickel
              1 mil or ½ mil 80 nickel

50–100 kHz: ½ mil 80 nickel
              ¼ mil 80 nickel

200 kHz and up, ¼ mil 80 nickel, ⅛ mil 80 nickel; square loop ferrites should be evaluated to see if the design can possibly be made to work with them; both cost and availability are much better than with ¼ mil to ⅛ mil square loop 80 nickel. The loop squareness, switching speed, temperature stability, and permeability of the nickel alloys are superior to those of ferrites, but cost rises exponentially with decreasing tape thickness. At very high frequencies it may turn out for a particular design situation that the high performance of a bobbin-core design (smaller size, less leakage inductance, etc.) will justify the cost. At extremely fast core switching speeds, the alloy thickness should be comparable to the skin depth of the frequency associated with the output rise time (core switching time). The skin depth in mils is roughly $4.4\sqrt{t}$, with t in microseconds.

9. A DESIGN EXAMPLE

Now let us look at a design example. Say we are designing a Jensen circuit DC-AC inverter and need to construct our own feedback transformer T_2. (See Figure 4-22.) We have decided from various criteria associated with the switch transistors that we want the fall time of t_2 (switching time) to be 0.1 μs. R_2 will provide the switching drive while not dropping too much of the feedback voltage due to the base drive current passed by R_1. Let us try a 1 : 1 : 1 turns ratio for t_2. We are given that the current to each transistor is 2 amps, and we decide that 5 volts on each winding of t_2 will make the base currents sufficiently independent of temperature for this application.

We need to determine now how much higher than 5 volts V_F must be, the value of R_2, and the design of t_2 itself. Our frequency is 33 kHz, so one-half period takes 15 μs. We have to try something and see where we are, because core area and turns determine volt-second support, while turns and the short-circuit current, through R_2, and the mean magnetic circuit length of the core determine switching speed. In a nutshell, we have too many variables and need to home in slowly on the design.

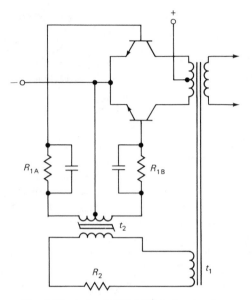

Fig. 4-22. A simple DC-AC inverter circuit.

Since cost rises rapidly with flux for thin gauge bobbin cores, try a modest core of 100 Maxwells and see where this leads us:

$$100ET = N\Phi \quad \text{gives} \quad 100 \times 5 \times 15 = 100N$$
$$N = 75.$$

This looks somewhat high from experience, but let us look into it further. For a $\frac{1}{8}$ in. groove width (consulting a tape-wound bobbin core manufacturer's catalog, we find that this amount of flux capacity can easily be put in a bobbin core of 0.375 nominal ID and finished dimensions of 0.335 ID, 0.478 OD, 0.175 height. We decide to wind the primary and both secondaries three-in-hand. With a 40% area utilization we have 112,000 circular mils multiplied by 0.4 for net copper area and divided by 225, the total number of turns, to obtain a wire area of 200 circular mils, which is #27. The mean length of turn is rougly 0.8 in., assuming the build over the core is one-third the inside diameter. The total length of each winding is then 5 ft., and the DC resistance of each winding is $\frac{1}{4}$ ohm, and at two amperes of load current we lose 0.5 volt each in the primary and the secondary. Power dissipation is two watts, a bit high.

If R_2 were 2.5 ohms and V_2 therefore 10 volts, our power dissipation in R_2 would be 10 watts plus the excess I^2R loss during the period the core was saturated, roughly the storage time of the transistors. If storage time turned out to be 0.5μs, then for 3% of the time power would be 40 watts, for an average power dissipation of about 11.2 W. The maximum current available from R_2 is 4 amperes, and for this core $l = 3.2$ cm, so $0.4\pi \times 75 \times 4 \div 3.2 = H$.

Our peak drive capability is therefore 117.8 Oersteds, and we estimate that even 1 mil 50 nickel would be fast enough. We are, however, asking for a $\Delta I/\Delta T$ of 40 amps per microsecond, and the saturation inductance of our primary winding will limit that. The saturation inductance we can estimate by unrolling the toroid into a long thin cylinder of mean radius 0.5 cm, length 3 cm, and use the air core inductance formula [Eq. (6.1), Chapter 6]. Assuming very optimistically that the saturated permeability is unity,

$$L \approx 4\pi \times 10^{-9} \times 0.5^2 \times 75^2/3.5 \approx 5\,\mu H.$$

For 10 volts available, $E = L(dI/dT)$ gives the rate of rise of current. $10 = 5(dI/dT)$, for a maximum of two amps per microsecond. At that rate, in 0.1 μs we would have at best 0.2 amps, for a drive level of a bit under 6 Oersteds. It would appear that a ¼ mil high nickel alloy would be required.

Going back to R_2, if it were ½ ohm and V_F were 7 volts, our average power dissipation in R_2 would be about two watts plus the additional losses during the core saturation interval. We note that saturation inductance will be much more of a problem, and our winding resistance is already too high. We could fit 200 Maxwells of flux capacity (80 nickel) in our original bobbin dimensions, so let us do that.

N now becomes 37 turns, and we have room for twice the wire area. Our winding resistance is now one-quarter of what it was before, or about 0.06 Ω. At two amperes we drop one-eighth volt in each winding. Our transformer primary voltage must then be 5.25 V, and if R_1 is ½ Ω, then V_1 is 6.25 V and the peak drive current becomes a bit over 11 amps. The peak core drive level is now 159 Oersteds. The saturation inductance is on the order of a quarter of the previous level, or 1.25 μH. For 6 volts available, dI/dT becomes 4.8 amps per microsecond. In 100 ns our current would have risen to only 0.48 amps, for a drive level of about 7 Oersteds. One-quarter mil 80 nickel will switch in about 0.075 microseconds at 7 Oersteds.

If we wished we could vary the design further; raising R_1 raises its average power dissipation. The point of diminishing returns will be where the power lost during the time when t_2 is saturated is equal to the power dissipation due to the load on t_2's secondary. If more flux is required the core cost will start to go up almost linearly with flux.

A low frequency, high power saturable reactor would be designed very much the same way, except for two items: first, the saturation inductance usually appears in series with the load, and the fraction of the input power dropped across it must be evaluated; second, the documentation available for determining switching speeds of 4 mil and 12 mil alloys is not good, and the user may have to personally evaluate a core material experimentally.

5
CURRENT TRANSFORMERS

1. INTRODUCTION

Current transformers, including most oscilloscope current probes, are most frequently used to measure the current amplitude and/or waveform in circuit locations where other measurement techniques are less efficient or less practical.

Measuring the voltage across a sampling resistor or inserting a moving coil AC ammeter is often not feasible with very high AC currents, pulse waveforms, or circuitry floating at high voltages. Most current sampling resistors have more inductance than a fast risetime current transformer. When the current is in a circuit floating at high voltage the oscilloscope would have to be floated at that voltage, and that is rarely desirable. In high current circuits the power loss associated with feasible resistor values is usually inconvenient or objectionable. Current transformers are superior in all respects except two: where the picosecond rise time of pulses being viewed is too fast for the rise time of even the fastest current transformers (in a transmission line geometry, of course; one does not obtain such speeds outside of matched transmission line systems), or where one needs low frequency response down to DC. There are DC "current probes." However, these are not current transformers strictly speaking, but Hall-effect devices which use a magnetic field to deflect a current flowing through a conductor, developing a transverse potential proportional to the magnetic field of the current being measured.

Current transformers can have remarkably low "insertion resistance," which is the effective resistance that seems to be in series with the current being measured due to their presence. Parasitic series inductance and stray capacitance can also be quite low, and are usually insignificant in a properly designed application.

Current transformers have a rise time limitation associated with the winding structure. There is also a "droop" on the observed waveform associated with the inductance of the winding (or other factors), and for current transformers with cores there is an ampere-second limitation corresponding to the volt-second limitation of conventional transformers with cores.

2. CHARACTERISTICS

The action of a current transformer is to attempt to force equal ampere-turns in both windings. Thus, for one turn in the current-sampling winding (primary) and 100 turns in the current-viewing winding (secondary) one views the current to be measured by the voltage the secondary winding develops across the load resistor. A schematic of a simple current transformer with a core is shown in Fig. 5-1. The inductance L (seen from the

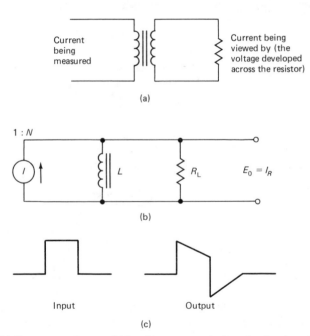

Fig. 5-1. (a) Current transformer; (b) low frequency equivalent circuit, where I is the current through the sense resistor, NI is the current being measured; (c) waveforms.

secondary, for convenience) clearly diverts an increasing current with time from the load resistor, so that for a rectangular input current waveform the output voltage is not rectangular, but droops.

The output may droop to zero following an exponential curve, or if the output volt-second product exceeds the volt-second support capability of the core the output waveform will collapse abruptly.

The high frequency performance of current transformers is limited by a leakage inductance associated with the less-than-perfect coupling between the secondary winding and the primary (often a wire threaded through the hole in the toroid), and by a shunt capacitance across the winding (Fig. 5-2).

The rise time for C' absent can be limited by the L/R time constant. The current source feeding the current transformer will of course attempt to develop whatever voltage is necessary to force itself through the inductance L', but stray capacitance and the properties of the source will usually limit the dv/dt the source is capable of generating. Hence, the current through L' may not be a faithful representation of the current being sampled unless L/R is much less than the rise time associated with the current pulse to be measured. A technique for raising the bandwidth of current transformers (distributed terminations) will be discussed later.

For small C' the rise time is approximately the same, but ringing following pulse transitions begins to be apparent. If C' becomes too large, the ringing will become quite large. For a highly underdamped design with unusually large shunt capacitance the rise time becomes limited by the $L'C'$ product.

3. APPLICATION

The action of a current transformer in forcing a secondary current equal to the primary current reflected through the turns ratio can be used to great advantage in measuring the current waveform in an AC power line going to some circuit. That signal can be fed to some logic or fault-sensing circuitry. In that case a unipolar (rectified) waveform is preferred as an input to the sensing circuitry. The current transformer forces its current through the rec-

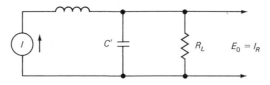

Fig. 5-2. High frequency equivalent circuit.

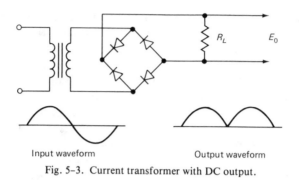

Input waveform   Output waveform

Fig. 5-3. Current transformer with DC output.

tifier bridge in Fig. 5-3 and develops a unipolar waveform across the sensing resistor.

Current transformers would be designed for the needed inductance and/or volt-second support on the secondary winding for the circuit of Fig. 5-1(a) or Fig. 5-3. Where rise time or wide bandwidth is a consideration they should be designed as pulse transformers, with attention to limiting the stray capacitance between the winding and the shield covering the entire secondary winding. Such a shield is usually necessary to keep noise on the primary wire from influencing the sampled voltage waveform.

4. AIR CORE CURRENT TRANSFORMERS

There is another type of current sampling device, used mainly in pulsed power research, where the currents involved are measured in kiloamperes, not amperes, and current rise times are on the order of 10^9–10^{12} amperes per second. These devices are called $\dot{I}$ (pronounced "I-dot") probes, a certain type of air core current sampling transformer is known as a *Rogowski coil*.

These devices are known as $\dot{I}$ probes because their output is proportional to the rate of change of current with time. The dot over a symbol denotes the time derivative of what the symbol represents. They are convenient for a number of reasons, and they also have their inconveniences.

They can literally be calibrated with a ruler. One configuration of an $\dot{I}$ probe is shown in Fig. 5-4. The coaxial transmission line is connected across the gap on the inside. The output voltage for such a configuration is

$$V_0 = \dot{I} \frac{\mu_0}{2\pi} w (ln \frac{r_2}{r_1}), \qquad \mu_0 = 4\pi \times 10^{-9} \, \text{Hy/cm}. \qquad (5.1)$$

In order to see the actual waveform, we must integrate it. Integrating the derivative of something gives us back that something, multiplied by some constant factors. One would take the output of Fig. 5-4, develop it across a resistance equal to the transmission line impedance, and then run that transmission line to a location where the measurement was to be done and integrate it there, as in Fig. 5-5. Radiated noise picked up by the cable in transit is filtered out by the integration. The restored waveform is then the original current waveform, with some droop present due to the finite time constant of the integrator.

$$E_0 = \frac{1}{RC} \int E \, dT \qquad (5.2)$$

where E is the V_0 of Eq. (5.1). Hence,

$$E_0 = \frac{\mu_0 w \left(ln \frac{r_2}{r_1} \right)}{2\pi RC} I. \qquad (5.3)$$

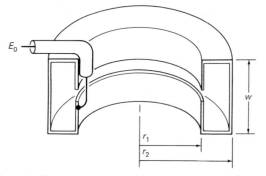

Fig. 5-4. Air core current transformer—cross-sectional view.

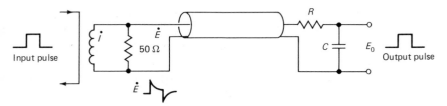

Fig. 5-5. I probe with integrating circuitry.

The rise time capability of such devices can be literally limited by the breakdown voltage capability of the transmission line, but the output voltages are quite low. A device of moderate dimensions ($w = $ ½ in., $r_2 = $ ½ in.) sampling a 100 ampere, 1 microsecond pulse, with a 100 microsecond time constant integrator (for 1%/microsecond droop) may only have a 10 millivolt output. That is a high impedance output, suitable for connection to an oscilloscope preamp input, but definitely not a 50 ohm transmission line. To measure nanosecond rise times one may have to use a sampling oscilloscope (assuming a repetitive waveform) and construct a cathode follower to buffer the 50 ohm sampling preamp input impedance from the integrator.

When measuring 10,000 amperes instead of 100 amperes, one can afford to divide the integrator output down with a fairly large resistance (perhaps 5000 ohm) and a 50 ohm resistor to obtain a voltage developed across an impedance which matches a transmission line impedance. This is vital if one wishes to propagate a fast rising pulse from anywhere to anywhere and maintain pulse fidelity.

Instead of a droop due to the inductance of a winding, the result is a droop due to the integrator time constant. For, say, a desired 1% droop, the integrator time constant would be selected to be 100 times the width of the pulse being measured. An electrostatic shield around the device is often necessary.

Where an output signal of at least millivolts is needed to measure tens or hundreds of amperes, one may take the (ruler-calibrated) nonmagnetic, nonconductive winding form of dimensions $\{r_1, r_2, w\}$ and wind upon it a winding of N turns. Such structures are called *Rogowski coils*. The output of a Rogowski coil is then N times the output of the single-turn structure of Fig. 5-4. However, there will be a rise time associated with the interwinding capacitance, the interturn capacitance, and the stray capacitance to the shield.

A rough estimate of the rise time of such a structure may be obtained from $t_r = 1.2ND$ nanoseconds, where D is the mean winding diameter in inches (the mean of the $r_1 - r_2$ difference and w).

For these types of i probes as well as conventional current transformers one may extend the bandwidth considerably by using a distributed termination. The winding of N turns may be divided into K portions, and the termination resistor divided by K. Each N/K turns is then terminated by R/K ohms, with all such terminated windings connected in series (Fig. 5-6). The total number of secondary turns is still N and the termination resistance is still R, but the rise time has been improved by a factor of K.

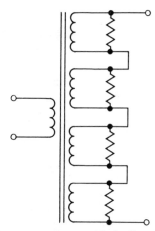

Fig. 5-6. Current transformer with distributed terminations.

In practice one rapidly runs into a nonlinear rise time improvement, because each winding rides on top of the pulse output of the previous windings, and the stray capacitance of each winding shunts high frequency components of the waveform. This minor detail can be handled by isolating each winding's stray capacitance with a common mode choke (Fig. 5-7). The common mode chokes would usually be a few turns on a very small ferrite toroid, since they need only support the voltsecond product associated with a fraction of the risetime and the output pulse amplitude.

It should therefore be possible to obtain increased outputs from single-

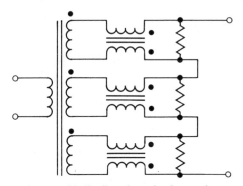

Fig. 5-7. Current transformer with distributed terminations and common mode chokes.

turn i probes (or multiturn coils with distributed terminations) by using pulse transformers to amplify the voltages associated with the i signal. Since the pulse transformers need only have a volt-second support to handle rise and fall time signals, they will usually be small ferrite toroids. Note, however, that their rise times must be much faster than the rise time of the i signal in order to faithfully reproduce it.

Some of the previously discussed techniques can be combined by constructing a single-turn i probe and distributing a number of pulse transformers around its periphery (Fig. 5–8). Here one has a number of small pulse transformers (usually on ferrite toroids, with modest step-up turns ratios, perhaps 1 : 5) with common mode chokes on the output of each, and series-connected terminated windings. It would usually be preferable to have the resistance terminations electrically closest to the transmission line output, with the common mode chokes between the windings and the terminations.

In Fig. 5–8, assuming five 1 : 5 pulse transformers, 25 times the output of a single-turn i probe would be obtained. With careful design the rise time should approach the rise times of the small step-up transformers, which

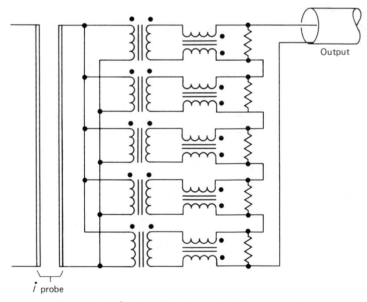

i probe

Fig. 5–8. An i probe with voltage gain before integration.

must have volt-second support capabilities sufficient to handle the $\dot{I}$ signal from the main current probe.

Larger amplitude signals can also be obtained with nanosecond rise times by summing the outputs of appropriate impedance transmission lines. Further information on $\dot{I}$ probes and Rogowski coils may be found in Baum (1978) and Sower (1981).

6
PULSE TRANSFORMERS

1. INTRODUCTION

If many electronics engineers regard magnetics design as a black art, then it is also true that many transformer engineers regard pulse transformer design in the same way. The reason is that a pulse transformer must faithfully reproduce the high frequency components of the leading edge of a rectangular input step, the low frequency components of the flat top of the pulse, and all frequencies in between. At the same time it must present to the source and load appropriately balanced parasitic components whose characteristic impedances match the source and load impedances.

The designer who has worked mainly with power transformers and inductors is usually ill-prepared to deal with the complex and demanding additional design criteria which a good pulse transformer design requires. It should meet broadband frequency response criteria while maintaining the proper ratio of leakage inductance to parasitic capacitance. It should also be simple to fabricate, and as small, light, and inexpensive as is possible within some given limit of available engineering time and budget!

It is true that a great deal of experience and involved calculations are needed to accurately predict the performance of a design. It is even more difficult to develop the best design for a given requirement.

It turns out that it is not too difficult to get within a factor of two on pulse rise time the first time out. If one knows what the relevant design equations are and can make an informed choice from among the several basic design types, almost anyone can design pulse transformers that work.

In order to introduce the reader to this area gently, we shall start with a consideration of air core pulse transformers, since these are the simplest. Later, we shall introduce ferromagnetic cores and their attendant advantages and disadvantages.

2. AIR CORE PULSE TRANSFORMERS

An air core pulse transformer has four major parasitic components: a total leakage inductance referred to the primary, an effective shunt capacitance referred to the primary, a series resistive component, and a primary inductance (Aslin, 1977). These are represented schematically in Fig. 6-1. In reality there is a primary leakage inductance between R_s and L_s this will reduce the voltage gain if that primary leakage inductance is not much less than the primary inductance.

For the case of a foil-wound pulse transformer, where each turn is the full width of the transformer, and each turn has insulation between it and the turn beneath, the geometric parameters are defined for a typical design in Fig. 6-2. The primary inductance of such a configuration is given by

$$L_p = \frac{\mu_0 \pi r^2 N_p^2}{r + b} \text{ Henries}$$ (6.1)

where μ_0 is the permeability of free space, $4\pi \times 10^{-9}$ H/cm. The total leakage inductance referred to the primary for an auto-transformer is given by

$$L'_p = \frac{2}{3} \pi \mu_0 \frac{r}{b} \Delta N_p^2 (1 - \frac{N_p}{N_p + N_s})^2 \text{ Henries}$$ (6.2)

and for a transformer with two separate windings

$$L'_p = \frac{2}{3} \pi \mu_0 \frac{r}{b} \Delta N_p^2 \text{ Henries.}$$ (6.3)

Note that Δ in Eqs. (6.2) and (6.3) above (see Fig. 6-2) is just $(N_p + N_s)(t + s)$.

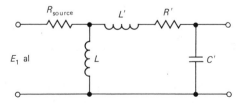

Fig. 6-1. The basic parasitic components of a foil-wound air core transformer.

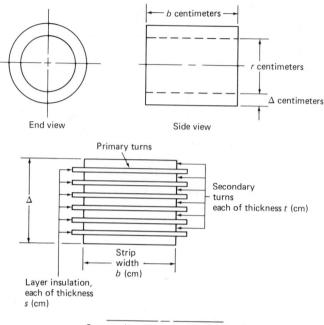

End view   Side view

Fig. 6-2. Typical winding section.

The effective shunt capacitance referred to the primary given by

$$C_p' = \frac{N_s^2}{N_p^2} \frac{\epsilon_r\epsilon_o 2\pi rb}{(N_p + N_s - 1)s} \text{ Farads} \qquad (6.4)$$

where ϵ_r is the dielectric constant (often called k) of the layer insulation, and ϵ_0 is the permittivity of free space, 8.85×10^{-14} Farads/cm.

The term $(N_p + N_s - 1)$ can usually be taken equal to $(N_p + N_s)$ with little loss of accuracy (e.g., more than 10 turns).

The radial (directed both outwards and inwards) pressure due to the magnetic field is given by

$$P = \frac{\mu_o N_p^2 I_p^2}{2b^2} \text{ Newtons/meter}^2 \qquad (6.5)$$

in which all dimensions are in meters, 6.9×10^3 N/m^2 = 1 psi, and $\mu_0 = 4\pi \times 10^{-7}$ Hy/m, for a two-winding transformer. The mechanical structure must be designed to take this into account. In high current applications it can be significant. For an auto transformer,

$$P = \frac{\mu_0 N_p^2 (I_p - I_s)^2}{2b^2} \quad N/m^2. \tag{6.6}$$

The resistance of a winding would be calculated from

$$R = \varrho l / A \tag{6.7}$$

in the usual manner.

The rise time (inductance-limited) is given by

$$T_r = \frac{1.1 L'}{R_L} \tag{6.8}$$

where R_L is the load resistance reflected via the square of the turns ratio to the primary and L' is the effective total leakage inductance referred to the primary.

The rise time (capacitance-limited) is given by

$$T_r = 1.1 R_L C' \tag{6.9}$$

where, again, both R_L and C' are referred to the primary (or the secondary—but they must both be as seen from the same side of the transformer).

The pulse droop (due to the current in the primary inductance) is given by

$$D = 1 - e^{-\left(\frac{R_s}{2L_p} T\right)} \tag{6.10}$$

where R_s is the source resistance, L_p is the primary inductance, and T is the pulse width.

One may view the air core transformer as a distributed or lumped L-C network, which has a certain characteristic impedance and rise time. Since we have equations which express those distributed parameters as lumped parameters referred to the primary, it is a simple matter to derive expressions for the input impedance of our "transmission line" and its rise time into a matched load (output impedance equals input impedance times the square of the turns ratio).

For a pulse transformer in which L' and C' determine the rise time, the rise time will be on the order of 0.35 resonant period,

$$T_r \approx 0.7\pi \sqrt{L'C'}$$

$$\approx 0.7\pi \left[\left(\frac{2\pi \mu_o r \Delta N_p^2}{3b} \right) \left(\frac{N_s^2 \epsilon_r \epsilon_o 2\pi rb}{N_p^2 (N_p + N_s)s} \right) \right]^{1/2}$$

$$\approx 1.4\pi^2 rN_s \left[\frac{\mu_o \epsilon_r \epsilon_o}{3(N_p + N_s)s} \right]^{1/2}. \tag{6.11}$$

Note that the rise time is linearly dependent on the mean winding radius, somewhat less than linearly dependent on the number of secondary turns, and independent of the winding width.

The input impedance Z_0 is given by $\sqrt{L/C}$:

$$Z_0 = \left[\frac{2\pi\mu_o r \Delta N_p^2 N_p^2 (N_p + N_s)s}{3bN_s^2 \epsilon_r \epsilon_o 2\pi rb} \right]^{1/2}$$

$$= \frac{N_p^2}{N_s b} \left[\frac{\mu_o \Delta s (N_p + N_s)}{3\epsilon_r \epsilon_o} \right]^{1/2}. \tag{6.12}$$

Note that the input impedance is independent of the winding radius and inversely dependent on the winding width.

3. CONSIDERING A DESIGN

Let us say that we were interested in designing a transformer with a 1 : 5 turns ratio, an input impedance of 8 ohms, and a rise time of 10 nsec (nanoseconds), to handle a 50 nsec wide pulse. Let us say that for reasons of dielectric strength we had selected $S = 0.05$ cm, and for current carrying capacity we had selected a secondary foil thickness of 0.012 cm. Then $\Delta \approx 0.3$ cm. Let us assume a single-turn primary winding and a five-turn secondary winding, since greater numbers of turns make the rise time more difficult to achieve. We find from rise time considerations that r must be on the order of 3 cm to meet our impedance requirements. The width b must be about 1.15 cm. The requirements for $L'C'$-limited rise time cannot be met if the mean winding radius is more than about 3 cm. We have not, however, considered whether the system rise time might be limited by the leakage inductance and the source impedance:

$$T_r = 1.1 \frac{L'}{R_S} , \quad \text{so} \quad \frac{10^{-8} \times 8}{1.1} = 72 \text{ nanohenries (nH)}$$

is the maximum leakage inductance allowable. For b of 1.15 cm, from Eq. (6.3), the actual leakage inductance is

$$L' = \tfrac{8}{3} \pi^2 \times 10^{-9} \times \frac{3}{1.15} \times 0.3 = 20.6 \text{ nH}$$

so we are safe on that count. The reader might wish to check whether the distributed capacitance and the source impedance limit the rise time.

Unfortunately, this value of L' means that the primary inductance will be about 86 nH, and for a 20 kV input the current drawn by the transformer primary inductance will increase at the rate of about 234 amperes per nanosecond of pulse width, becoming about 12,000 amperes in 50 nanoseconds.

It is for this reason that people put cores in transformers—to increase the primary inductance from otherwise unacceptably low levels while maintaining specified leakage inductance and shunt capacitance levels.

4. INTRODUCING THE FERROMAGNETIC CORE

Let us consider the results of placing 20 square centimeters of ferrite cross-sectional area in the coil, such that we have a complete magnetic circuit with negligible air gap. Typical ferrite materials will have a permeability of several thousand times that of a vacuum ($4\pi \times 10^{-9}$ Hy/cm), and for this geometry a single-turn inductance of perhaps 10 microhenries. As we will see later, this inductance, measured well below one megahertz, may not be valid for very short pulses.

The current drawn by our primary inductance is now no more than one-hundredth the previous level and possibly less, or around 100 amperes for a 20 kV, 50 nsec input pulse. There is of course a disadvantage. Ferromagnetic materials have a limited volt-second support capability, and after a while they saturate and the primary current is once again limited only by the air-core primary inductance.

If we assume our ferrite has a full flux reset, so that the full height of the

B-H loop is available, we can assume a maximum flux density of 8000 gauss. The total flux capacity of our core in Maxwells is $\Phi = BA$ where B is in gauss and A is in square centimeters.

For $A = 20$ cm^2, our flux capacity is $\Phi = 8000 \times 20 = 160,000$ Maxwells, and the volt second support is given by $100\,ET = N\Phi$, where T is in microseconds. For 20 kV input,

$$T = \frac{1 \times 120,000}{100 \times 20,000} = 60\,\text{nsec.}$$

So the design looks viable if our magnetizing current is acceptable and our estimate of primary inductance is accurate. The primary load current (20 kV into 8 ohms) is 2500 amperes, and diverting 4% of that to L_p at pulse end gives 4% droop.

We see that we can meet the design requirements with the possible exception of an adequately high primary inductance. At this point, if the designer is unfamiliar with core properties at these pulse widths he should build two units: one with the ferrite core protruding about one-quarter of the coil length from each end, and the other with a complete magnetic circuit. He should carefully compare the performance of the two at various pulse widths. Ferrite U and I cores with 1 sq. in. (6.45 cm^2) of cross-sectional area are available, and several cores can be assembled together to give the required area.

5. WINDING AND EDDY-CURRENT LOSSES*

Resistance losses in the main windings of saturable inductors in short-pulse modulators are markedly influenced by skin and proximity effects in the winding conductors. In consequence, both the amount of the loss and the optimum winding form are dependent on the duration of the inductor current pulse. This dependence is complex; only a few general principles of special importance are discussed here.

Let l_w be the average length per turn, a_c the conducting-material cross section in one turn, and ϱ the resistivity of the wire used in the inductor winding. The resistance of an N-turn winding is then $N\varrho l_w/(\eta a_c)$, where η is a coefficient < 1 introduced to account for skin and proximity effects—the

*This section has been reprinted from Godfrey T. Coate and Laurence R. Swain, Jr., *High-Power Semiconductor-Magnetic Pulse Generators* (Cambridge: MIT Press, 1966), Research Monograph No. 39, Section 2.1.5, pp. 17–21.

crowding of fast-changing currents into portions of the conducting area nearest the core so that only the portion ηa_c of the cross section a_c is utilized. In terms of the total useful conductor cross section in the winding,

$$a = N\eta a_c. \tag{2.25}$$

Therefore,

$$R = N^2 \varrho l_w/a \tag{2.25a}$$

and the energy loss per cycle of operation is

$$J_w = N^2 \varrho(l_w/a) \int_{T_r} i^2 \, dt \tag{2.25b}$$

where i is the winding current and T_r is the repetition period. Inasmuch as saturation-period currents are many times switching and resetting currents, the integral in Eq. 6.13b can be evaluated as $I_{max}^2 T_{sat}/2$, where I_{max} is the peak value of a half-sine current pulse of duration T_{sat}. (For other saturation-interval current waveforms, T_{sat} is adjusted to an effective value; for example, twice the saturation time if the current pulse is rectangular.) Evaluating $I_{max} = H_{max}l/N$ in terms of

$$H_{max} = (B_{max} - B_s)/\mu_e = (B_s/\mu_e)\,(\Delta\lambda/\lambda)^*$$

yields, with the aid of Eq. 2.3, 2.11, and 2.13,

$$J_w/J_{st} = (\varrho T_{sat}/4)\,(l/a)\,(l_w/A).^{**} \tag{2.25c}$$

Inasmuch as ϱ is determined by the conductor material (copper) and T_{sat} by the required operation of the inductor, minimizing the inductor loss ratio J_w/J_{st} requires maximizing:

1. The ratio a/l. This ratio may be considered an effective winding depth—when multiplied by l (somewhat more than the winding length) it equals the useful conductor cross section a.
2. The ratio A/l_w—or the area-to-perimeter ratio for the core times the ratio (less than one) of core perimeter to l_w.

---

*$\Delta\lambda/\lambda$ is the pulse compression ratio of the inductor, expressed in terms of volt-time integrals.
**J_{st} is the peak energy stored in the inductor.

For DC or very low-frequency inductors, the procedure, once a specific core has been chosen, is to maximize a/l by selecting a wire size so that, with minimum space lost to insulation, the core window is filled. In short-pulse inductors, skin and proximity effects (through their effect on η) place a practical limit on a/l that is often far less than that corresponding to a full core window. In consequence the inductor losses are increased, and this increase cannot be avoided by using more copper in the winding. In fact, adding conductor cross section is doubly undesirable because it increases both the losses and the weight of the inductor.

One useful form of short-pulse inductor winding comprises a single layer of turns arranged to fill the inner perimeter of the core-plus-insulation toroid as completely as possible. For some high-voltage windings it maybe possible to choose a wire size such that the required number of turns N exactly fills the available space. More often, however, to avoid requiring impracticably large wire, it is necessary to employ a K-filar winding, where K is an integer chosen so that KN strands of convenient size wire fill the space available. For example, if $K = 4$, four strands of wire are laid side by side to form a four-element "tape," and N turns of this "tape" are wound on the core. The four strands are connected in parallel at each winding terminal; interstrand insulation is necessary only between strands in different turns. It is essential that all strands of all turns fit in a single layer against the core insulation, and that the available winding space be filled as evenly and completely as possible. The wire diameter d (of a single strand) is immaterial, provided only that it is appreciably greater than the skin depth δ—which may be estimated as $2.6/(f_{mc})^{1/2}$ mils, where f_{mc} (in megacycles per second) is taken in the range of important frequency components of the current pulse. For current pulses of several microseconds duration, δ is of the order of 5–10 mils, and $d >> \delta$ for wire strands of moderate size convenient for winding.

Losses in a single-layer winding are high (relative to losses in a well-designed low-frequency winding) because a in Eq. (2.25c) is δ times the conductor-occupied part of the core-plus-insulation inner perimeter, which is somewhat less than l. Thus a/l is at most a few mils for current pulses in the several-microsecond range. Unfortunately, the losses cannot be reduced by increasing or decreasing wire size in the single-layer winding or, if the condition $d > \delta$ is maintained, by providing more conductor cross section in a multiple-layer winding. In a $d > \delta$ multiple-layer winding, the layers of conductors function much as the walls in a waveguide. Each interlayer space forms (with the conductors on either side) a substantially isolated small inductor in which losses (associated with skin-depth conduction in layers on both sides) are high.

For short-pulse inductors the only useful alternative to the single-layer winding is a multiple-layer construction in which the condition $d < < \delta$ is satisfied for individual strands. Then

$$\frac{\text{Loss in one turn of any one strand}}{\text{DC loss in same turn of same strand}} \approx 1 + \frac{1}{3} \left[\frac{H}{\Delta H} \left(\frac{d}{\delta} \right)^2 \right]^2 \tag{2.26}$$

where H is the magnetic-field intensity in the vicinity of the strand of wire considered (caused by currents in all the layers of strands wound over it) and ΔH the increment of H from one side of the strand to the other (caused by current in that strand). For strands nearest the core, $H/\Delta H \approx n$, the total number of layers of strands in the winding. Thus, if $t \approx nd$ denotes the portion of the total depth of winding occupied by conductors, then $H/\Delta H \approx t/d$, and Eq. (2.26) indicates that losses in the innermost layer of strands increase rapidly if t/δ is allowed to increase much beyond δ/d. In effect, t/δ must be limited to a few times δ/d to avoid a decrease of η in Eq. (2.25) that would more than offset the increase of a_c provided by increase of t.

The most practical means of producing multiple-layer windings with $d < < \delta$ is use of very finely stranded litz wire, in which many insulated strands are twisted together so that each occupies all possible positions to approximately the same extent. Thus the total flux linkages are the same for each strand, and, with the strands connected in parallel at the winding terminals, the current divides equally among the strands. For current-pulse durations of the order of 100 μs or more, litz-wire windings can be constructed to provide significantly less loss than single-layer windings. The advantage of the litz wire becomes less for shorter pulses and disppears altogether when the skin depth becomes only a little more than the smallest available strand diameter. For example, suppose $\delta = 5$ mils, and that 2-mil conductors with 2-mil-thick insulation are the smallest strands available. For $\delta/d = 2.5$, a winding with $t/\delta = \delta/d$, or $t = 12.5$ mils, provides an AC loss substantially equal to its DC loss, so that a in Eq. (2.25c) can be taken to be equal to $12.5/3 \approx 4$ mils times the core-plus-insulation inner perimeter; here the factor $\frac{1}{3}$ enters because the perimeter is occupied about $\frac{2}{3}$ by insulation and only about $\frac{1}{3}$ by conductors. In comparison, for a single-layer winding a can be very nearly δ (5 mils) times the perimeter. Slightly less loss may be possible in a litz winding of somewhat larger t, but the improvement is sharply limited by the increased AC/DC loss ratio for $t/\delta > \delta/d$.

The fast-changing flux that causes skin and proximity effects in saturable-inductor windings also causes saturation-interval eddy-current losses in the magnetic-core material (as already mentioned in connection with Eq. 2.20) and in the protective case for the core if the case is metallic. It has been found convenient to account for these losses in connection with winding loss rather than with the major core loss; like winding loss, they are determined by saturation-interval current, whereas the major core loss is a function of switching and resetting voltage waveforms. Metallic-cased cores should be avoided in short-pulse modulator applications (plastic cases are available), because eddy-current losses in them are large, sometimes larger than the loss in a single-layer winding. This loss occurs despite the fact that metallic cases are always split to avoid short-circuiting the inductor. For pulses of a few microseconds duration, skin depths in the case metal (usually aluminum) are much less than the metal thickness, and the loss associated with skin-depth conduction in both sides of the case is a large fraction of the skin-depth conduction loss in a (copper) single-layer winding. The same changing flux that causes these losses penetrates the saturated magnetic core material. The eddy-current losses produced, though not negligible, are usually relatively small because the magnetic tape is thin.

6. SOME HIGH FREQUENCY EFFECTS

There are further considerations affecting core loss in laminated materials. One example is an interesting version of the wire proximity effect, related to skin effect.* Some tape wound C-cores had anomalously low high-frequency core losses, in that the material resisitivity appeared higher than it actually was. The induced eddy currents in thin laminations close to each other tend to repel each other, since the currents are flowing in opposite directions on each side of the insulation between adjacent wraps of the core material. Skin effect would force the eddy currents to the outer surface of the laminations. This is similar to the proximity effect, in which the current is forced away from itself by neighboring currents flowing in the same direction in a conductor or in several adjacent conductors. The net result in both cases is to reduce the conductor area the current flows through. In the case of induced eddy currents, the actual current is determined by the induced voltage in an increment of core area and the material resistivity. Since the current here is

*Frank Allen, National Magnetics; personal communication.

confined to a smaller area, the apparent resistivity is higher, the eddy current less, and the eddy current I^2R loss is lower. This means that the eddy component of core loss, as frequency increases, should change from increasing linearly with frequency to increasing somewhat less than with the square root of frequency, as the skin depth at that frequency becomes comparable to, first, the alloy thickness and then the thickness of the interlaminar insulation. In any case it would not scale up from 10 kHz to 500 kHz the same as it would from 10 Hz to 500 Hz.

The charging of the interlaminar core capacitance on the leading edge of the pulse draws a small but noticeable current spike when rise times are below 50 nsec, as does an energy absorption phenomenon associated with electron spin in ferromagnetic materials.

7. PHILOSOPHICAL CONSIDERATIONS

From thermodynamic considerations it has been argued (Glascoe & Lebacqz, 1964) what the maximum efficiency of a pulse transformer would be if, at the end of the pulse, there were no temperature difference between the coil and the core and minimum (and equal!) stored energy in both coil and core. Similarly, in a power electromagnetic transformer it has been argued that the transformer would be in its lowest energy state if the power dissipated per unit volume of both coil and core were equal. Früngel (1965) maintains that, within this criterion, minimum total power will then be dissipated in the transformer when the transformer volume is a minimum, and that the volume will be a minimum when the perimeters and volumes of coil and core are approximately equal. This, then, is a criterion for equal total iron and copper losses.

There are thermodynamic plausibility arguments for the rate of heat generation per unit volume being equal in both core and winding.

How low an energy state the unit might be in might be judged by the temperature *difference* between core and winding, since thermodynamics has the concept of the potential energy level being related to the temperature difference between two terminals, and the theoretical work that can be produced by a flow of thermal energy from a hotter to a colder terminal is the thermal energy generated multiplied by a theoretical maximum conversion efficiency factor (one minus the ratio of absolute temperatures).

Quantum mechanics has the concept that the lowest energy state of a system is not a zero energy state, but rather that there is a certain irreducible amount of energy associated with this state. The lowest energy state is called the *ground state,* and its energy level is called the *ground state energy.*

It may very well be that in a transformer which has been designed for most efficient volume utilization that there will be an inherent differential rate of thermal energy generation between core and winding. If this turns out to be the case, the transformer designer should not not be upset, for the quantum-mechanical plausibility argument makes it all right. On the other hand, if our best designs lead to essentially no differential rate of thermal energy generation, the thermodynamic plausibility argument makes *that* all right. The point here is that there may be circumstances in which either of these concepts will apply, and one viewpoint or another may give useful insight into a particular problem. The primary focus of the designer's attention should not be on these questions, but rather on applying design tools which work to produce a transformer which works. Philosophical considerations may make one feel better about a particular design tool or use thereof, but are only useful insofar as they lead to useful tools.

Let us pick one of our transformer designs, optimized for winding loss only, and see how it corresponds to this concept. Take Case I, $P = 1.8$, $Q = 2.7$. The core volume is calculated to be 2.68 cu. in. and the coil volume is 0.59 cu. in. The coil and core volumes are clearly not equal.

Now let's take a look at a Case I design optimized for total loss (core and winding combined), and see how it does. Consider a fairly efficient design, $P = 4.7$, $Q = 6.8$. The core volume is calculated to be almost 20, and the coil volume about 2. We know that the core loss is about 45% of the total loss, so the loss per unit volume of the core is about one-tenth that of the coil. The coil perimeter is calculated to be about 7, whereas the core perimeter is about 22, almost three times the coil perimeter.

A comment on the contrast between these results and the earlier philosophical considerations of other authors may be in order. The loss mechanism associated with the interaction of a magnetic field with ferromagnetic matter is not the same as the loss mechanism associated with the interaction of an electric field with a conductor. The energy densities of the magnetic and electric fields (per unit volume of transformer) are probably not equal, either. It should therefore be not at all surprising that in such an (optimized for minimum volume) transformer the core and coil volumes or perimeters or loss densities are disparate.

On the other hand, it should be noted that these considerations may very well apply to pulse transformers, when the effects of high frequency coil parasitics (leakage inductance and coil capacitance) or high frequency core parasitics (magnetic field penetration depth into laminated materials, etc.)

are taken into consideration. Some discussion of this topic may be found in vol. 5 of the M.I.T. radiation laboratory series.

8. SOME COMMENTS ON MATERIALS
AND SUPPLIERS

Various manufacturers of magnetic materials, including Ferroxcube, Indiana General, Stackpole, Micro Metals, TDK, Magnetics Inc., Magnetic Metals, Arnold Engineering, and National Magnetics, have not only varying degrees of applications information but large amounts of technical data on the performance of their various materials. The reader is strongly urged to contact all possible sources, obtain all available data, and study everything carefully. These companies have many knowledgeable engineers on their staffs and they are familiar with many of the more subtle characteristics of their materials. Some of these manufacturers make unique materials (the 12 mil Z-type silicon steel available from National Magnetics is a good example), and the designer must be aware of the performance advantages each company offers.

The reader is also cautioned that not all similar-appearing items from different manufacturers are the same. Many people offer 2 mil tape-wound C-cores for pulse applications, but the pulse permeability may vary by a factor of 4 or more depending on the supplier. Pulse permeability is not necessarily the permeability as measured under DC, low frequency AC excitation, or even high frequency AC excitation. It is the apparent permeability as measured under pulse conditions.

Further, many manufacturers can provide material performance equal to special selected grades or apparently proprietary items offered by others. Some manufacturers can work more closely with the engineer than others, and can tailor special core characteristics to a specific application.

The difference between two designs can easily be the familiarity of one engineer with more types of materials and the capabilities of more suppliers. The following very general guidelines are suggested, at the risk of some repetition. For pulse transformers which handle pulse widths of several hundred microseconds to several milliseconds, 4 mil Z-type silicon steel is an appropriate core material. For pulse widths of 1-2 μsec to a few hundred microseconds, 2 mil silicon steel is appropriate. For pulse widths of a few tenths of a microsecond to a few tens of microseconds, 1 mil 50 nickel C-cores or tape-wound toroids are often appropriate. For pulse widths of a few tens of nanoseconds to a few tens of microseconds, ferrite cores are often appropriate. For pulse widths of a few tens of nanoseconds to a few microseconds, air core pulse transformers are often appropriate, especially where very high peak powers (megawatts to gigawatts) are involved.

9. MATHEMATICAL MANIPULATIONS

Let us look at a few of the relationships we have introduced so far and see if we can find a relationship between the given parameters of a pulse transformer application and the transformer design parameters such as turns, core and coil dimensions, etc.

$$T_r = 1.4\pi^2 r N_s \left[\frac{\mu_0 \epsilon_r \epsilon_0}{3 (N_p + N_s) s} \right]^{1/2} \qquad (6.15)$$

$$Z_0 = \frac{N_p^2}{N_s b} \left(\frac{\mu_0 \Delta s (N_p + N_s)}{3 \epsilon_r \epsilon_0} \right)^{1/2}. \qquad (6.16)$$

For units with cores only,

$$100 E T_{pw} = N\Phi = NBA \qquad (6.17)$$

$$A \cong [\frac{\pi}{4}(2r - \Delta)]^2 = \frac{\pi^2}{16}(4R^2 - 2r\Delta - \Delta^2)$$
$$\cong 2r^2 \quad \text{for} \quad \Delta << R \qquad (6.18)$$

$$100 E_p T_{pw} = N_p B(2r^2). \qquad (6.19)$$

E_p and T_{pw} are given. The designer selects a B appropriate to the particular core material. T_r is given and Z_0 is either given or can be inferred from the circuit application. The ratio of N_p to N_s is also given.

Our knowns are now r, N_p, b, s, and Δ. Now Δ is just $N_p(t_p + s_p) + N_s(t_s + s_s)$, and the designer often will choose the foil thickness based on winding convenience rather than allowable resistance. The foil thickness t for the moment will be assumed to depend only on N_p and s and hence will not be an independent unknown parameter. We will look at winding resistance later.

Note that r, N_p, b, and s make four unknowns. Unfortunately, we have only two equations for air core designs or three equations for iron core designs, and cannot solve for four unknowns. It appears we have to look at winding resistance now.

$R = \varrho 2\pi r N / bt$ for foil wound geometry:

$$R = \frac{\varrho 1}{A} = \frac{2\pi r N_p}{bt_p} + \frac{\varrho 2\pi r N_s}{bt_s} = \frac{\varrho 2\pi r}{b} \left(\frac{N_p}{t_p} + \frac{N_s}{t_s} \right) \quad (6.20)$$

If we pick t_p and t_s to be in the same ratio as N_p and N_s, then we have one more unknown and one more relationship.

We now have five unknowns and four relationships. Perhaps we can decide that we care what inductance the primary has, and that may help. For an air core,

$$L_p = \frac{\mu_o r^2 N_p^2}{r+b} \text{ Henries,} \quad (6.21)$$

which seems to give us one more relationship among our existing unknowns but naturally only replaces Eq. (6.18) for core designs. Can we now uniquely determine an air core design from the information given? The unknowns are r, N_p, b, s, and Δ. Our relationships are Eqs. (6.14), (6.15), (6.16), and (6.17).

Equation (6.18) spoke of flux density, which is irrelevant for an air core design. So, for air core design we are short one relationship.

10. MORE DISCUSSION AND EXAMPLES

The situation is actually not so bad. We *can* design an air core transformer. The layer insulation thickness must support the voltage between turns, and this gives us a connection between the input voltage, turns, and layer insulation thickness via a working stress level which the designer is free to choose. The same conceptual relationship exists in iron core designs, so we do not necessarily need to invoke a primary inductance.

Beware of specifying too much. If we have four unknowns and we specify six parameters which are functions of those, it will usually be impossible to meet simultaneously more than four of those six specified parameters.

We haven't said anything about the permeability of the core material for iron core designs, or whether the effective permeability is determined by a gap in the magnetic circuit. Nonetheless, these are minor items. The designer can specify these and tweak the design later. For iron core designs we do have enough relationships to specify a design uniquely.

It may (and often does) turn out that the transformer rise time is limited not by its own $L'C'$ product, but by the time constant of the load impedance with the transformer leakage inductance, or by the time constant of the source impedance and the internal shunt capacitance of the tranformer (including any capacitive component of the load).

One may use the idea that putting a core into a transformer which meets all design requirements (except primary inductance) as an air core will raise the primary inductance by a factor of anywhere from around 3–10 for a simple rod core inside the coil only to perhaps 10^3–10^4 for high permeability alloys in gapless construction.

For typical configurations of tape-wound C-cores or ferrite cores one can assume the primary inductance will increase over an air core by about one hundred to a few thousand. Depending on materials and physical size, the range can go anywhere from 10 to 10^5.

Between resistance considerations, which may vary from significant to insignificant, and volts per turn or volts per layer considerations, which may vary from irrelevant to so critical that turns per layer and/or layer insultation thickness is determined by allowable dielectric stress (which may conflict with capacitance or leakage inductance requirements for wire wound units, forcing a foil-wound design approach, and the myriad core materials and attendant permeability characteristics over a pulse width range of nanoseconds to milliseconds, it begins to look like we will not be able to come up with a simple cookbook criterion for pulse transformers as we did for power transformers or inductors.

In some cases we may need a pulse transformer with high voltage isolation between windings. The need for distance between windings, combined with the need for tight coupling between windings and minimal delay time for the pulse transmitted through the transformer impose conflicting and often exasperating design requirements. They not only make it difficult to obtain a viable design at the first attempt, but may tax the ingenuity of the designer.

11. METHODS OF STARTING A DESIGN

Very well, how do we start a pulse transformer design? The answer to this depends on which specifiable parameters are the most important. Say we know that our design is going to have its rise time limited by leakage inductance and source impedance, and we have an allowable leakage inductance number. Let us say that this is a high voltage, high peak current, high repetition rate application, so that we care what the copper losses are. Given the turns ratio and voltage per turn, we can choose s to hold a given level of voltage stress. Assume the simplest possible geometry, a single primary turn, and the foil wound multi-turn secondary beneath. For the single-turn primary, use $100ET = N\Phi$ to get the flux density in the air core to the B at which we expect to run the core (20,000–30,000 gauss or less for 2 mil silicon steel with full flux reset, or some other as appropriate).

We now have a winding radius; calculate resistances and set allowable conductor thicknesses for some assumed coil width. We now have a winding build. Increase the coil width b until the leakage inductance requirement is met. Go back and recheck winding resistance, reduce conductor thickness as appropriate. These two parameters can be traded off for any specified resistance. Now calculate the primary inductance. It is usually too low. Place a simple rod core inside the coil, expect a primary inductance increase of about a factor of 10. If that is much too low, a C-core can be inserted around the coil and the inductance recalculated.

What we just did in the previous example was to design a pulse transformer according to the following procedure:

1. Assume a foil-wound design.
2. Assume a given level of volts per turn.
3. Pick a layer (turn) insulation thickness, assuming a high volts-per-turn design.
4. Determine the winding radius based on the fixed flux density limit which follows from the voltage per turn and the pulse width.
5. Estimate the winding width.
6. Pick a conductor thickness for that winding width such that copper loss requirements are met (with due attention to skin effect).
7. Calculate leakage inductance, then vary the winding width b until the leakage inductance is met for the assumed coil build.
8. For that coil width, recheck the new winding resistance. Vary conductor foil thickness as appropriate.
9. For the new coil build (due to the new foil thickness) pick a new coil width such that leakage inductance still meets specifications.
10. Now all parameters should be in line except primary inductance. Calculate the air core inductance, and add some type of core to the design such that we have adequate inductance and a pulse droop less than the specification limit.

In one particular case where this procedure sequence was used, the requirement was for a pulse transformer to deliver a quarter megavolt at a gigawatt to a resistive load. Pulse width was in the low microsecond range and repetition rate was below a kilohertz.

If an iron core had not been used, the transformer size (less than 2 cu. ft) would have been around ten times the volume. Meeting both resistance and leakage inductance (less than 100 nanohenry) requirements would have been improbable.

As we said previously, how we start a pulse transformer design depends on which specifiable parameters are the most important. If leading edge pulse fidelity is important and unusually high voltages are not present, one could start by making a ballpark estimate of the turns on the winding with the greatest number of turns from $T_r = 1.2ND$ μsec, where D is the mean winding diameter in inches. Guess a winding diameter. From that go to step 4 above and follow through to step 10. If a viable design is not obtained it will be obvious whether to raise or lower N or D, whether the rise time, pulse width, winding resistance, or other parameters are mutually incompatible.

If pulse width capability is most important and rise time not too significant, it may be simplest to design the unit as a conventional transformer, as outlined in Chapter 2.

If magnetizing current is important, and high voltages do not complicate the situation excessively, determine the primary inductance required, and design it as a choke using the techniques developed in Chapter 3. Then go to step 7 above and rework the design as necessary. If a viable design is not obtained for this case, go to a gapless core (tape-wound toroid) and modify the circuit for full flux reset in the core.

If both leading edge and magnetizing current are important, set up the application for full flux reset, then obtain a trial number of turns from the rise time estimate and a winding diameter guess.

It may seem unscientific to start a design procedure with guesswork. A few design attempts will convince the reader that this design algorithm homes in very rapidly on core and coil dimensions. More complex and exact design tools are available in the references cited at the end of the chapter, but unfortunately the more exact techniques take more time and require a working familiarity with advanced mathematics. This is simpler and gets the result faster. The reason is that this is not really guesswork. What we do is pick the most important parameter or parameters and design for that. Then we modify the design to obtain a close approach to conformance with the remaining requirements *in order of their importance.* You should take the performance requirements and rate them in order of relative importance. If everything is totally important, nothing can be secondary to anything else, the needed performance requirements drown in an overwhelming sea of importances. When one can differentiate the relative significance of the various parameters, the really hard part has been accomplished.

12. THE DESIGN STARTS WITH THE CIRCUIT

Where we start with a pulse transformer design is often a complex interaction with the circuit. Some parameters are the most critical ones, and others are ones that fall out as they may and we have to get them to some acceptable limit or find a way to live with them. Many high frequency converter transformers should be treated as pulse transformers where one starts the design with leading-edge or rise-time considerations, and plays these off against pulse-width capability. Let us take as an example a common situation, that of a gate pulse coupling transformer for the power mosfets in a half-bridge PWM inverter operating at 100 kHz. The first step is to find out what the load really requires. As Ed Oxner of Siliconix showed in his *Powercon 9* paper, the load on the transformer (the mosfet gate) is a highly nonlinear capacitance. We rig up a breadboard gate driver and measure the *average* gate current drawn by the gate when driven by, for example, a 12-volt 100-kHz signal, and the mosfet drain switching a small load. Let us say that we found this average gate current to be 30 mA for this particular mosfet. Let us further say that we want the mosfet switching time to be on the order of 50 nanoseconds. Since there are two switching transitions per cycle, the peak gate current pulse would be about 50 ns wide, twice per 10 μs period. The duty cycle of this waveform is therefore 1%, and for 30 mA average the design peak current would be 3 A. For a 12-volt signal, we should therefore treat the load as having an impedance of 4 ohms. If we assume that the pulse transformer rise time is inductance-limited, then we would want the current rise time (L/R time constant of transformer leakage inductance and the 4 ohm load impedance) to be on the order of 20 nanoseconds. This gives us a leakage inductance limit of 80 nanohenries. The load is too complex for a rigorous analysis to be simple. This approach gives an approximation which is usually close enough.

We now juggle coil dimensions against core flux capacity. When these factors line up we then check core loss, and if there are no thermal problems we then look at primary inductance and see if we can live with it.

Let us assume that our primary and secondaries all have equal turns.

Leakage inductance:

$$80 \times 10^{-9} = \tfrac{2}{3} \ \pi\mu_o \ \frac{r}{b} \ \Delta N_p^2$$

$$3 \cong \frac{r}{b} \ \Delta N_p^2.$$

Take a guess:

$$\Delta = 0.3\,\text{cm}, \qquad r = 0.5\,\text{cm}, b = 2\,\text{cm}$$

$$3 \cong \frac{0.5 \times 0.3}{2} N_p^2$$

$$\frac{6}{0.15} = N_p^2 = 40.$$

$N = 6$ would do.

Flux capacity:

$$100ET = N\Phi.$$

Maximum pulse would be at 100% duty, or $T_p = 5\,\mu\text{s}$:

$$100 \times 12 \times 5 = 6\Phi.$$

Our core must therefore have a flux capacity of 1000 Maxwells.

We decided from an examination of the core loss curves for Magnetics, Inc., type S ferrite (we want the core loss to decrease with increasing temperature) that a flux density of 3000 Gauss would probably give an acceptable core loss for the size core we would end up with. The required core area is then $\Phi/B = A = 0.3\,\text{cm}^2$.

A coil of mean radius 0.5 cm clearly will have enough room for a core of area 0.3 cm².

We need a core with a window length of 1 in. (a bit more than the actual winding length) and a window height of a bit more than ⅛ in., say ³⁄₁₆–¼ in. The core area of 0.3 cm² (≈ 0.05 sq. in.) could be met with a ¼ in. × ¼ in. core area.

We discover from the Magnetics, Inc. catalog that there is a standard E-E core which would do the job.

We can now proceed to design the coil.

Let us say that we want our windings spread over the 0.8 in. winding length. Foil would be the most convenient way of controlling this parameter. Let us further say that 22 mils of insulation between each winding and the next will give us an adequately low level of interwinding capacitance. Two wraps of 0.010 Nomex 411 interleaved with one wrap of 2 mil Kapton will give adequate dielectric strength and adequate wicks for impregnation. Impregnation into the structure of each winding is not important, as the winding has only 12 volts, and a mil or two of turn-to-turn spacing gives us a voltage far below corona inception.

Use 2 mil copper foil slit to 0.800 wide, interleaved with 2 mil Kapton film slit to 0.950. Our coil build is then:

tube	0.030
6-turn primary	0.024
interwinding insulation	0.022
6-turn secondary	0.024
interwinding insulation	0.022
6-turn secondary	0.024
outside wrap	0.012
	0.158

The actual build of the coil itself is 0.116 in., or 0.29 cm. This winding structure will fit comfortably within a window height of 0.25 in., so we have a physical fit.

A few more things need to be checked in order to have a complete design. The reader might find it an interesting exercise to calculate the copper loss, the exact core loss now that we know the core volume, and the primary inductance.

13. MISCELLANY

Interleaving of windings is useful as a means of reducing leakage inductance where interwinding capacitance can be allowed to increase. Interleaving is also useful in some high frequency inverter transformer designs, and in resonant charging reactors which have a second winding to return excess stored energy to the supply (this being the basis of many resonant charging and capacitive load voltage regulation functions). Interleaving was briefly discussed in Chapter 1.

The design of resonant charging reactors is covered in the chapter on power reactors (Chapter 3).

Shields can be useful in controlling interwinding capacitance. The application of shields in power transformers is considered in the chapter on field gradient control (Chapter 7).

Shields do, however, have their drawbacks. The designer is cautioned to calculate the capacitance between windings and shields in pulse transformers, and to ensure that the dielectric thickness is great enough, or the area low enough, so that the capacitance from windings to shields is sufficiently low to preclude its interfering unduly with pulse transformer performance.

If one is not certain of the magnitude of the capacitance between two windings or a winding and a shield, these may be easily calculated from Eq. (6.4), assuming a single primary and a single secondary turn.

In point of fact, one may specify an initial capacitance or leakage inductance, go through the numbers, and come up with a winding width of 2.5 cm (1 in.), a conductor thickness of 0.001 cm (0.00025 in.) and an insulation thickness of 0.0001 cm (0.000025 in.).

One could hardly pick up something so thin and fragile, much less wind with it.

This indicates that, in this case, one should go to round magnet wire of equivalent area. At N turns per layer, S becomes multiplied by N and so becomes more manageable. The wire diameter plus S should now be comparable to the original N $(T + S)$ and performance will be somewhat similar, although definitely not identical. Since we no longer have each turn shielding the stray capacitance of the turn above it from ground, each turn of the layer looks directly at the core or the previous winding or layer. If this is critical the successive layers of a layer-wound coil can all be wound in the same direction, left to right, left to right, etc. In a toroid the capacitance usually becomes higher and more random for the same design compared to a single coil implemented on a C-core.

14. THE CASE OF A COMPARATIVELY HIGH IMPEDANCE CAPACITIVE LOAD DRIVEN FROM A FAST-PULSED VOLTAGE SOURCE

Assuming a pulsed voltage source of low impedance, and a load capacitance which is much larger than the internal capacitances of the transformer, in order to obtain a clean pulse it will be necessary to insert damping resistances in series with the load.

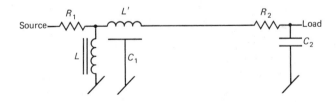

L' is the equivalent lumped leakage inductance. C_1 is the distributed shunt capacitance across L'. C_1 is presumed to be much less than C_2.

R_1 is a resistor (small, but always present) in series with the transformer primary. It can aid in damping of the $L'C_1$ resonance, but if too large it will give excessive pulse droop due to the pulse current which flows through it and the primary inductance L. R_2 damps the $L'C_2$ resonance. It is the essential component for obtaining a clean pulse response where the rise time of the transformer is determined by the leakage inductance and the load capacitance.

As C_1 becomes larger compared to C_2, it will be necessary (for proper damping) for R_1 to become larger compared to R_2, putting more severe demands on the transformer design, as L must become higher. The proper design procedure is to keep C_1 much less than C_2, perhaps 10–20% at most. If $C_1 < C_2$ the impedance of $L'C_1$ will always be higher than that of $L'C_2$, implying that R_1 should always be greater than R_2. If R_1 is greater than R_2, $L'C_1$ will be damped and $L'C_2$ overdamped. It is therefore not practical to damp both resonances to obtain the fastest rise time. If one damps $L'C_2$, there will be an oscillatory component of pulse magnetizing current on the leading edge due to the undamped $L'C_1$ resonance. In order to obtain the proper (desired) rise time (that associated with L', C_2, and R_2), the driver must be able to tolerate this behavior in its load. This circumstance is seen with fast pulse transformers driving power mosfet gates, as well as with grid pulse coupling transformers for power vacuum tubes. One of the more common methods of dealing with these situations is to slow down the rise and fall times of the input pulse to make them comparable to the rise and fall times that would be obtained with faster pulse edges and sufficient damping. In effect, the transformer acts as a low pass filter and we are restricting the signal bandwidth to the pass band of the $L'C_2R_2$ filter.

15. COPING WITH THE REAL WORLD OF TRANSFORMER SPECIFICATIONS

A common occurrence is that a circuit designer will request a transformer and not give sufficient data. For example, let's say we have the following specification:

1. One primary, one secondary, 1 : 1 turns ratio
2. 1000 V in, 1000 V out
3. 100 V maximum droop
4. pulse width ¼–1½ μsec

5. 40 kV isolation between windings
6. 40 nsec rise time

We can make the following observations:

1. It was not specified whether the 1 : 1 turns ratio was an inverting or non-inverting application. The pulse feedthrough capacitance may (depending on construction) be in phase or out of phase with the output pulse, and may markedly affect the rise time.
2. A source impedance was not specified, so one does not know what primary inductance is required, or what level of pulse magnetizing current is tolerable, or what the impedance level of the parasitic inductance and capacitance of the transformer should be.
3. No allowable limits on pulse overshoot or ringing were mentioned, nor is any mention made of pulse fall time or backswing.
4. A load impedance and a duty cycle are not specified, so we do not know what peak or average currents will pass through the transformer.
5. Nothing was said about allowable size, weight, whether a solid, liquid, or compressed gas dielectric is expected or allowable, the operating temperature, other environmental requirements, connections, etc.
6. The circuit in which the transformer is to operate was not given. Even a simplified circuit can be helpful to the designer in understanding the transformer requirements.

Assuming it is not possible at the moment to go back to whoever requested such a device and obtain the needed data, one could do well to apply a bit of deductive reasoning to the problem and perhaps discover some of the design requirements.

Let us say that the quote came from an experimental scientist working in a laboratory, so we assume a laboratory environment and we assume that a liquid-immersed unit will be acceptable. Liquid immersion allows the connections to be in the liquid as well, which at 40 kV is very desirable for safety reasons.

We further assume that because load current was not mentioned, the load current was not large enough to appear significant to the experimenter. Similarly, we assume the duty cycle to be low, probably less than 1%.

If the foregoing assumptions are correct, it is plausible to assume that our experimenter has a pulse generator at ground which can deliver at least a few amperes of pulse current, and probably not more than 20 amperes.

A noteworthy item is the 40 kV isolation requirement. We should start by assuming that however we construct the transformer, the stress between windings will be moderate, say 100 volts/mil. We then have a distance between windings of 0.4 in. At this point, the mean diameter of our winding is guaranteed to be more than 0.8 in.

Let us assume that we will need a core, and that for this short pulse width we will use either a ferrite or a ½ mil tape-wound nickel alloy core, either 50 or 80 nickel.

The volt-second support of our core is given by

$$100 \, ET = N\Phi \; (T \text{ in } \mu\text{s}, \; \Phi \text{ in Maxwells})$$

and a very rough estimation of the rise time of the transformer is $T_r = 3ND$ (T_r is ns, D is in.). Let us see whether we have any conflicts or conclusions with the data and assumptions so far.

Convert the rise time estimate to the CGS system:

$$T_r = 7.5 \, ND \, (\text{cm})$$
$$40 = 7.5 \, ND$$
$$ND \cong 5.$$

$100 \times 1000 \times 1.5 = N\Phi = NBA$, where B is the flux excursion in Gauss and A is the core area in square centimeters. For a ferrite, high nickel, or low nickel core we can assume 7,500, 15,000, and 30,000 Gauss for cores with full flux reset. (Oh yes, remember to tell the customer he should arrange his circuit to provide reset current for the core, so that the full flux capacity can be utilized.) Since the tape-wound alloys will have stacking factors on the order of 50% (including the core case for larger cores), the effective flux densities for low and high nickel become 7,500 and 15,000 Gauss. D is the mean winding diameter, which is not the diameter of the core (see Fig. 6–3).

If we assume for the sake of a simple analysis that all cross sections are round, then our core area A may be expressed as πr^2, where r is found in Fig. 6–3 to be $(D - 1)/2$.

Then

$$\pi r^2 = \frac{\pi}{4} (D - 1)^2$$

and

$$100 \times 1000 \times 1.5 = NB \frac{\pi}{4} (D - 1)^2.$$

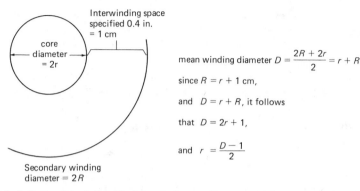

mean winding diameter $D = \dfrac{2R + 2r}{2} = r + R$

since $R = r + 1$ cm,

and $D = r + R$, it follows

that $D = 2r + 1$,

and $r = \dfrac{D - 1}{2}$

Fig. 6-3. Inferring the relationship between the core diameter and the mean winding diameter.

We have from the rise time a relation connecting turns and winding diameter, $ND = 5$, or $D \cong \%$. Substituting,

$$\frac{4}{\pi} \times 1.5 \times 10^5 = NB\left(\frac{5}{N} - 1\right)^2$$

where B is a constant to be selected. This equation has only one unknown, and can be solved for N.

For 15,000 Gauss it is approximately the equation

$$N^2 - 22N + 25 = 0,$$

while for 7,500 Gauss it is approximately the equation

$$N^2 - 35N + 25 = 0.$$

Both of these equations have real solutions, but in both cases N turns out to be slightly negative. That means either that this combination of parameters is impossible to physically realize, or that our rise time estimating parameter is not appropriate to this situation, with a large interwinding space and relatively thin (we presume) windings. Unfortunately, this rise time relationship was derived primarily for this type of geometry, although it is a rough approximation. The results so far indicate that our closest approach to the requirements is with very small N. It is to be expected that very small levels of interwinding capacitance will be required. Just for example, assume an effective shunt capacitance (including the load capacitance, which the experimenter also neglected to specify) of 50 picofarads. To charge that to 1000 volts in 40 nanoseconds requires (from $IT = CE$) 1 ¼ amperes. That implies that our source impedance must be much less than 800 ohms.

If the RC rise time is equal to the L/R rise time, that implies our leakage inductance must be on the order of 32 μH. That implies N much more than a few turns for coil diameters of a few centimeters. If N were one turn, based on volt-second support considerations, a core area of at least 10 square centimeters would be required. We would have a primary inductance on the order of 10–50 μH, and a pulse magnetizing current of about 30–150 amps, which is far too much. For 10 turns we would need only 1 cm^2 square of core area, and could expect a primary inductance of 100–500 μH, for a pulse magnetizing current in the 3–15 ampere range. If the least primary current were to give the maximum allowable droop of 100 volts, our source impedance would need to be about 30 ohms. For 10 turns and a 1 cm^2 core, what might our leakage inductance be?

$$L' = \tfrac{2}{3}\ \pi\mu_0(\frac{r}{b})\Delta N^2 \quad \text{(all dimensions centimeters)}$$

Assuming a fairly short winding length b to keep the interwinding capacitance down, use $r = 1.5, b = 1, D = 1, N = 10$. Then $L' \cong 4\,\mu$H, and for a 30 ohm source impedance the L/R rise time is about 130 ns. We can calculate the interwinding capacitance from the air core pulse transformer formula [Eq. (6.4)], or obtain a better result for this case by using the formula for the capacitance between concentric cylinders. For short, stubby cylinders the end fringing effects will be significant, unless the length is at least several diameters. To reduce the L/R rise time to about 40 ns will require the leakage inductance to be about one-third of its present value, which can be done with $b \cong 3$ cm. For that case we can calculate the interwinding capacitance using a relationship for the capacitance between coaxial cylinders (Radio Engineers, 1970):

$$C = 0.24\,\epsilon_r/\log_{10}\frac{r_2}{r_1}\ \text{pF/cm}$$

where r_2 is the radius of the outer cylinder and r_1 is the radius of the inner cylinder.

For our case of a 3 cm length, assuming $\epsilon_r = 4$,

$$C = \frac{0.24 \times 4}{\log_{10} \dfrac{1.5}{0.5}} \, pF = \frac{0.96}{0.48} \cong 2 \, pF.$$

This indicates that the interwinding capacitance will probably not be significant, as even with fringing effects is seems unlikely it would exceed 6 pF. Intrawinding capacitance may be a problem, but with only 10 turns one could space them a millimeter or so and hold the intrawinding capacitance to perhaps a few picofarads. The customer's load capacitance, however, is going to be the dominant component. Depending on that and the source impedance, we might or might not meet the required rise time with our tentative geometry (length of 3 cm and leakage inductance of about $1\,\frac{1}{3}$ μH). For higher source impedances the rise time might be RC limited, while for a switched voltage source the rise time will be LC limited unless we increase the winding length considerably.

You may have noticed that we did not take some initial requirements, go through a few neat formulas in a totally mechanical manner, and somehow come out with a perfect design. This is the real world, and not all examples have happy endings. We now have an idea of what the tradeoffs are going to be, and can discuss with the customer the tradeoffs between dielectric stress, peak drive power, source impedance, droop, and rise time for this particular situation. Once the customer has made a few decisions, deciding what he is willing to give in order to get what he needs, then we can design a transformer.

7
FIELD GRADIENT CONTROL

1. INTRODUCTION

Physics is the study of this universe, its constituents, and their behavior. To the extent that we understand clearly some part of the subject and can accurately predict the methods to be used for the acquisition of new data, it is a science. To the extent that we do not know what results a particular experiment will have, and rely on intuition to govern the choice of the next experiment or "suddenly realize" that a particular datum is missing or not yet discovered, it is an art.

Many of the basic concepts of physics are an attempt of the physicist to explain something which was observed. A good example of this is the electric field. I personally have never seen one, and I have never met anyone who has. Yet, it has been observed that matter acts as if it were composed of particles, some of which attract or repel each other. We call one type "positive" and the other "negative," simply because they are different. The opposite types attract each other, and two of the same type repel each other. There seems to be no physical thing which we can see moving between our particles, yet there is a force there. It is as if each were "aware" of the other, the awareness becoming fainter with distance.

This property of awareness we call the *electric field*. The characteristic we call *charge*. The electric field is not a thing. It is a manifestation in this universe of a concept; it has direction and magnitude, and fits very nicely into some natural laws we have discovered.

Electrons and protons are particles which carry one unit of charge each. The measure of how much charge an item has is its excess or deficiency of electrons. When the number of electrons and protons in an item do not match, or when they are not neatly arranged very close to each other, an external electric field is manifested.

Conductors are materials on or through which electrons may move freely or easily. Insulators are materials over or through which electrons move only with great difficulty. As the temperature increases, conductors tend to become poorer conductors and insulators tend to become poorer insulators. None of the ordinary materials used in transformer construction will cross over from conductors to insulators in normal circumstances.

2. THE ELECTRIC FIELD GRADIENT

The potential (voltage) of a particular conductor is measured (defined) by the amount of work it takes to transport a unit charge from a great distance up to the surface of the object. That seems to say that we are measuring the charge stored on that object. In fact, we are measuring the electric field which results from that charge, and that also depends in a special way on the physical geometry of the object. This geometric relationship between charge and voltage is defined by $CV = Q$. The greater the capacitance, the less voltage is present when a given amount of charge is stored.

Referring to Table 7-1, look at columns B and F. Column B is the diameter of a sphere or wire. Column F shows the intensity of the electric field (the gradient) at the surface of the conductor. Note that the units are volts per mil of distance. As the diameter of the conductor gets smaller (approaching a point or sharp edge), the intensity of the electric field rises to truly absurd levels. Imagine what it would be at a really sharp edge!

Table 7-1. Electric Field Gradient Versus Radius and Voltage.

(A)	(B)	(C) VOLTAGE FOR $E = 20$	(D) VOLTAGE FOR E	(E) VOLTAGE FOR $E = 500$	(F) E FIELD NEAR CONDUCTOR AT 15 kV
AWG	DIA. (IN.)	VMIL	$= 200$ V/MIL	V/MIL	(V/MIL)
	0.25	2,500	25,000	62,500	120
#10	0.100	1,000	10,000	25,000	300
#14	0.062	620	6,200	15,500	484
#16	0.050	500	5,000	12,500	600
#20	0.031	310	3,100	7,800	962
#24	0.020	200	2,000	5,000	1,500
#26	0.015	150	1,500	3,750	1,923
#30	0.010	100	1,000	2,500	3,000
#40	0.0031	31	310	780	9,620

A useful equation: E (volts per mil) $= 10^{-3} V$ (volts)$/R$ (inches), where R is the surface radius of curvature.
NOTE: At field gradients above about 500 volts per mil in most dielectric media we would expect corona and eventual breakdown starting at the conductor surface and extending outward as the corona breaks down the dielectric and develops sharp-tipped paths of carbon through the dielectric.

3. BREAKDOWN

This is very important, because any gas, liquid, or solid will break down if the electric field gets too intense. What happens is that a stray electric charge (there is always a random electron or ion around) is accelerated by the electric field. It hits an atom, knocking an electron loose from it. These electrons in turn are accelerated, hitting other atoms and knocking more electrons loose. All the positive ions produced are likewise accelerated in the opposite direction, knocking other electrons loose and creating more ions and electrons. You can see how there can be an avalanche multiplication of the number of free charges.

When about as many free charges are generated as are losing their energy through collisions after a short distance, a small blue glow is usually visible around a sharp point in air. Sometimes there will be a short burst of avalanching charges that will die off and then repeat. This is commonly called *corona.* When the number of charges grows and multiplies rapidly, finds or creates a path, and flows rapidly along it, a flash of light is usually visible. This is called an *arc,* a *spark,* or a mistake.

4. THE ELECTRIC FIELD IN A VOID
WITHIN A DIELECTRIC MEDIUM

If we now look at the situation where a solid dielectric medium contains a void, we find something interesting (see Fig. 7-1). Electric fields in the three shapes of voids are quite different. If the dielectric constant of the medium is *K,* the electric field in void A is essentially the same as that in the dielectric. The electric field in void B is *K*/3 times that in the dielectric. The electric field in void C is *K* times that of the electric field in the dielectric. This is true even if the void is in contact with one of the electrodes. The only assumption we have made is that the width dimension of void A is small compared to the length, and vice versa for void C.

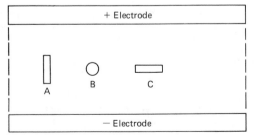

Fig. 7-1. Voids in a dielectric medium.

The gas in a small void (in a solid dielectric) is likely to be either air or low pressure vapor from the potting compound, which will break down at 50 volts per mil or less. If we have designed a working stress level of 50–300 volts per mil there will be, at the very least, internal corona wherever there is such a void. Furthermore, if our solid dielectric consists of something like epoxy resin with a layer of something which epoxy may not bond well to (such as untreated Teflon or polypropylene or even mylar), under thermal cycling a thin crack (void C) may appear. We will have corona. If the void (due to material orientation) is of shape A, then a current can creep along the surface (surface creep dielectric strength is much less than bulk breakdown strength). The remaining bulk thickness of dielectric at each end of that slot would be subjected to nearly the full electrical stress.

5. THE ELECTRIC FIELD GRADIENT AND CORONA

If one electrode in a solid dielectric medium is a sharp edge or corner, we would have a potential situation for corona. It does not matter whether the sharp edge is at high voltage or at ground.

Commonly, designers will very carefully radius and shield all terminals on a high voltage winding and ignore the fact that they now have a smoothly curved electrode facing a sharp one. This is not the best design practice.

Corona is objectionable for several reasons. First, there are military specifications governing the performance of transformers. In some situations a transformer must meet those specifications or the customer will not buy it. It also turns out that the electron and ion activity which is corona will, if allowed to continue, cause a progressive degradation of the dielectric material. It turns into carbon, silicone slime, or other partly decomposed materials. Electrical breakdown will inevitably follow—the only question is when.

Corona is also objectionable because the electron and ion activity, being somewhat random, induces small, high frequency currents in external circuitry. This electrical noise can interfere with the proper functioning of some systems.

You can see now that it is important: (a) to use not-too-fine leads coming out of transformer windings; (b) to interpose a larger radius conductor between any sharp edge at high voltages (such as the edge of a shield) and ground, or other surfaces at largely different voltages; (c) never to leave an exposed sharp point or edge of any kind around a high voltage assembly; (d) to avoid incompatible dielectric materials; and (3) to avoid incomplete impregnation or potting.

One other noteworthy item—never leave a conductor (such as a field-shaping electrode) without a definite conductive path to somewhere. Conductors "floating" in a high voltage region can occasionally cause problems. For example, it may turn out that the insulation resistance from the conductor to the physically nearer electrode is much higher in proportion than the insulation resistance to the farther electrode. In this case the capacitance of the conductor charges up to a voltage determined by the ratio of insulation resistances. At this potential, the conductor may then break down the shorter length of insulating material. If the energy involved is not enough to establish a permanent arc path, an intermittent discharge will ensue. This will generate electrical noise, much as corona does, and in some cases can lead progressively to total failure of the insulation system. Dielectric uniformity and positive control of the potential of all conductive materials play an important role in the design of reliable high voltage components and systems.

6. A PHENOMENON AT A DIELECTRIC INTERFACE

At a dielectric interface between air or some gas and a solid dielectric, some very interesting things can happen. Let us look at the situation shown in Fig. 7-2. An electron (there is always a free electron floating around) is accelerated by the electric field toward the positive electrode. It hits the dielectric surface and knocks another electron loose, leaving a positive ion behind. Both electrons continue on, hitting the dielectric surface again, knocking another two electrons loose, and leaving two positive ions behind. All four electrons continue on their way, and in this way we can build up an avalanche of

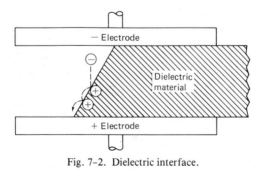

Fig. 7-2. Dielectric interface.

electrons, leaving a path of positive ions behind on the dielectric surface. This creates a conductive path on the dielectric surface, which electrons from the negative electrode can then flow across. This is called *breakdown*.

Depending on the dielectric constant of the material, the minimum breakdown potential will be when the dielectric surface between the electrodes makes roughly a right angle with the negative electrode. Now look at the slightly different situation illustrated in Fig. 7–3.

The electron, as it is accelerated by the electric field toward the positive electrode, moves *away* from the dielectric surface. No positive ions are formed. The breakdown potential of this geometry may be as much as a factor of two or three higher than that of the previous situation. This phenomenon has actually been used to make three-terminal spark gaps, wherein a trigger electrode is embedded in the dielectric and biased in such a manner as to pull the electric field into or out of the dielectric surface. Electrons are thus caused to be accelerated either toward or away from the surface.

The significance here is that the transformer designer, being aware of this phenomenon, can make an effort to shape the insulating surfaces in high voltage designs to minimize the likelihood of electrical breakdown.

7. MECHANICAL FIELD GRADIENT CONTROL TECHNIQUES

Many high voltage windings are wound with relatively fine wire. While the field gradient inside the winding structure may be quite uniform, the wire itself is far too small to bring out as a self lead due to the intense electrical field associated with a wire of that small diameter in free space. It may be desirable to terminate the fine wire to a larger one. This can be done in the following manner.

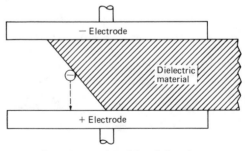

Fig. 7–3. Another dielectric interface.

Strip an appropriate length of each wire (¼ in. or more of the smaller one, ⅛ in. or more of the larger); wrap a few turns of the smaller around the larger (see Fig. 7-4). Solder the wires together in the normal manner. Take a pair of flush-cutting diagonal cutters and cut off all but about ⅛ in. of the soldered part. Touch the soldered part to the oxidized shank of the hot part of a soldering iron (hot enough to melt solder, but where the solder will not wet the surface). Add a little solder to the joint. As it melts it will wet the already soldered wire and form a neat ball.

When it becomes necessary to have smoothly rounded inner edges of the core which face the coil, there are a number of ways to accomplish this end. Some methods are more suitable for a single unit than for a production run, and some are more suitable for C-cores than stacked laminations or ferrites.

To radius the edges of a C-core one may most expediently use a fine file (fairly new and sharp). It is vitally important that every stroke of the file be *toward* the edges of the laminations. That way the cutting face of the file tends to hold the laminations in place. If the motion of the file were toward the middle of the core, the cutting force of the file would tend to lift up and peel back the laminations. That would destroy the core in short order.

If this same end must be accomplished on a ferrite there are two options. First, the ferrite manufacturer may be able to provide a special core with the appropriate edges already radiused. This is preferable for production runs. In other cases the ferrite cores, along with a drawing of the desired shape modifications, may be sent out to a local company which does form grinding with diamond grinding wheels. Such grinding can also be done in house. Obtain a small bench grinder, and ask a local supplier of grinding wheels for a *soft silicon carbide* grinding wheel, fine grit. The supplier will know what that means. Also get a tool for dressing grinding wheels and use it as you do the grinding, because soft grinding wheels wear somewhat faster than the hard-bonded wheels.

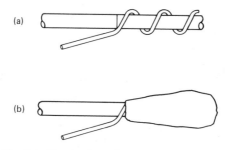

Fig. 7-4. Terminating a fine wire to a larger wire.

This operation can be performed on a core made from a stack of E–I laminations. The laminations must be carefully and exactly assembled in the position in which they will be reassembled around the core, and held in place with a few small C-clamps. The stack of laminations may then be filed, taking care always to file toward the edge of a lamination. This is a fairly dicey operation, not too suitable for high volume production. Where the coil may be spaced away from the core on two sides, the following techniques are usually more suitable.

Various items of rounded conductive material may be bonded to the core to control the electric field gradient (see Fig. 7–5).

For physically larger structures and higher voltages it may be appropriate to take a piece of hard copper tube, remove a 90–120° segment, and bond it over the corner of the core. Note that since a voltage gradient exists along the edges of the laminations, it is important to place some insulating material (tape, paper, etc.) between the tubing and the exposed lamination edges. Otherwise a partial shorted turn may result [Fig. 7–6(a)]. Note that for most efficient utilization of space the segment should be cut out, and perhaps the tubing deformed, so that the corner of the core rests on the inside surface of the tube. In many cases it may be simpler to form strips of copper foil as indicated in Fig. 7–6(b), and bond them to the inner portions of the core window facing the high voltage coil. With radii of curvature appropriate to the potentials involved, a design with moderate and uniform electric fields can be obtained.

The phrase *corona ring* or *corona roll* has come to mean a smoothly curved piece of conductive material, often in the shape of a ring, which shields the sharp points or edges of the assembly behind it from the high

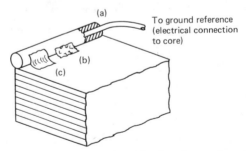

Fig. 7–5. Shielding the edge of a core. (a) Wire soldered to a larger wire or tube; perhaps #10–14 tinned copper wire, or copper tubing as appropriate. (b) Epoxy or cyanoacrylate adhesive to hold in place. If conductive epoxy is used here, (a) would not be necessary. (c) 3M #1170 aluminum foil tape with conductive adhesive.

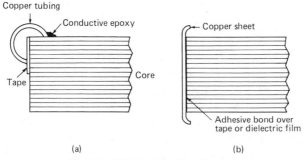

Copper tubing

Conductive epoxy

Copper sheet

Tape

Core

Adhesive bond over
tape or dielectric film

(a)   (b)

Fig. 7-6. Shielding the edge of a core.

voltage field between that assembly and elsewhere. The corona ring would be tied to the high potential source (to which the assembly to be shielded is referenced) by a direct low impedance path such that the discharge of stray capacitances does not pass through passive or active circuit components. Fifty picofarads of stray capacitance charged to 20 kV stores 10 millijoules of energy. If allowed to discharge through a small rectifier diode, that energy could quite easily destroy it. Take, for example, the circuit illustrated in Fig. 7-7.

Let us say that our load requires a maximum current of 20 mA. This implies relatively small components for the rectifier/filter and regulator. All the components could be assembled on 10 square inches of board area. The Zener and the resistor may dissipate a few watts, and small heat radiators would probably be sufficient. All these things have points and sharp edges. Such an assembly would perhaps be shielded with corona rings in the manner shown in Fig. 7-8. The top and bottom surfaces are conductive planes, smoothly joined to wire or tubing at the edges. The edge of the component assembly is located in from the corona rings a distance on the order of the gap between the corona rings. The ring diameter in this example is on the order of the gap between rings. For these relative proportions, most of the

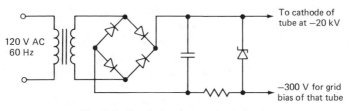

To cathode of
tube at −20 kV

120 V AC
60 Hz

−300 V for grid
bias of that tube

Fig. 7-7. Simple floating regulator circuit.

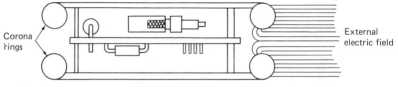

Fig. 7-8. Shielded assembly.

electric field goes to the corona rings as shown. Very little goes to the assembly. The proportions indicated above are guidelines only.

In another example, a corona ring could be used to shield edges of shields and fine wires associated with a transformer winding. In most cases, simply making the width of the shield equal to that of the high voltage winding does a fairly decent job of hiding the sharp shield edge from the field while sacrificing a minimum of shielding efficiency. The overall shape of the winding, including its shield, is still that of a rectangular cylinder, and the ends of the cylinders are relatively sharp edges. The dielectric constant of the layer insulation may not be sufficiently higher than that of the surrounding medium to adequately distort the electric field. In this case one could either cover each end of the winding with a corona roll of relatively high resistance material (to avoid a shorted turn) but with a low enough resistance to act as a field-shaping electrode when its potential is referenced to the outside of the winding or the shield. Alternatively, one can progressively reduce the widths of the last few layers of the winding so that the winding layer edges approximate the desired radius of curvature. The former technique allows most of the winding to be shielded, while the latter technique requires less space in the core window, hence gives a smaller design at the expense of shielding efficiency. The latter technique is also very difficult to do neatly without an automatic traverse winding machine.

8. ELECTRICAL FIELD GRADIENT
CONTROL TECHNIQUES

For a high voltage output transformer with many tens of layers and many thousands of turns, space may be at a premium. Let us say that the end of the winding near the primary is floating at 16,000 volts DC, while the other end of the secondary winding has 16,000 volts AC superimposed on 16,000 volts DC. (Just what kinds of rectifier circuit give rise to this sort of situation will be discussed shortly.)

In this case we are looking at the amount of insulation required between the edge of the coil and the core, as well as the insulation above and below the winding, and how best to utilize that space. Let us say that we choose working stress levels for our insulation system of 200 volts DC per mil and 100 volts per mil AC. At the bottom edge of the coil we would require 80 mils of insulation, and 240 mils at the top. The most efficient method of space utilization would be to smoothly taper the winding width to obtain the desired margin profile. This would best be done on a winding machine with automatic traverse. Adding a small shim on alternate sides of the traverse mechanism at the end of each layer would progressively narrow the winding width. On very large, hand-wound transformers, this would be done by hand. Various widths of insulating material could be used to guide the winder.

Rectifier configurations in which one end of the transformer effectively operates at a DC potential, and the other end has all the AC voltage of the winding, are often desirable because they minimize the electrical stress on the transformer.

Take, for example, a transformer that feeds a full wave bridge which delivers 28,000 V DC. Each end of the winding must be insulated for the peak AC voltage which appears there. If one DC terminal of the bridge is grounded, the stress on *each* end of the winding is 14,000 V DC ± 14,000 V AC (peak).

Now let us look at a full-wave doubler. Admittedly, the ripple amplitude may be higher, but it has some interesting consequences in transformer design. In Fig. 7-9, the end of the winding connected to the midpoint of the two capacitors floats at half the output voltage. The end connected to the two rectifiers goes from ground to 28,000 V, and so its operating level is seen to be 14,000 V DC ± 14,000 V AC (peak). Clearly, less insulation would be required in this situation—it would allow the physical size of the transformer to be reduced. In a given volume, more conservative stress levels could be employed. (It does, however, require identification of the winding terminations in relation to the external rectifier circuit.)

Another very useful configuration is the full-wave quadrupler shown in

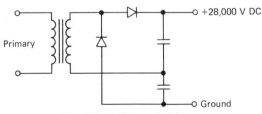

Fig. 7-9. Full wave doubler.

Fig. 7-10. Note that one end of the winding floats at half the DC output voltage, and the other end has an AC voltage superimposed on that DC voltage. Since the full-wave quadrupler may be seen to be two full-wave doublers, that AC voltage must have a peak-to-peak value equal to half the output voltage. So our transformer secondary winding would have to develop 14,000 volts P-P or (for sine wave excitation) about 5,000 volts RMS. The RMS secondary current is, of course, twice that of the full-wave doubler (as found from Schade's curves in Appendix C).

There you have it. Minimal working stress on the insulation system of the transformer, relatively few turns for the output voltage, and one end of the winding tied to a DC potential, with the tradeoff being higher ripple amplitude or larger output filter components.

The case of the three-phase transformer with a wye secondary is extremely interesting (see Fig. 7-11). Wye secondaries are usually used for high output voltages, so that no one winding has to develop the entire output voltage. The number of turns of wire required is thereby minimized. On the other hand, for high current, low voltage secondaries a delta is usually used, so that all three secondaries may share the current for better thermal design and smaller conductor diameters.

Note that the midpoint of the wye secondary shown above will float at half the DC output voltage.

If the end of each secondary nearest the primary winding is connected to the wye center, a minimum of insulation is required between primary and secondary.

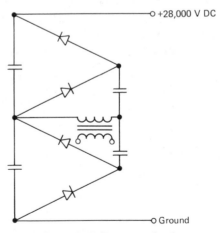

Fig. 7-10. Full wave quadrupler.

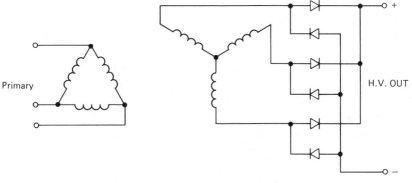

Fig. 7-11. Three-phase Δ-Y circuit.

The insulation on the outside of each coil between it and the core must of course be half the DC output voltage plus the induced winding voltage. The insulation between coils (assuming sine wave excitation, 120° phase shift between coils) would be that necessary to hold off 1.5 times the peak AC voltage per leg.

9. USE OF SHIELDS IN HIGH VOLTAGE TRANSFORMERS

I am going to discuss a very important concept relating to the use of electrostatic shields in high voltage transformers. This will lead in to general concepts regarding use of shields in general-purpose transformers. Referring to Fig. 7-12, let us look at a low-power supply (the bias supply) floating at high

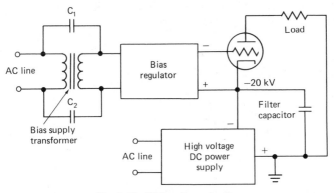

Fig. 7-12. High voltage circuit.

voltage as part of the larger circuit including some of the load components and ground returns and showing some stray capacitances.

Here we have a simplified schematic of some sort of high voltage regulator, controlling the power from the power supply to the load. Note the capacitors C_1 and C_2. Let these represent the stray capacitance from the finish of the primary winding to the start of the secondary winding (C_1), and from the finish of the secondary winding to the start of the primary winding (C_2). Usually one (C_1) will be relatively large and the other very small by comparison. A plausible value for the larger would be in the range of 30–300 pF or so, depending. *That capacitor is charged to essentially the full potential of the high voltage DC power supply.* Let us say that the regulator tube arcs internally, some external part of the system flashes over to ground, or a triggered spark gap inside the high voltage power supply may fire. The 20 kV potential at the cathode of our vacuum tube goes essentially at once to ground.

Now, what happened to C_1? A moment ago it was sitting there, blissfully charged to 20 kV. Let us look at the equivalent circuit of our new situation (Fig. 7–13).

What is going to keep C_1 from discharging through this external circuit? Certainly not the components of the bias regulator–that is only a 200 volt regulator. The components certainly were not designed to withstand 20 kV. What else is left? The AC power line? Notice we have not drawn any explicit connections from the AC power line to ground. Do you think the AC power line is going to float up to some high voltage and nothing else connected to that power line would complain? No, indeed.

We presume that one of the two AC input power terminals is usually pretty close to ground, but what about the other one? It goes to other things, maybe through a push button switch on the front panel. Is that switch rated to hold off 20 kV? What about the guy pushing the button?

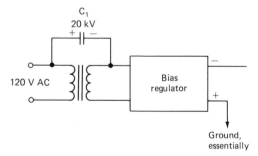

Fig. 7–13. Part of the circuit of Fig. 7–12 after a high voltage arc.

You can see that at best, the energy stored in the interwinding capacitance of the transformer and other stray capacitances (there are *always* stray capacitances to *somewhere* from *everything*) will discharge only through the bias supply components, most likely blasting some component. At worst, the discharge path will include the AC power line, the switch on the front panel, the president of the company who bought your transformer to go in his equipment, and his customer (who happened to be touching both the president and the cabinet containing the transformer). How does business look now?

If there is a way, high voltage will find it. It is up to you to make it foolproof. Here is how:

1. Control the stray capacitance between windings with electrostatic shields, so that the stray capacitance is to the shield and much less to other windings or the core.

2. Provide absolute return paths for all stored energy, so that the discharge path of that energy is controlled by design.

3. Ensure that stored energy is returned to its source.

4. If such a discharge path must pass through an area containing sensitive components, then install energy-diverting components such that it passes around, not through.

Let us look at a start on doing this (Fig. 7–14). Here we have a basic concept. Notice how the energy stored in the capacitance of the winding to the right-hand shield is only that associated with the output voltage of the bias supply itself.

Notice how the right-hand shield is at the high voltage source point for the circuit.

Notice how the left-hand shield becomes a return path for the energy stored in the capacitance of the high voltage shield.

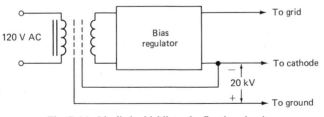

Fig. 7-14. Idealistic shielding of a floating circuit.

Notice how the primary winding is now shielded from high voltage transients.

On paper this works very nicely, but in the real world there is a minor detail which deserves our attention. The wire from the left-hand shield in Fig. 7–14 to "ground" has some inductance, associated mainly with its length. A high voltage arc is a fast rising spike, with a rise time in nanoseconds. The discharge current will therefore develop a voltage spike of some magnitude along that ground return wire, thereby elevating the potential of the left-hand shield to, in some cases, an appreciable fraction of the high voltage supply. In this case a moderate transient would be coupled into the AC power line. The solution is to add a third shield (Fig. 7–15).

There is another reason for the third shield. "Ground" is sometimes a nebulous location. The exact point at which the low potential point of the high voltage energy storage capacitor is located is rarely the point at "ground" to which a high voltage arc will leap. The high frequency impedance between these two points may also develop a voltage due to the arc, and this voltage can also appear on the middle shield of Fig. 7–15. The left-hand shield of that figure serves to conduct those transients into power line ground. Since any properly designed high voltage system contains a low impedance path from the prime power ground reference to high voltage ground, this concept provides for positive control of high voltage transients.

The implementation of this technique in most transformers requires that the two shields associated with the high voltage circuitry (the two right shields of Fig. 7–15) be located both between the primary and secondary windings and on the outside of the secondary winding. This is shown explicitly in Fig. 7–16.

In addition to controlling the energy stored in the stray capacitance between the external surface of the high voltage winding and nearby components, it has the convenient advantage that the external surface of the unit is now largely at ground. The spacing to nearby components is thus made less critical and the tendency for dust pickup and corona is reduced.

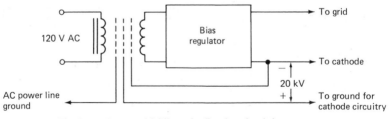

Fig. 7–15. Proper shielding of a floating circuit in some cases.

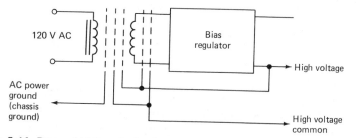

Fig. 7-16. Proper shielding of a floating circuit for best control of stray capacitance.

You may note that I haven't said what to connect the core to. It varies somewhat. In high voltage transformers it is usually connected to chassis common. In high frequency inverter transformers (usually for reasons of noice control, but perhaps on occasion for other reasons) it may need to be connected to high frequency AC common, or perhaps power line common, or it may even be appropriate to allow it to float. I'm not going to give you a pat answer to this because there isn't one. Look at each individual situation in a new unit of time, use your common sense, and do what seems right. If the results aren't optimum, repeat the above steps.

There is an additional consideration peculiar to high frequency inverters operating off the AC power line. If high frequency noise spikes on the primary winding are giving electromagnetic interference problems, it would be appropriate to place an additional shield on each side of the primary winding (above and below) and return that to the power common point of the inverter transformer drive circuitry. This serves to return the currents due to those voltage transitions to their source. In this implementation, our transformer would now appear as shown in Fig. 7-17.

There is one more consideration, relating mainly to transformers used in

Fig. 7-17. Primary shields added to Fig. 7-16.

low noise power supplies. It may be, when all appropriate shields have been implemented and properly connected, that a certain very small amount of AC ripple exists on the load. All attempts to remove it fail, and the regulator does not appear to be generating it. There may be some indication that it acts like common mode noise at the transformer operating frequency, and that it seems to be coming somehow from the transformer. It could very well be stray capacitance from the winding edges, coming out past the shields and going to the core or elsewhere. The cure for that is to make the shields wider than the winding, bend the edges inwards, or make total box shields or some such. This is extremely difficult to implement on high voltage transformers.

There is another possibility which should be evaluated first. Assume that we are still talking about the transformer of Fig. 7-17, and that we are looking at the last layer of the secondary winding and the shield right above it (the first shield to the right of the secondary winding in Fig. 7-17—see Fig. 7-18).

You can see that there is some stray capacitance from each turn of the winding to the shield. There is a (different) AC potential on each turn, and different AC currents will flow between each turn and the shield. The source of those currents is the turns of the windings themselves. Those currents *will* return to those sources. The only question if how. If the shield were returned to the last turn of the winding, then that total current would just circulate back through the turns of the winding to their sources. However, that shield is returned to the output of the regulator. The current will therefore circulate back through the external circuitry and its impedance (current sensing resistors, wiring inductance, rectifier diodes, etc.) to get back to its source. That current, flowing through that impedance, gives rise to a voltage which modulates the output voltage of our regulators.

The solution in this case is to implement first and last layer shields on each winding where needed. Our transformer (and a rather general one it is) now would look like Fig. 7-19, with the shields numbered for reference.

Now that you have learned to do it the hard way, we can take a few short cuts.

Very rarely would all these shields be needed in every circumstance. Some

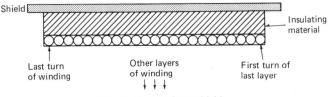

Fig. 7-18. Last layer shield.

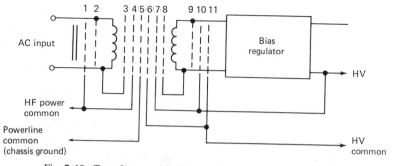

Fig. 7-19. Transformer with all necessary shields for any occasion.

circuit configurations can greatly simplify the implementation of needed shields. You will see that some become phantom shields, their function merged into that of another shield. Take the full-wave doubler of Fig. 7-20, for example. The following comments would apply equally well to the balanced quadrupler of Fig. 7-10. Let us look at what we have here. Our experience and the customer's application have (presumably) told us that primary first and last layer shields (2 and 3 of Fig. 7-19) are not needed. We have the AC power line and not a high frequency inverter for an AC power source, so shields 1 and 4 of Fig. 7-19 are not needed. Shield A of Fig. 7-20 corresponds to shield 5 of Fig. 7-19.

Shield C is beneath the first layer of the secondary, and shield D is above the last layer. Shield C clearly corresponds to the first layer shield 8, but it also fulfills the function of shield 7 even though it is not connected to the high voltage output. Being connected to the junction of the two filter capacitors (which are, or should be, identical and of reasonably low inductance) it has a low impedance path to the high voltage output. Note also that the insulation required between it and shield B at ground must only hold off half the output voltage.

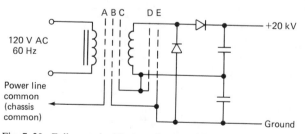

Fig. 7-20. Full wave doubler transformer with simplified shielding.

Shield D clearly corresponds to shield 10. Being above the last layer and returned to one end of the winding, one might think that it would perform some of the function of the last layer shield (9). As it turns out, it does. Not as well as a discrete last layer shield might, but fairly well in most cases, and the benefits are free. Note that the insulation between shield D and the last layer of the winding must hold off only the AC voltage developed by the secondary winding.

Shield E corresponds to shield 11. Note that here also the insulation required between shields D and E is only half the DC output voltage.

A similar line of reasoning may be used for the primary shields to be associated with the primary winding of a high-frequency inverter transformer when the drive circuitry is a half bridge.

You see that once you understand the theory behind shields in transformers, you can look at a particular situation and determine which of these shields are necessary and, from their function, specify the proper location for connecting them. Some manufacturers of transformers will install the five shields of Fig. 7–20 but not tell the customer exactly what they did. To all external appearances, the transformer may have no shields whatsoever if shields A, B, and E are connected internally to the mounting surface of the transformer. The performance will be noticeably different from one without any shields.

There is a very important detail relating to shield construction in certain types of high voltage transformers. In certain applications, a high voltage transformer may have a shield which is held at the DC isolation potential—such as a transformer which supplies low voltages (heater, etc.) to a load floating at a high potential—or our transformer may be a doubler or quadrupler high voltage output transformer with a shield floating at half the output potential. In either case, the application may be one in which the load can arc or be crowbarred. In that case, the energy stored in the capacitance of the shield to ground will discharge very rapidly. Most designers will construct the shield by connecting the external lead to one end of the foil, taping the start of the shield down, wrapping the foil around the coil, taping down the finish, and proceeding with the construction and winding of the coil. Let us look at the charge distribution of such an arrangement (Fig. 7–21).

Consider what happens when there is an arc and the high voltage shield is abruptly shorted to ground. A high peak current flows out of the shield lead as all the charge stored in the high voltage shield capacitance flows rapidly to ground. Note, however, that some of the charges of Fig. 7–21 are on the opposite side of the core from the shield lead. That charge must flow through the hole in the middle of the core in order to reach ground. *That charge tran-*

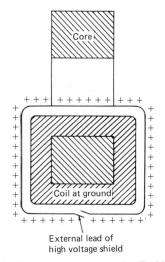

Fig. 7-21. Cross section of core and coil with shield.

sient induces a high voltage transient in every turn of every winding on the transformer. In a 20,000-volt isolation transformer it can theoretically approach 20,000 volts per turn! In practice, core losses and the stray capacitance of other windings will damp it considerably, but an induced voltage spike of even 200 volts per turn can create a very nasty transient, which can not only cause external components to fail, but also can cause dielectric failures within the winding structure.

The way to virtually eliminate this problem is to balance the stored charge on the shield so that half of it discharges in each of two opposite directions through the core, and a minimum of the charge passes through the core in the first place. Such a balanced construction is shown in Fig. 7-22.

Notice that the shield connection is at the *outside center* of the coil, so the charge stored on the shield inside the core flows symmetrically in both directions away from the center line; thus in theory the net induced voltage is zero if the stored charge on each half of the shield portion inside the core window is zero. Notice also that most of the charge stored on the shield is discharged externally, without ever flowing through the core at all. Constructing a shield in this manner requires no more effort than constructing it the wrong way, and one creates not only a vastly more reliable transformer but imparts much more reliability to the circuitry the transformer serves. System reliability is thus raised, and that's really what it's all about.

The attentive reader will notice that the correctly implemented shield, hav-

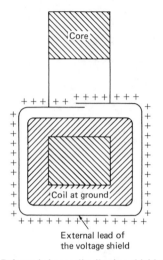

Fig. 7-22. Balanced charge distribution shield construction.

ing no flux linkages to the core, has in effect very little equivalent series inductance. It thus acts as an effective shield in high frequency or low noise applications. Contrast this with the worst possible shield implementation, that in which copper foil laminated to mylar tape (mylar slightly wider than the foil) is wrapped around a toroid to effect a "shield" between windings on a toroidal transformer. Each wrap of the shield adds another large increment of inductance in series with the shunt capacitance of the shield, and the first part of the shield is almost entirely isolated from the last part for high frequencies! The virtual impracticality of implementing an effective high voltage shield on a toroidal transformer is a very good reason for not designing high voltage toroidal transformers if any other implementation is possible.

An awareness of these concepts will greatly facilitate the design of reliable compact high voltage transformers. As the designer is now aware, it requires not only clever theoretical design but meticulous attention to the physical details of transformer fabrication and processing to ensure that the designer ends up with what he thought he was designing.

8
HEAT TRANSFER

1. INTRODUCTION

There is one concept, the most basic and important, which the engineer must understand about heat in transformers or in any component.

It does not vanish. It does not go away. It goes *somewhere*—and you have to make sure that it can go wherever you want it to, and easily.

Heat is energy. It flows, whenever there is a temperature difference, from a region of higher concentration to a region of lower concentration. If an energy source is generating heat in a certain volume at a constant rate, the temperature of that volume will rise until the temperature difference between that volume and its surroundings is sufficiently high that some heat transfer mechanism will remove heat as fast as it is being generated. (There is always *some* heat transfer mechanism available. There is no such thing as a perfect insulator.)

The stuff of which this universe is made, if left to its own devices, appears to behave in a certain repeatable fashion. We call this behavior *natural laws*. The conservation of energy is one of these laws. The fact that we call these observed phenomena *laws* does not mean that nothing will ever happen that violates them—it just means that over a great period of time many individuals have made careful and extensive observations, and things seem to have behaved in a certain consistent manner so far. In order to express our confidence (and our desire) that things continue to behave as they seem to, we call some physical phenomena *laws*. It is well to bear this in mind, as sometimes natural laws turn out to be nothing more than someone's fixed idea of how they suppose things are. At one time "everybody knew" that heavier-than-air machines could not possibly fly, and that the sun revolved around the earth.

In this subject we are dealing with some fairly basic, well known phenomena. Heat can be transferred by three basic means—conduction, convection, and radiation. There are also other phenomena involving heat transfer. The Peltier and Seebeck effects involve the conversion of heat to electricity and vice versa by the passage of an electric current through the junction of dissimilar conductors. Thermocouples and thermoelectric coolers operate on these principles. The Nernst effect involves an electric potential developed by the flow of heat across a magnetic field. Certain salts can be cooled by application and removal of a magnetic field. These phenomena have very little to do with conventional transformer design. The heat absorbed by a liquid when it boils, called the *heat of vaporization*, is very important in some liquid-cooled designs and will be discussed later in this chapter.

The manifestations of energy most relevant to this subject are the motion of masses and the exchange or transfer of quanta of electromagnetic energy, called *photons*. The transfer of photons is called *radiation*.

Energy can be transferred by the coupling of motion from one mass to another, or the transmission of a photon from a particle to free space or to another particle. When a particle of mass transmits a photon, it loses some of its energy of motion. When it absorbs a photon it gains energy of motion. When the particles that comprise a mass have more (random) energy of motion, the mass is said to have a higher temperature.

2. CONDUCTION

In the heat transfer mechanism called *conduction*, two masses are physically coupled together. The energy of motion of the particles comprising one portion is coupled to the other portion across the area of contact.

3. CONVECTION

In the heat transfer mechanism called *convection*, some portion of our system is a liquid or gas in which the particles have attractive forces much weaker than those present in solids. In liquid or gases the particles are free to move independently under the influence of attractive or external forces. A hotter liquid or gas is less dense than a cooler one and the greater energy of motion of the individual particles keeps them further apart. Reduced density of some portion of a liquid or gas relative to some other portion causes the lower-density portion to experience a buoyant upward force. Bulk physical motion is thereby coupled into the fluid, and the heated material moves upward and presumably away from the heating surface. A cooler portion of the fluid is pulled in from beneath to take its place, and we have a net transport of thermal energy away from the hotter region. The reverse effect takes place when a hot fluid meets a cooler surface or portion of the fluid.

The convection phenomenon requires the presence of acceleration or a gravitational field. In the absence of gravity or acceleration there is no convection.

4. RADIATION

Heat transfer by radiation is the absorption or emission of photons by particles of mass. Particles of mass, since they have electromagnetic properties, radiate energy when they are in motion. They are surrounded by other similarly behaving particles, and so an equilibrium condition is attained where the emission and absorption of photons is equal, and on the average our particles neither gain nor lose energy.

Nothing can be totally isolated from radiative heat transfer. At any temperature above absolute zero $(-273.16°C)$ particles radiate energy. Deep space (between galaxies) is filled with background radiation at about 3° above absolute zero.

The intensity of radiation from an object depends on the fourth power of its absolute temperature. The intensity of radiation from one object to another (the rate of radiative heat transfer per unit area) depends on the difference in the fourth powers of their absolute temperatures. It also depends on a property of the surface called emissivity. For a perfect "black body" radiator the emissivity (ϵ) is unity. For everything in the real world it is less than one, and greater than zero.

ϵ is about 0.9 for most electrical nonconductors, and for metals coated or painted at least 0.002 in. thick with an externally nonconducting material. The decision of whether or not to coat a metal part to enhance heat transfer through radiation but slightly inhibit heat transfer through convection, can be made for a particular case depending on the temperatures and the temperature differences involved. For metals ϵ varies roughly from 0.04 (copper) to 0.25 (polished stainless steel). The lower range relates to materials of lower electrical resistivity, with ϵ increasing roughly as the resistivity.

5. HEAT TRANSFER

Radiative heat transfer is not always a significant heat transfer mode in transformers, since it presumes a fairly "toasty" design (or a high altitude or space application). The case of (for example) 55% convective heat transfer,

45% radiative heat transfer presumes a particular transformer-to-ambient temperature relationship. Fig. 8-1 shows the radiative heat transfer coefficient for various temperatures. A comparison of this graph with the one for convective heat transfer (Fig. 8-5) will indicate the magnitude of the difference between these mechanisms at the operating temperature of a particular design. One can definitely not make the blanket assumption that some particular fraction of the heat will be transferred by radiation and the balance by convection.

When uncertain about what the operating temperature of a unit will be, ignore the radiative component (it can only help lower the temperature) and calculate the temperature based on convection off the exposed surface and conduction through the mounting surface, and see what the temperature rise is. The radiative component can then be allowed for, new convection and conduction characteristics calculated, and the actual temperature arrived at by successive approximation.

In calculating heat flow through the mounting surface, remember that the heat, once transferred to the chassis, has to go *somewhere*. There is a certain

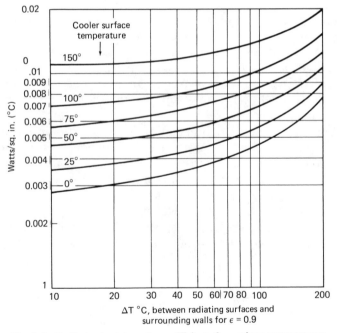

Fig. 8-1. Radiative heat transfer coefficients for various temperatures.

resistance to heat flow through matter, so the chassis temperature at the heat source will be higher than elsewhere (unless, of course, something else which generates a lot of heat is mounted on the chassis right next to the transformer—but that is not usually wise).

The concept of thermal resistance to heat flow is directly analogous to the opposition a resistance offers to the flow of a current. The thermal energy corresponds to electrical current, and the voltage drop corresponds to the temperature difference.

6. HEAT FLOW AND THERMAL RESISTANCE

Just as we use $R = \varrho l/A$ and $E = IR$ to calculate the resistivity and voltage drop of an electrical conductor, so we use $T_{max} = T_{surface} + Q(b/Ak)$ to calculate temperature drop. Q is heat input in watts, b is the conduction length, A is the area the heat flows across, and k is the thermal conductivity of the material (the reciprocal of bulk thermal resistivity). b/A corresponds to the term l/A used to calculate resistance, and $T_{max} = T_{surface} + Q(b/Ak)$ corresponds to $E = E_0 + IR$, Ohm's law with a zero offset term to correct for the potential of the surroundings.

Take the case of a simple slab of material, of area A and thickness b, with a heat input on one surface of Q watts (Fig. 8-2):

$$T_{max} = T_{surface} + \frac{Qb}{Ak}. \qquad (8.1)$$

If Q were expressed in watts per square inch our equation would become

$$T_{max} = T_{surface} + \frac{Qb}{k}. \qquad (8.2)$$

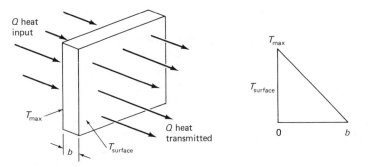

Fig. 8-2. Temperature versus distance for a simple slab.

You may have seen the thermal resistance of a transistor to its mounting surface or the thermal resistance of a heat sink expressed in degrees C per watt. Each has a particular area, and so for each item there is a particular thermal resistance which we can add just as with electrical resistors in series. The temperature rise of a composite of several slabs (assuming zero thermal resistance at the interfaces)* is given by Fig. 8–3:

$$T_{max} = T_{surface} + \frac{Q}{A} \left[\frac{b_1}{k_1} + \frac{b_2}{k_2} + \frac{b_3}{k_3} + \frac{b_4}{k_4} \right] \qquad (8.3)$$

When we have a uniformly heated volume (such as a winding on a core or the core itself) the temperature rise is shown in Fig. 8–4.

With the heat uniformly generated at q watts per cubic inch,

$$T_{max} = T_{surface} + \frac{\frac{1}{2}qb^2}{k'} \qquad (8.4)$$

where k' is the thermal conductivity determined by the volume percentage of metal filling the dielectric medium of thermal conductivity k.

With reference, to Fig. 8–4, Table 8–1 gives the thermal conductivity enhancement of a dielectric of thermal conductivity k uniformly filled with wire, heat flow transverse to the wire cross sections.

In the case where our material is anisotropic, we would calculate the thermal conductivity in each direction separately. The thermal resistance to heat flow across the turns of a foil-wound coil or across the laminations of a tape-wound core would be found by starting at the midpoint (or wherever the axis or plane of no heat flow is located) and, using the method of Fig. 8–3 or the data of Table 8–1, summing the thermal resistance of each layer of metal and insulation. In the case of a coil wound with layers of round wire, the thermal resistance along a layer is the thermal resistance of resin, 70% metal-filled, plus the thermal resistance of the resin-filled margins. In the case of the ends of a foil-wound coil, the thermal resistance of dielectric in the margins would be added to the (very low) thermal resistance from the center to one side of the winding. Conduction or convection heat transfer coefficients can then be added on to obtain a total thermal resistance in each direction and a total temperature rise.

Convective heat transfer may be via either natural convection, in which

---

*This, by the way, is not a trivial assumption.

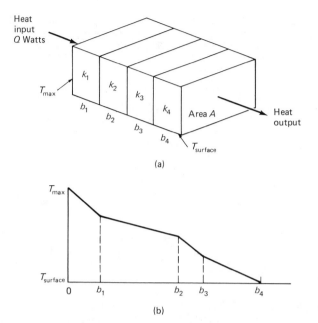

Fig. 8-3. Temperature versus distance for a composite of several materials.

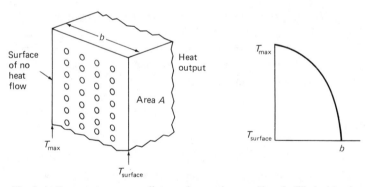

Fig. 8-4. Temperature versus distance for a volume uniformly filled with wire.

density gradients and gravity cause physical motion of the fluid, or via forced convection, in which fans or pumps are used to move fluids at much higher velocities than those associated with natural convection—typically 100–1,000 linear feet per minute for air—past or through heat sources or sinks.

Table 8-1. Thermal Conductivity Enhancement of a Dielectric.

PERCENT METAL BY VOLUME	k' (APPROXIMATELY)
10	$1.2k$
15	$1.35k$
20	$1.5k$
25	$1.6k$
30	$1.8k$
35	$2.0k$
40	$2.3k$
45	$2.6k$
50	$3.0k$
55	$3.5k$
60	$4.2k$
65	$5.3k$
70	$7.0k$

SOURCE: General Electric Company, 120 Erie Blvd., Schenectady, New York (1982).

Depending on the particular situation, *the boundary layer thickness associated with natural convection in air is on the order of* ¼-1 in. This tends to round off the surface contours of objects. Therefore, even though one may calculate the exact surface area of a transformer in excruciating detail, the effective surface area for natural convection is closer to the surface area of a plastic bag pulled tightly around the transformer. For vertical fins, as height increases the fins become somewhat more effective due to the chimney effect (i.e., the greater the column of low density air, the greater the buoyant force), imparting greater convective velocity to the fluid, and thereby reducing the boundary layer thickness. For transformers molded in a high thermal conductivity material one could enhance the surface heat transfer efficiency either with molded fins, bonded metal fins, or molded holes in the unit.

Figure 8-5 gives the convective heat transfer per unit length for either a thin vertical slab *or* a cylinder of diameter equal to the slab height for various heights (diameters). Note that the same curves represent both shapes.

One could evidently obtain the convective heat transfer coefficient for the transformer of Fig. 8-6 in the following manner: Take the convective heat transfer coefficient of a cylinder of length Z and diameter the average of X and $Y;$ add to that the convective heat transfer coefficient of two slabs of height the average of W and Y and length XV (the two slabs approximate the ends). The sum of those two convective heat transfer coefficients will give a fair approximation of the total convective heat transfer coefficient. To this

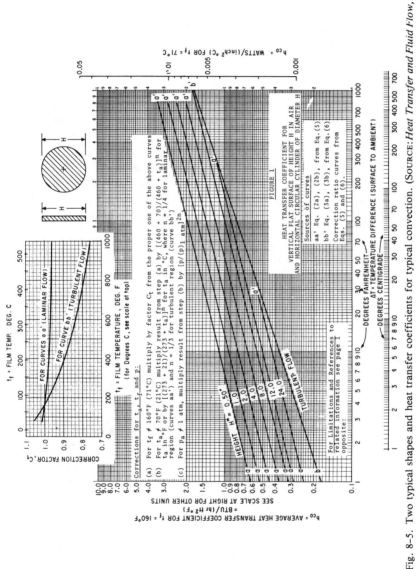

Fig. 8–5. Two typical shapes and heat transfer coefficients for typical convection. (SOURCE:*Heat Transfer and Fluid Flow*, Section 504.2, p. 2, December 1971. G. E. Data Books, General Electric Company, 120 Erie Blvd., Schenectady NY 12305.)

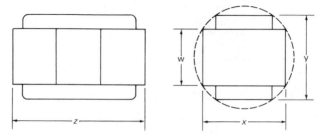

Fig. 8-6. Approximating the convective heat transfer coefficient for a transformer in air.

may be added, if desired, the conductive heat transfer coefficient represented by heat flow across the core laminations, across the mounting surface interface, and the thermal conductivity of the chassis to the heat spreading out into it. Keep in mind (as will be shown later) that heat from the core is not necessarily transferred uniformly from all surfaces. Laminated cores transfer their heat mostly at the lamination edges.

In the case of forced convection of air, given some number of linear feet per minute of air flowing parallel to the transformer surface (or an estimate of the air velocity being stirred past the whole surface), the simplest solution to calculating forced convection heat transfer is to determine the effective heat transfer surface area of the transformer, smoothing the contours to within $\frac{1}{4}$ in. or so, and then (having in hand the catalogs of various semiconductor heat sink manufacturers) look up various heat sinks which have similar surface areas. Take the published thermal resistance of that heat sink as the thermal resistance from surface to ambient for the particular case. This gives a fairly reasonable approximation to reality.

Exact calculation of forced convection heat transfer coefficients is an ability which can take longer to learn than transformer design. A number of references on the subject are given at the end of this chapter.

The heat of fusion (melting or freezing of a solid) or heat of vaporization (boiling of a liquid—heat given up when the vapor condenses) are special heat transfer situations in which the material exhibits thermal energy storage characteristics at some point as the temperature increases or decreases. Once the fluid has absorbed its heat of vaporization, it then boils and the thermal energy is carried off as vapor. The heat of vaporization of liquids used as boiling coolants can be quite significant. Water absorbs about one calorie per gram for every centigrade degree of temperature increase, but when it boils it carries off 540 calories per gram. Fluorochemical liquid FC-77 ab-

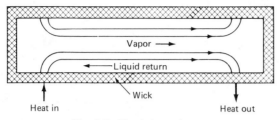

Fig. 8-7. Simple heat pipe.

sorbs 0.25 calorie per gram for every centigrade degree of temperature rise, but when it boils it carries off 20 calories per gram.

A very interesting device called a *heat pipe* is sometimes used where one must remove a moderate amount of heat with very little temperature rise and deposit it elsewhere. A heat pipe basically consists of a sealed tube lined with a wick, the wick saturated with a fluid, and a clear space down the middle (Fig. 8-7).

A heat input at one end vaporizes the fluid out of the wick. The vapor flows to a cooler portion of the heat pipe and condenses back into liquid, which flows back up the wick by capillary action. Since evaporation and condensation take place at virtually the same temperature (there is only a miniscule temperature difference between the ends of the heat pipe, associated with the pressure drop of the vapor flowing through the central space) the heat pipe transfers heat with almost no temperature drop. By way of comparison with conventional materials, a simple heat pipe an inch in diameter can have the same thermal resistance as a bar of solid copper a foot in diameter.

7. FLUOROCHEMICALS

Typical properties of some fluorochemicals are shown in Table 8-2. It is noteworthy that the dielectric strength of the vapor is comparable to that of the liquid, which makes fluorochemicals ideal for the cooling by boiling of components in a high voltage environment.

While water solubility may be on the order of a few parts per million and oil solubility may be quite low, the presence of such contamination or of dust or dirt of any kind can adversely affect the dielectric strength of fluorochemicals, as can dissolved gasses.

A rough indication of comparative heat transfer coefficients is given in Fig. 8-8.

Table 8-2. Typical Properties of Fluorinert Liquids.*

Property	FC-88	FC-78	FC-77	FC-75	FC-40	FC-43	FC-48
			Physical Properties				
Nominal boiling point, °F	88	122	207	216	320	345	345
Vapor pressure at 77°F, mmHg	570	260	42	30	3	<1	3
Density at 77°F, lbs/ft³	101	106	111	110	117	117	—
Viscosity at 77°F, cs	0.3	0.4	0.8	0.8	2.4	2.6	3.1
Pour point, °F	-150	-135	-150	-135	-60	-58	-80
Heat of vaporization at boiling point, Btu/lb	37	41	36	38	31	30	34
Heat capacity at 77°F, Btu-lb/°F	0.24	0.24	0.25	0.25	0.27	0.27	0.24
Thermal conductivity at 77°F, Btu/hr-ft²-°F/ft	0.032	0.036	0.037	0.037	0.038	0.039	—
Coefficient of expansion, ft/ft³-°F	0.0009	0.0009	0.0009	0.0009	0.0008	0.0008	0.00067

Electrical Properties

	42	43	45	55	55	55	46
Dielectric strength at 77°F, kv/0.1 in.							
Dissipation factor at 77°F,							
1 KC	<0.0003	<0.0003	<0.0003	<0.0003	<0.0003	<0.0003	<0.00025
1 KMC	0.0007	0.0017	0.0019	0.0036	0.0050	0.0055	0.008
3 KMC	0.0013	0.0023	0.0074	0.0065	0.0061	0.0065	0.009
8.5 KMC	0.0035	0.0050	0.0181	0.0090	0.0038	0.0036	0.008
Dielectric constant at 77°F,							
1 KC	1.72	1.81	1.86	1.86	1.89	1.90	1.94
1 KMC	1.77	1.83	1.91	1.87	1.91	1.92	1.95
3 KMC	1.73	1.84	1.88	1.86	1.91	1.92	1.93
8.5 KMC	1.75	1.84	1.89	1.90	1.89	1.90	1.90
Insulation resistance at 77°F, megohms	2×10^5 minimum						

*The "Fluorinert" liquids are clear, colorless, perfluorinated fluids, relatively dense and of low viscosity. The major difference between them lies in their respective boiling points, ranging from 88°F to 345°F. Characteristically, their pour points are quite low, in most cases being well below −100°F.

SOURCE: 3M Company, St. Paul, Minnesota.

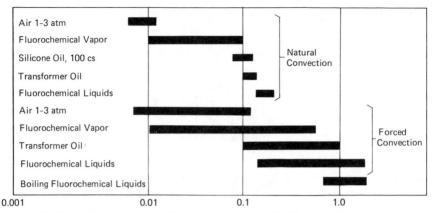

Fig. 8-8. Typical heat transfer coefficients for various fluids. (Source: Ahearn, J. H., et al., "Cooling of Electronic Equipment by Means of Inert Vapors," Natl. Conf. on Aero. Electronics, May 15, 1957. After 3M Fluorinet Data Book.)

8. BOILING LIQUID HEAT TRANSFER

In the event that the heat flux from the component being cooled is sufficiently high to cause the fluorochemical to boil, the component temperature tends to stabilize somewhat above the boiling point of the coolant. This is shown in Fig. 8-9 for some materials. The heat flux transfer capability does not increase without limit, however. There comes a "boiling crisis" at a critical heat flux where the liquid cannot contact the surface because the hot vapor forces it away faster than the vapor can leave and allow liquid to contact the surface. At this point the component temperature rises abruptly, the superheated vapor next to the component carries enough heat from the component to boil the liquid before it even touches the component, and the component gets hot enough to transfer a fair amount of heat by radiation. The heat flux associated with this sudden, tremendous temperature rise at the boiling crisis is referred to as the *maximum heat flux* or the *burnout heat flux,* for an obvious reason.

One may calculate (or at least obtain a fair estimate) of the burnout heat flux from the following equation:

$$q_c = F_f \left[\frac{\sigma a (\varrho_F - \varrho_g)}{\varrho_g^2} \right]^{0.25} h_{fg} \varrho_g, \qquad (8.5)$$

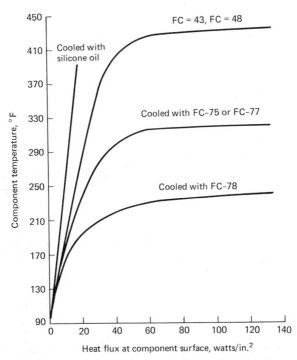

Fig. 8-9. Typical temperature of components cooled with the fluorinert liquids or silicone oil. Heat sink of 77°F.

where q_c = critical heat flux in watts per square meter;

σ = surface tension of liquid, Newtons/meter;

a = ratio of the acceleration of the fluid to the normal acceleration at the surface of the earth, due to its gravitational field;

ϱ_F = fluid density in kilograms per cubic meter;

ϱ_g = vapor density in kilograms per cubic meter;

h_{fg} = latent heat of vaporization in watt-seconds per kilogram; and

F_f = futz factor which seems to range between 0.1–0.2, depending on the fluid, the exact geometry involved, and the patterns of convection currents in the liquid which may happen to form from time to time.

In other words, sometimes it will burn out at a lower heat flux than at another time. I would suggest that when using boiling liquids for heat transfer, one stay well below the calculated burnout heat flux (by at least a factor of three or four).

Visual observation of the unit in action in order to ensure that it is well below the burnout heat flux is also recommended in applications where one is pushing to the limit.

9. SOME COMMENTS ON MATERIALS AND PACKAGING

Commonly used liquids for dielectric and heat transfer requirements include refined mineral oil such as Shell Diala-AX, silicone fluids such as Dow Corning's, or the fully fluorinated liquids offered by 3M under the trade name Fluorinert. High thermal conductivity potting materials are available from (among others) Bacon Industries, Emerson Cummings, and National Beryllia.

Silicone fluids have excellent dielectric strength but are somewhat hygroscopic, i.e., they absorb water from the air and hold onto it, somewhat to the detriment of their ultimate electrical capabilities. They also tend to polymerize in an arc and are not necessarily self-healing.

Liquid-filled systems must have some means for accommodating the coefficient of volumetric expansion associated with the change in temperature of a heated liquid. This is usually either a flexible membrane inside the case with the dry side vented to ambient pressure, a bellows system, or a gas cushion, wherein the gas pressure is allowed to increase as the liquid expands and compresses it. Sulfur hexafluoride is preferred as the gas because it has very good insulating properties (about 120 kV/in. at 1 atm.).

Sulfur hexafluoride is also used as a dielectric and heat transfer gas in hermetically sealed transformer and power supply designs. With a density about five times that of air and a specific heat of about half that of air, respectable convective heat transfer rates can be achieved.

10. A CALCULATION OF THERMAL RESISTANCE FOR AN ANISOTROPIC OBJECT

Just for an example, let us look at the thermal conductivity of a tape-wound C-core. Assume 12 mil silicon steel, with a 95% space factor. That means that 5% of the volume is taken up with interlaminar insulation. If our core cross section is 2 × 2 in., then 5% of the 2 in. E dimension is insulation. Whether we calculate it as 1.9 in. of steel and 0.1 in. of insulation or 12 mils of steel and 0.6 mils of insulation, etc., we will obtain the same result.

The thermal conductivity of steel is ~ 0.1 cal/sec/cm/cm^2/°C, and that of our insulation (usually a varnish) is estimated to be 0.0004. Since both steel and varnish have equal areas, our composite thermal resistance per unit length would be the sum of each length divided by the conductivity of that length:

$$\frac{b}{k} = \frac{b_1}{k_1} + \frac{b_2}{k_2} = \frac{1.9}{0.1} + \frac{0.1}{4 \times 10^{-4}}$$

$$\frac{2}{k} = 19 + 250 = 269$$

$$k = 0.0075.$$

This is less than one-tenth of the conductivity of the steel in the perpendicular direction, where the heat flows through the steel laminations out to the edges of the core. What we did here was to add the thermal resistances of two items to get the thermal resistance of the two items in series.

From Fig. 8–3, we see that this is the technique used for calculating heat flow through composite structures. If we wanted to calculate the temperature rise of our C-core, which generates its heat as core loss throughout its volume, we would use Fig. 8–4, except that we would use for k' our k number determined above. Since the thermal conductivity is different in each direction, more heat will flow in one direction than another. Our total temperature rise would be (extending the idea of Fig. 8–4 from the reasoning of Fig. 8–3 and expressing the heat flow parallel and perpendicular to the laminations separately):

$$T_{max} = T_{\perp \text{ surface}} + \frac{1}{2} q_\perp \frac{b_\perp}{k_\perp}$$

$$T_{max} = T_{\parallel \text{ surface}} + \frac{1}{2} q_\parallel \frac{b_\parallel}{k_\parallel}$$

T_{max} will be the peak temperature at the inside of our C-core. T_{max} is equal for the above two equations.

We can observe that $q_\perp + q_\parallel$ is the total q representing the heat generated by the core loss in a particular volume and we know that more heat will flow parallel to the laminations than perpendicular. In this case we have a symmetrical situation, and while we have some nonlinear effects at the corners, we will simply say that there is, to a first order, no interaction between the components, so $b_\perp$ and $b_\parallel$ are equal. (For our 2 × 2 in. cross section, the planes of no heat flow are perpendicular to the outer surfaces and intersect in the middle, so both b's are equal to 1 in.) We now have two equations and four unknowns (the two surface T's and the two q's). If our core is potted in a high thermal conductivity epoxy (much higher than $k_\parallel$) then equal temperatures will be forced at both surfaces. We can then solve for the heat flow in each direction.

If our C-core is convection cooled, we must add the thermal resistance associated with the convective heat transfer coefficient off each surface for an assumed temperature of each surface. Add the q's thus obtained and see if they add up to the given total q. Make a new guess at the surface temperatures and repeat the above procedure. By a successive approximation procedure we can find the heat flow in each direction and the temperature rise over ambient for each surface. As the reader will find when he does this, with a thermal resistance in one direction over ten times that in the other direction, very little heat flows perpendicular to the laminations and the temperature rise over ambient at the surface where the edges of the laminations are exposed (the surfaces that get the heat which flows parallel to the laminations) is much greater than at the two perpendicular surfaces. This is why some effort must be made to cool the edges of tape-wound C-cores where significant heat is generated in the core. The same reasoning holds true for cores assembled from stacks of E–I laminations.

The problem can be particularly severe for three-phase transformers constructed with tape-wound E-cores, as coils are wound on all three legs and very little of the core edges are exposed. It is therefore important that some type of heat spreader be considered, or that the coil be spaced away from the core on those two surfaces to allow convection of a heat transfer fluid along the core edges.

9
MATERIALS AND
FABRICATION METHODS

1. INTRODUCTION

The preceeding chapters showed how to design various types of magnetic components and discussed their function in some applications.

It is the purpose of this chapter to familiarize the novice designer with basic fabrication techniques that he can understand, apply, and modify as needed, using materials suitable to his specific situation. Many of the fabrication techniques and considerations developed over the last century or so are included. I have not made an attempt to cover every facet of the subject in detail. After the designer has become familiar with the fabrication methods in this chapter, variations will suggest themselves from particular circumstances which arise in the course of design work.

The mention of one or a few materials does not imply that nothing else exists. A tape manufacturer such as 3M has hundreds or even thousands of products, many of which are suitable for use in the fabrication of magnetic components. It would be impossible to discuss all types of materials for all applications in a book of easily readable length. A few items are mentioned here for the purpose of developing a basic familiarity with enough different materials to suit a fairly wide variety of applications.

A particular manufacturer or his product may be mentioned for convenience of reference. This does not imply that other manufacturers have inferior products. They may have superior products. When a manufacturer is mentioned as having a unique material or capability, it is to be understood that the uniqueness is only within the experience and awareness of the author.

The reader is urged to talk to his local suppliers of magnet wire and materials, to discuss his specific requirements with several core manufacturers, and generally make his own determination among his various sources. With that out of the way, we begin with the concept of thermal classes.

2. THERMAL CLASSES OF DIELECTRIC MEDIA

Insulating materials are classed by the manufacturer according to how well they retain their insulating properties after long exposure to elevated temperatures. This is done by picking a set of temperatures and assigning each a designation (for convenience of reference). For a life expectancy of, say, 10,000 hr at a "maximum rated" temperature, the subject material is tested at each temperature. The probability of failure within the (10,000 hr) life expectancy will be lowest at the lowest temperature, and higher at higher temperatures. If an unacceptable failure rate is 1% at 10,000 hr, it is a simple matter to test materials and assign them thermal classes. It is all ultimately based on the chemical stability of various kinds of polymers, minerals, or other chemical compounds with time and temperature.

MIL-T-27 references certain thermal classes which magnetic components may be specified to meet. The thermal classes of dielectric media are derived from an AIEE standard. To some degree these are related (see Table 9-1).

MIL-T-27 specifies a 12-week life cycle test (2,016 hr) for units with a 10,000 hr life expectancy. A number of samples from the production lot (a qualification lot) are tested in a specified manner. No failures are permitted.

The *dielectric loss* of a particular dielectric at the maximum working temperature can have a great deal to do with the reliability of a pulse trans-

Table 9-1. Thermal Classes of Insulation.

AIEE	MIL-T-27D	MODERN THERMAL CLASS DESIGNATION: MAXIMUM OPERATING TEMPERATURE
O	Q	85°C
A	R	105°C
B	S	130°C
F	V	155°C
—	T	170°C
H	U	180°C
—	U	200°C
C	U	220°C

former or high frequency power magnetic component. One dielectric failure mode is related to the self-heating of the dielectric in regions of intense high frequency electric fields by the high frequency loss of the dielectric, which increases as the temperature increases. This leads to thermal runaway in the dielectric in microscopic regions around defects or impurities. Dielectric media should be chosen only after consultation with the manufacturer regarding their loss properties at frequency and temperature.

The corona threshold of a finished unit is an excellent test for assessing the reliability of a high voltage transformer. The test parameters should be chosen so as to bear a realistic relationship to the actual working conditions of the unit. The most severe corona threshold test is with square wave excitation; this test should be used when that is the actual working condition. The rise time of the square wave in the corona test should of course be comparable to the actual rise time of the circuit in which the transformer operates. Information on corona threshold testing may be found in MIL-T-27 and its references, as well as from manufacturers of corona testing equipment such as Biddle.

3. INSULATING PAPERS

Several common types of insulating papers are available for the combined functions of electrical insulation and mechanical support. The choice of which to use is governed by several factors. The mechanical stiffness of the material must be appropriate to the wire size. For foil-wound coils, a high degree of cut-through resistance is often necessary because the copper sheet, when slit, may have a sharp burr on one or both edges which can puncture some sheet materials.* A soft, porous material may be appropriate for ensuring impregnation of layer-wound coils of many turns of fine wire with high dielectric stress.

For class 105 or 130 applications, kraft paper, a specially processed, acid-free paper (also called blue-line neutral kraft), is appropriate. It is made in two forms—an easily impregnated paper ranging from ½ mil to 5 mils, and a stiff, dense paper which does not impregnate too well (called Fibrelec) in thicknesses of 7 mils or more.**

Copaco, a stiff 100% rag paper, is available from Lenni Products. It is particularly convenient for heavy wire (#10–20) as it comes on rolls with the

* Thin Sheet Metals Co. can provide degreased copper foil slit to width with a burr less than 5% of material thickness.
** The Schweitzer Division of Kimberly-Clark in Lee, Massachusetts, is an excellent source of a wide variety of kraft papers.

edges folded over about ⅛ in. This material is called *cuff,* and the paper is said to be *cuffed.* It not only gives greatly enhanced stiffness at the coil ends where it is most needed, but serves as a convenient control of margin spacing. These materials do not impregnate all that well. However, over the last five years over 1000 units of one design using 15 mil Copaco and vacuum impregnation with a Class 155 epoxy resin have been operating in the field at 500 volts per mil for a 1 microsecond pulse, with most of the units running 8 hours per day, 250 days per year, at 15 pulses per second. The units are guaranteed to have a life of 10^9 pulses at 15 Hz. To date there have been no electrical failures.

For Class 155 or 180 applications, or for higher reliability at lower temperatures, several interesting materials are available. DuPont offers a polyimide paper called Nomex. It has exceptional high temperature capabilities, and is available in a soft, porous paper called Nomex 411 and a stiff, dense paper called Nomex 410. The combination of Nomex 411 impregnated with 3M's epoxy #280 (an oven-curing Class 155 system) has shown excellent reliability in a variety of applications.

Chase-Foster (and others) offer a variety of composite dielectric papers. Kraft paper or a polyester fabric may be laminated on either side of Mylar; Copaco may be laminated to Mylar; or Nomex may be laminated to Kapton. Kapton is a polyimide film made by DuPont, to which epoxies bond readily. Epoxies do not bond well to Mylar (Mylar film is often used as a release film). The composites of Mylar are treated with adhesion-promoting intermediate materials in order to provide structural integrity.

Nomex-Kapton-Nomex laminate, interleaved with Nomex 411 (the only available laminates at present being with the more dense Nomex 410), such that about 15% to 20% of the thickness is Kapton, seem to do well even at 500 volts per mil. They should have exceptional life, as the Kapton film stops the "corona trees" which develop with high electrical stress and can lead to dielectric breakdown. If one were to decide to use this system an interleaved dielectric of, say, 2 mil Kapton and 10 mil Nomex 411 could be manually wound. Chase-Foster may make Kapton/Nomex 411 laminates available in the future, but they report little demand to date.

A West Coast supplier of insulating papers and films who does not object to selling modest quantities is Fralock in Los Angeles [(213) 873–6665]. If they do not have a needed material in stock both Chase-Foster [(401) 434–2340] and DuPont [(800) 441–7515] have indicated a willingness to assist in development programs. Fralock has the capacity to make custom laminates for specialized applications.

4. TAPES AND WIRE INSULATION

Tapes hold something in place, temporarily or permanently. An example of temporary use is to hold a wire in place at the end of a layer while one's hands are occupied cutting a piece of layer insulation. Another example would be holding layer insulation in place until the next layer of wire is wound, when the second layer of wire holds the layer insulation beneath it in place.

An example of permanent use is securing and providing strain relief for the start or finish of a winding. Another example is to provide electrical insulation of a specific amount in a specific location. Tape is also used to provide mechanical strength in a particular situation. The latter example is best illustrated by a glass cloth tape in which the glass fibers, when embedded in a resin matrix, give a composite of exceptional strength.

Tapes may be adhesive coated on one or two sides, or may have no adhesive coating. The adhesives are usually thermosetting, so that when the fabricated unit is baked the adhesive cures and forms a permanent adhesive bond.

Non-adhesive coated tapes are most commonly a fibrous mat of paper or synthetic material, or glass fiber cloth woven in a narrow width. The latter is frequently used for impregnatable interwinding insulation and outer wraps on toroids.

Adhesive coated tapes may in general be classified as either film or fabric.

Film tapes are thin, offer high dielectric strength with minimal physical build, and have a slick surface over which wire may slide.

Fabric tapes will impregnate from one side and will trap air voids if wound overlapping. Fabric tapes are thicker than film tapes and can result in much higher physical strength due to their fibrous nature.

It is important to use only electrical grade tapes. High-strength tapes are used for securing packages for shipping, but they are not acid-free and can cause long-term electrical failures. Equally obvious but common mistakes include using a pencil to mark the layer insulation before cutting it to length. The conducting graphite trace is an obvious risk to the life of the unit. Corrosive materials are likewise a liability. Not all paper is acid-free, for that matter. Use only materials specified for electrical applications.

Temporary materials which will not be left in the unit when it is finished need not have the thermal class of the unit. They need only be sufficiently strong and sticky to do the job of the moment. All materials which become part of the unit must have at least the thermal class for which the unit is rated.

Other temporary tapes need not have a thermosetting adhesive, as something else ultimately holds in place what the temporary tape held. Permanent usage tapes should have a thermosetting adhesive if at all possible.

The choice between film or fabric may be made on the basis of elementary physical attributes. Very high temperature applications require a tape with a silicone adhesive. Kapton film tape coated with silicone adhesive is somewhat less convenient to use than others since the adhesive "gum balls" easily when the tape is cut.

Fabric tapes such as 3M #28, an acetate cloth tape with excellent electrical properties and a Class 105 temperature capability are available. Others are 3M #27, a Class 130 glass cloth tape, and Mystik #7001, a Class 180 glass cloth tape with silicone adhesive.

Mystik #7367 is Kapton tape with an acrylic adhesive. It is a Class 155 material and extremely satisfactory film tape with epoxy resin compatibility.

Most polyester film tapes are Class 130. No adhesive bond should be expected unless the manufacturer specifically indicates the surface has been treated in some manner to give adhesion.

Teflon tape with a silicone adhesive is available where a Class 180 material with releasing properties is required.

For specific fire retardant requirements Kapton, glass cloth, creped Nomex paper, Tedlar, and other materials are available from a variety of manufacturers.

A heat bondable Kapton film is available from Fralock. Called T-183 or 184, it has a 1 or 2 mil thick phenolic butyral adhesive applied to one or both sides of any standard thickness of Kapton. It is handled as a dry film, but when it is heated the adhesive melts, reflows, and cures, bonding the film to the adjacent surface.

5. WINDING CONDUCTORS

The most common winding conductor is round copper wire with a coating of a film insulation. Copper foil or sheet may be used, slit to width and interleaved with a paper, film, or composite material for interturn insulation. The latter is desirable where relatively high currents and/or high frequencies are involved, or where a better space factor is desired than that obtainable with magnet wire (the general term for round or square, aluminum or copper, film-coated or fabric-served wire). Copper foil may not exhibit a good adhesive bond to epoxy impregnants. Under thermal cycling cracks will develop, degrading heat transfer and allowing moisture intrusion. A black oxide finish on the copper will enhance adhesion considerably, thereby obviating this failure mechanism. Magnet wire is available in square cross section in sizes from 0 to 14, but is extremely difficult to wind neatly and is rarely necessary. Aluminum conductors are lighter than copper, but for prototype work they present problems in obtaining reliable terminations. I would recommend that square conductors or aluminum conductors be considered only when the design will not fly any other way.

Of the various film insulations available, some have advantages over others in the same or nearby temperature classes. Where is it desired to bond the turns of a coil together, a butyral adhesive coating is available over both a polyurethane film (Class 105) and a nylon overcoated polyurethane film (Class 130). Both of these are solderable films, in that the film dissolves in hot solder. A Class 155 polyester film wire is available with or without a nylon overcoat. Both of these are solderable film insulations. It would be expected that the nylon overcoat would raise the moisture resistance of the base film. In critical applications the manufacturer should be consulted, as he should have the greatest familiarity with the capabilities of his own formulations. The cross reference chart (Table 9-2) gives generic insulation type, applicable NEMA standards, and brand names various wire manufacturers assign to these products.

6. THE USE OF CURE-IN-PLACE DIELECTRIC RESINS IN COIL FABRICATION

Cure-in-place dielectric resins serve many functions in the fabrication of winding structures and finished transformers. They displace air voids between conductors and within layers of electrical insulation or mechanical conductor supports. This replacement of air with a solid medium has many consequences. The electric field is more homogenous, as dielectric constants are now closer to each other. Thermal conductivity through the coil is improved, aiding heat transfer. With the outside air and its attendant moisture and contaminants excluded, the probability of electrical breakdown is vastly reduced. With adjacent turns bonded together and held in place by a highly adherent solid, the repulsive force between adjacent turns in high current applications will not blow the end turns out of the coil. The coil can now survive handling during assembly, and thermal cycling over its life, without unravelling. In many cases, the coil and core can be held firmly together for the life of the unit without extra mechanical assembly operations. Impregnating the core and coil as a unit can provide a good mechanical bond and improved thermal path between coil and core.

Table 9-2. Cross Reference of Magnet Wire Trade Names.

FILM INSULATION TYPE	NEMA STANDARD MW 1000/1973	ANACONDA WIRE & CABLE CO.	BELDEN CORP.	CHICAGO MAGNET WIRE CO.	ESSEX GROUP, INC.	HUDSON WIRE CO.	PHELPS DODGE CORP.	REA MAGNET WIRE CORP.	VIKING WIRE CO., INC.
Oleoresinous (105°C)	MW1-C	plain enamel	Beld-Enamel	plain enamel	plain enamel	plain enamel	enamel	plain enamel	enamel
Polyvinyl formal (105°C)	MW15-A MW15-C MW18-A MW18-C	Formvar	Formvar	Formvar	Formvar	Formvar	Formvareze	Formvar	Formvar
Polyvinyl formal Nylon (105°C)	MW17-C	Nyform	Nyclad	Nyform	Nyform	Formvar N	Nyform	Nyform	Nyform
Polyvinyl formal butyral (105°C)	MW19-C	cement coated Formvar	—	bondable Formvar	Bondex*	Formvar AVC	Bondeze	—	F-Bond
Solderable acrylic (105°C)	MW37-C	—	—	—	Ensolex*	—	—	acrylic	—
Solderable acrylic nylon (105°C)	MW39-C	—	—	—	Ensolon*	—	—	Nylon acrylic	—

Polyurethane (105°C)	MW2–C	Analac	Beldure	Solder-Brite	Soderex*	urethane	Sodereze	Solvar	poly-urethane
Polyurethane butyral (105°C)	MW3–C	cement coated Analac	—	bondable Soder-Brite	Soderbond*	urethane SB	—	Reabond-A Solvar	P-Bondall
Polyurethane Nylon (130°C)	MW28–C	Nylac	Beldsol	Nysod	Soderon*	urethane N	Nyleze	Nysol	poly-Nylon
Polyurethane Nylon butyral (130°C)	MW29–C	cement coated Nylac	Beldbond	Soder-Brite Bondable/N	Soderbond N*	urethane N–SB	Sy Bondeze	Reabond-A Nysol	—
Polyester solderable (155°C/ 180°C)	MW26–C	—	Celemid	F/Sod	Solidex*	solderable polyester	—	Thermsol	Vipoly
Polyester Nylon Solderable (155°C/ 180°C)	MW27–C	—	Celenon	F/Sod N	Solidon*	solderable polyester-N	—	Amid-Thermsol	Vipoly N
Polyester (180°C)	MW30–C	—	Isonel	polyester	Thermatex* 200	polyester	Thermaleze 200	Isonel 200	Isonel 200, Isonel 200–P

(continued)

Table 9-2. (Cont.)

FILM INSULATION TYPE	NEMA STANDARD MW 1000/1973	ANACONDA WIRE & CABLE CO.	BELDEN CORP.	CHICAGO MAGNET WIRE CO.	ESSEX GROUP, INC.	HUDSON WIRE CO.	PHELPS DODGE CORP.	REA MAGNET WIRE CORP.	VIKING WIRE CO., INC.
Polyester Nylon (155°C–180°C)	MW24-A MW24-C	Anatherm-N	Beldtherm N	polyester Nylon	Nytherm* 180°C	—	Thermaleze N	Amidtherm	polyester Nylon
Polyester polyamideimide (180°C–200°C)	MW35-A MW35-C MW36-A MW36-C	Anaclad-A	armored poly-Thermaleze	polyester	Thermelex* GP-200	polyester A1	armored poly-Thermaleze 2000	Therm-Aimid	Isonel 200-R
Polyester Polyamideimide thermoplastic overcoat (155°C–180°C)	—	cement coated Anaclad-A	—	—	poly-Bondex* 155/180	—	armored poly-Thermaleze Bondeze	Reabond-S Therm-Aimid	—
Polyimide (220°C)	MW16-C MW20-C	ML**	ML**	—	Allex*	polyimide	ML**	Pyre ML**	ML**

Reproduced courtesy of Essex Magnet Wire & Insulation Division, Essex Group, 1510 Wall Street, Fort Wayne, IN 46804.
*Essex trade name.
**DuPont trade name.

192

The impregnation of coil and core as a unit is contraindicated in several cases:

1. The unit may be so large or heavy that it is inconvenient or impractical to impregnate coil and core together, in which case the coil would be impregnated first, and then the core assembled onto the coil.

2. The core may be a tape-wound nickel-iron C-core. These cores are extremely strain-sensitive, and must be carefully mounted and isolated from mechanical shock or strain if the maximum magnetic properties of the core are to be obtained and preserved. Most epoxy systems shrink on curing and impart too much mechanical strain to these cores. Polysulfides such as PRC's 1201Q (liquid) or 1201 HT (paste) are often suitable for bonding coils to such cores, or for mounting them.

3. The core may be a ferrite. Most ferrites have some degree of porosity, and vacuum impregnation with thermosetting resins will reduce the ferrite permeability. This can cause unacceptable degradation of the core magnetic properties. Some ferrites are available with a parylene coating. This may help to seal the ferrite.

In lieu of or following impregnation, the unit may be placed in a mold or potting shell. (A potting shell is a thin dielectric shell to which the potting material adheres, the shell becoming the exterior surface of the finished unit.) A potting material is then introduced around the unit, surrounding it and filling voids between coil and core. Additional environmental protection and enhanced heat transfer may be obtained by judicious choice of the potting compound and the mold geometry.

As a compromise between impregnation and molding, the unit may be dipped in a thixotropic resin, which results in a thick shell of resin (a conformal coat) on the unit when cured.

The electrical consequences of voids in high voltage transformers are severe. The presence of an air void in a winding or in the insulation between a winding and ground can cause corona, which will eventually lead to dielectric breakdown and local conversion of the dielectric material to electrically conductive carbon and unsightly tars. This was discussed in the chapter on field gradient control (Chapter 7).

Even with perfect impregnation, voids may be caused by the liquid impregnant draining from the high points to low points during oven curing. Curing the coil with one end up can exacerbate this problem, as the resin not only drains from the top but pools at the bottom and can trap an air bubble in the coil margin at the bottom of the coil when it cures. When a high voltage coil is impregnated and then assembled onto a core there will be a gap between the coil ends and the inside of the core. Corona can occur in this gap unless it is potted. When epoxy resins cure there is a volumetric shrinkage *due to the change in molecular structure of the material.* In some cases this shrinkage can cause cracks in a coil. The cracks are voids and corona can then start.

All materials have a coefficient of volumetric expansion. A perfectly cured impregnated coil can, on cooling down from its curing temperature, develop cracks *due to thermal shrinkage.*

Under thermal cycling some parts of a transformer may expand more or less than others. If the resin used can not accommodate the strains due to differential coefficients of expansion, cracks can occur.

Under thermal shock the outer part of a unit may expand or contract faster than the interior, causing mechanical failure.

These kinds of failures can be prevented by adequate care in processing and the choice of suitable materials.

Voids can also develop when the potting material does not adhere to the conductors or insulating materials.

When copper or brass foil is used as an electrostatic shield and not adequately degreased, voids between the foil and the impregnated coil will develop. The thermal conductivity of such an air gap is rather low. Several of these within a coil can cause the temperature rise to be far beyond design limits, leading to catastrophic failure. Most silicone or urethane impregnating resins have little or no adhesion to magnet wire or core materials, and under extremes of temperature can separate from their surrounding surfaces as a consequence of their large coefficients of volumetric expansion.

There are primers available for obtaining adhesion of urethane or silicone systems to various substrates. These primers are not suitable for impregnation. For most applications, epoxy resin systems offer an adequate range of performance characteristics and adhesion.

In some cases the unit may be designed to operate in pressurized gas or a liquid (such as oil or fluorochemical). In these cases, potting or molding is usually not required, and corona due to voids between coil and core is largely eliminated. Thermosetting tapes may be relied upon to hold the coil together if tests show the thermoset polymer to not be affected by the gas or liquid over the temperature range of concern.

Dielectric media such as Teflon, polyethylene, or polypropylene are very useful in transformer design, since they have low dielectric constants, high dielectric strength, and are available in both block and sheet form. Virtually nothing bonds to these materials without special treatment, so they are not suitable for most epoxy-impregnated designs.

When expecting adhesion between a resin system and a cured polymer substrate, the designer may find a few rules of thumb helpful: Urethanes and silicones will adhere to epoxies (primers are required with some systems), but epoxies will not adhere to either. Some silicones can adhere to urethanes (primers may be required), but urethanes will not adhere to silicones. Epoxies bond nicely to Kapton but not to Mylar, although Mylar is commonly used in transformers. In that usage one either has a noncritical commercial grade application or takes a (hopefully calculated) risk.

Where it is desired that the insulation resistance between windings be extremely high, in a relatively low-voltage application, a wrap of kraft-Mylar-kraft laminate, Kapton, or a Kapton laminate will do wonders. Epoxy-impregnated kraft paper may give an IR of perhaps 10^8–10^{10} ohms, but a single wrap of Mylar or Kapton embedded in the dielectric can easily raise it several orders of magnitude, and the difference at elevated temperatures will be even greater. The reason this works is that the volume resistivity of Mylar or Kapton is many orders of magnitude higher than that of a thermosetting impregnating resin.

The insulation resistance of a resin-impregnated structure may be easily calculated if the cured impregnating resin is the controlling element.

One would use $R = \varrho l/A$, where ϱ is the volume resistivity of the cured resin, l is the insulating thickness, and A is the interwinding area.

7. SECURING CONDUCTORS

There are several commonly used methods of securing conductors. For the start of a winding, a piece of tape may be used, as shown in Fig. 9–1. This method is useful for wire sizes of less than approximately #33, or whatever size it is practical to wrap a piece of tape around. When the wire is large enough that additional strength is needed to hold down the tape securing the first turn, an additional piece of tape may be applied [Fig. 9–1(b)].

When sleeving is needed on the start (or finish, or tap) leads of the winding, strain relief for the sleeving may be effected by splitting the end of the sleeving and securing with a piece of tape in either of the manners shown in Fig. 9–2.

The finish of a winding is best secured by placing a piece of tape down a few turns before the last turn of the winding, *sticky side up,* and using the last few turns to hold the tape down. After the last turn the tape is folded

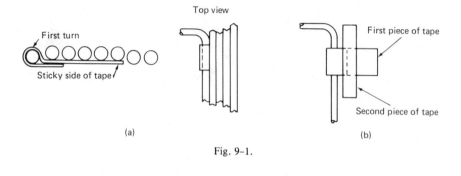

Top view

First turn

Sticky side of tape

First piece of tape

Second piece of tape

(a)

(b)

Fig. 9–1.

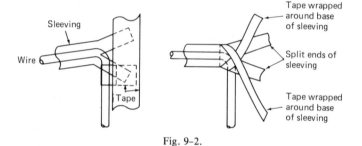

Sleeving

Wire

Tape

Tape wrapped around base of sleeving

Split ends of sleeving

Tape wrapped around base of sleeving

Fig. 9–2.

over the winding. A second piece of tape, wrapping all the way around the coil if (rarely) needed, may be used to secure the finish tape, as shown in Fig. 9–3.

Taps are a little more difficult than a finish to secure but are very similar. The tap at the end of a layer is the simplest tap to secure. Anticipate the tap a few turns before the tap point and place *two* pieces of tape, sticky side up and 4–10 wire diameters apart [Fig. 9–4(a)]. The first piece of tape is folded over the free end of the winding to secure one lead of the tap [Fig. 9–4(b)]. The free lead is looped out as far as needed for the length of the tap, and then returned to the coil. It is secured by the remaining piece of tape folded over it. A piece of tape may be placed over both of the first two tape ends to hold them down [Fig. 9–4(c,d)].

A tap in the middle of a layer must be secured very much as in Fig. 9–4, but the additional complication of the following turns is present. The tap leads must not be allowed to rest on the adjacent wires but must be protected or cushioned, usually with a piece of tape or two, or a piece of the layer insulation. If the tap is taped down and the rest of the layer wound on top of it,

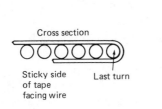

Cross section

Sticky side
of tape
facing wire

Last turn

Top view

Wire

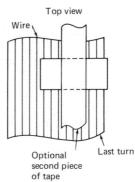

Optional
second piece
of tape

Last turn

Fig. 9-3.

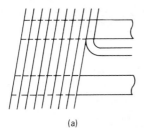

(a)

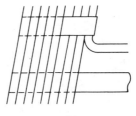

(b)

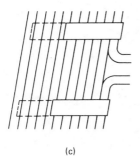

(c)

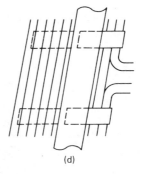

(d)

Fig. 9-4.

then the protection (and electrical insulation!) must be placed on top of the
tap lead.

If the tap is held in the air until the remainder of the layer is wound and
then laid on top of the remainder, or laid back down over the portion of the

layer wound before the tap, then of course the cushioning is placed under the tap lead. Tape is used as required to hold the various items in place (Fig. 9–5).

It is difficult to wind further layers over the tapped layer because the coil has become tapered, lumpy, and irregular. Fig. 9–6(a) shows a cross section of Fig. 9–5. Placing some cut pieces of layer insulation opposite the tap lead for the remainder of the layer (a width and thickness comparable to the tap lead), gives the coil a symmetrical lump, no taper, and succeeding layers may easily be wound [Fig. 9–6(b)].

Taps on high current windings are sometimes needed. It may be desired to vary the load voltage, in which case the tap must carry the full winding current. When the wire size is inconveniently large to pull out a tap as in Fig. 9–5(a), this can be accomplished by stripping the wire for a short length, soldering on a tab of copper foil, insulating the solder joint and tab, and continuing the winding. It is often possible to put taps on the primary (which would have a smaller, more convenient wire size) than on the secondary of a step-down transformer.

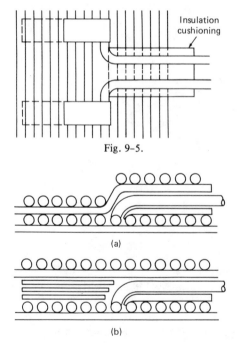

Fig. 9–5.

(a)

(b)

Fig. 9–6.

When the secondary is a multifilar winding (several strands in parallel) and a center tap is required (the center tap carrying a relatively low current, such as the center tap on vacuum tube heater transformers) a tap of only one of the strands may be pulled out in the manner shown in Fig. 9–4 or 9–5, and accomplish all necessary electrical functions.

When fine wire is wound by hand it will not always feed evenly in the desired direction but may jump back and wind over several previously wound turns before jumping forward again and continuing to wind in the desired direction. The causes of this are vibration and runout in the shaft of the winding machine or (more commonly) muscle tremors in the hands of the winder. This must not be allowed to happen. The machine must be backed up, the wire unwound and rewound properly. The reason is the same as the reason why insulation must be placed above or below a tap in the middle of a layer. When round wires lie on top of each other at an angle, they make contact at a very small point. The pressure is quite high at that point and the thin film of insulation on the wire can be damaged, leading to a shorted turn and failure of the transformer or inductor.

In high voltage transformers the layer insulation serves several functions. One is to cushion the wires against mechanical damage of the film insulation. Another is to provide a wick of controlled thickness into which an impregnant may be introduced. The third function is mechanical support of the winding during the fabrication process. The thickness, stiffness, and porosity of the layer insulation must all be considered in the selection of layer insulating media. It may be necessary to specify a composite (kraft paper laminated to Kapton, or whatever) to obtain a desired combination of physical and electrical properties.

8. WINDING MACHINES

Coils for transformers or inductors may be wound in a variety of manners. For something small with many turns, especially when one does not need to count the exact number of turns as they are wound, a bench-mount hand-cranked grinder may be used. These devices have a simple gear arrangement that speeds up the shaft rotation by 10 or 20 to 1 over the turning speed of the hand crank. One hand turns the crank, while the other hand feeds the wire onto a bobbin affixed to the shaft in place of the grinding wheel. This is extremely crude, but simple, and I have seen circumstancs where it was expedient.

A step up from this arrangement is to C-clamp a variable speed electric drill to a table, operate the trigger control with one hand and feed the wire

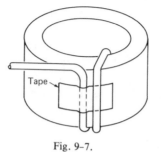

Fig. 9-7.

with the other. A simple turns counter may be arranged by gluing a small magnet to the side of the chuck and using a reed relay capsule (mounted on the bench, suitably near the magnet) to couple signals into an electronic impulse counter.

Commercial winding machines are readily available for as little as a few hundred dollars. They consist of a headstock shaft supported by bearings for minimal shaft rotation eccentricity, a variable speed motor drive, a foot operated speed control (leaving both hands free for winding), and a mechanical or electromechanical counter. High torque, low speed versions are available for winding heavy wire or foil. Semi-automatic winding machines which feed the wire at controlled turn-to-turn spacing are available as one's budget moves upward from hundreds to thousands of dollars.

Winding a toroid of few turns may most expediently be accomplished by holding the toroid between the thumb and one or more fingers of one hand and passing a calculated length of wire through the hole and around the cross section of the core. Each succeeding turn is held down with the thumb as the winding proceeds around the core. If needed, a piece of tape may be used to secure the start to the outer periphery of the core (Fig. 9-7).

Succeeding turns of wire hold down the right side of the piece of tape. If the layer is 360° the last turns hold down the left side. The finish may be secured as the finish of a layer-wound coil. A piece of tape may be wrapped around 360° of the outer periphery of the core to further secure the flying leads. On very small toroids no tape may be needed, the wires staying in place by themselves adequately for further processing such as an epoxy dip and cure.

9. MAKING A MANDREL FOR A COIL OR BOBBIN

Most winding machines will be configured so that the headstock (drive shaft) is terminated in a three-jaw chuck, typically of ½ in. capacity. Simply sketch a ½ in. diameter aluminum cylinder, an inch or so long, with a square or rectangular shape extending from one end and coaxial with the cylindrical part. A local machine shop can then machine the square section from that part of a round shaft. For larger winding forms (over ½–1 in. square) it is more convenient to machine a shaft and have the machine shop press a rectangular block (the winding form with a precisely bored hole through the center) onto the shaft. The hole should be true to the rectangular winding form to such an extent that the wire may be wound at the maximum coil diameter with wobble or runout of less than one-half wire diameter.

In some instances the shaft may be a threaded rod, with a set of coil mandrels, all of which have clearance holes. The blocks are held to the threaded shaft with nuts and washers. This is not suitable for heavy wire (too much torque) or very fine wire (mandrel will not be coaxial with the chuck of the winding machine).

In some cases it will be necessary to make a tube or winding form for the coil. With a mandrel machined to fit, a layer of nonadhesive plastic film (Mylar, Teflon, or polyethylene, 1 mil thick) may be wrapped and sprayed with dry film mold release, and a narrow strip of glass fiber cloth wound on the mandrel. This structure is then heated, saturated with epoxy resin, and cured, and then the mandrel is carefully driven out. The finished winding form is then trimmed to length and the exterior surfaces sanded. A coil may then be wound. This approach is time consuming and is only recommended where time does not permit an outside fabricator to make a length of rectangular epoxy fiberglass tubing to order. I have used Dorco Electronics of Paramount, California for almost 20 years and found them to be an excellent supplier of such custom tubes.

Pot cores have bobbins available to match the pot core sizes. They must be supported on both ends or the coil flanges will bulge out from wire pressure during winding and the finished coil will not fit the core. A mandrel design such as that shown in Fig. 9–8 has been found adequate.

When the bobbin is assembled on such a mandrel, it is held by compression and the end flanges adequately supported. The washers may be notched to match the bobbin flange notches. A similar design may be used for square nylon bobbins.

10. HOW TO WIND A BOBBIN

When the bobbin has holes in the side flange simply pass the start through the hole, bend it at a right angle once inside the flange, secure it with tape as needed, and proceed to wind. In some cases it may be necessary or desirable

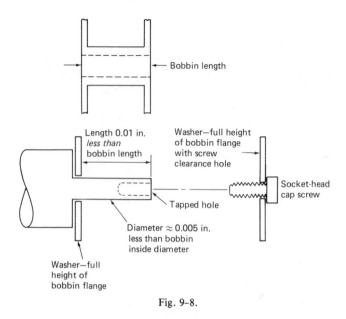

Bobbin length

Length 0.01 in. *less than* bobbin length

Washer—full height of bobbin flange with screw clearance hole

Socket-head cap screw

Tapped hole

Diameter ≈ 0.005 in. less than bobbin inside diameter

Washer—full height of bobbin flange

Fig. 9-8.

to secure the wire so that adequate initial winding tension may be applied (Fig. 9-9).

When the bobbin has no side holes or it is not desirable to use them, the start (or tap or finish) must be secured to the winding surface (or portion of the coil already wound) and then laid over to a flange, run up the flange and over. Tape or insulating material must usually be used to protect the exiting lead from the abrasion of succeeding turns being wound and whatever potential differences may be present.

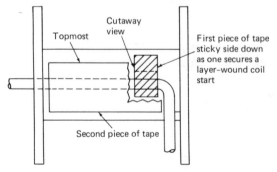

Cutaway view

Topmost

First piece of tape sticky side down as one secures a layer-wound coil start

Second piece of tape

Fig. 9-9.

11. HOW TO WIND A COIL WITHOUT A BOBBIN

Bobbins are usually used to support the turns of a winding which may not have layer insulation to support the layers of wire. They are also used where the greatest space utilization factor is desired and the film insulation on the magnet wire is adequate electrical insulation for the potential developed by two layers of winding.

If the coil is to be layer wound with insulating support sheet between layers, or is to be a single layer, epoxy–glass fiber or kraft paper tubing may be used. Kapton, Nomex, and a wide variety of materials are available in tube form from various specialty manufacturers.

If no space is available for a winding form or no winding form is desired, an aluminum mandrel can be machined to the desired coil diameter, etc. (Be sure to radius the corners on square or rectangular mandrels!) After mold release, the wire is wound on the form while liquid epoxy resin is fed onto the coil (wet winding). The wound coil is then placed in an oven for curing.

As a highly desirable alternative to wet winding, bondable magnet wire can be wound on a mold-released mandrel and the wires bonded together by passing a current through the wound coil for sufficient time to cause the bondable layer to melt. On cooling the coil stays together as a unit, and may be wrapped with glass fiber cloth tape and vacuum impregnated with epoxy resin if desired.

12. HOW TO WIND A LAYER-WOUND COIL

Layer-wound coils are specified to avoid placing turns of succeeding layers directly against each other. This may be for electrical reasons, to ensure that the placement of turns is regular, to wind a coil of several layers such that the layers are of equal length and the end turns do not fall off the coil during or after winding, or where multiple coils are being wound on a production multiple winder. Where interlayer insulation is important in a bobbin-supported coil, layer insulation may also be used.

To wind a layer-wound coil, choose a winding form (round, square, or rectangular tubing as required, cut to length) from a material and shape which is available, appropriately easy to fabricate, and has the desired electrical and mechanical properties. Spiral wound kraft paper tubing is often used, as is the more expensive glass-epoxy tubing. Molded bobbins of epoxy, diallyl phthalate, or Nylon are available, although the latter usually require a winding form with end plates to support the flexible Nylon bobbin flanges.

Choose a layer insulation based on the degree of mechanical support required, its thermal class, its electrical properties after impregnation, the ease of impregnation, the ability of the impregnant or encapsulant to form an adhesive bond with the layer insulation, and the ability of the layer insulation surface to hold the turns in place so they do not slide around during winding and handling, as well as availability and cost. In some cases a laminate or composite material may be required. Select the tape or tapes to be used to hold things together while winding the coil. Some of the more commonly used tapes are paper, Mylar, and glass cloth, or Kapton with a thermosetting adhesive.

Begin by securing the start of the winding as previously discussed. Then rotate the mandrel while guiding the wire onto it so that adjacent turns end up next to each other with very little space between them, but so that succeeding turns do not pile up on top of previous turns. The wire being wound meets the mandrel at an angle slightly less than 90°, lagging the advance of the winding by about 0–10°. With fine wire the angle may be only a few degrees off perpendicular. Much practice will be necessary to develop the fine touch necessary to hand-wind neat coils with wire sizes in the range of 30–38 AWG. The first few layers on square or rectangular coils are more difficult to wind than succeeding layers because of the sharp corners. Coils with large aspect ratio (cross-section dimensional ratio over 3 : 1) are more difficult to wind than nearly square coils. The beginner might practice by winding 18 AWG wire on a 2 in. diameter round form. This is easy. Then try a 1½ in. square form, then a ¾ × 2¼ in. form, then wind #28 AWG on a 1 in. diameter round form and a ½ in. square form. Machines with automatic wire traverse feed, while expensive, are well worth the price when many coils of hundreds of turns of 24 AWG or finer wire must be wound.

When the prescribed number of turns of the first layer has been wound, you are ready to put in the sheet layer insulation. Place a piece of tape on the coil covering the last few turns. This may often be only a temporary item to hold the winding in place and free the operator's hands. Then cut a piece of layer insulation (precut in strips of the proper width) to the required length. Wrap the strip around the coil in the winding direction. Allow a bit for overlap and cut with scissors. That gives the correct length. Then take a piece of tape (width ¼–1 in. or so as appropriate—start with about one-fourth the layer insulation width and change if needed to what suits the particular job best)—and stick it to the layer insulation as shown in Fig. 9–10(a). Then place it on the coil, slipping the wire between tape and layer insulation [Fig. 9–10(b)]. Press the tape entirely down. Wrap the layer insulation around the coil, maintaining a moderate tension (practice will teach you the proper amount). Place a second piece of tape across the lap, parallel to the coil axis

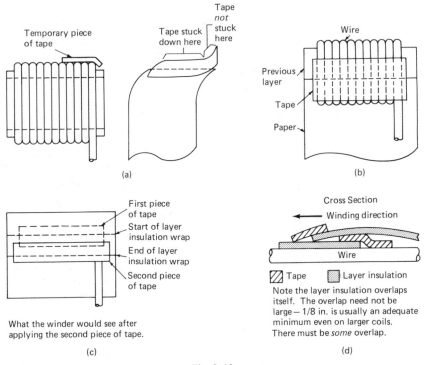

Fig. 9-10.

(perpendicular tape placement can build up lumps and allow the layer insulation to twist and swim in the winding direction, leading to skewed coils, offset wraps of layer insulation, and a sloppy product). After final tape placement the coil is ready for another layer to be wound. It will look like Fig. 9-10(c). Then wind another layer. Taps and finishes are handled as previously described.

When a layer-wound coil is wound in a bobbin and the layer insulation width exactly matches the inside bobbin width (or perhaps one-fourth wire diameter less), the layers may be wound from one bobbin wall to the other and a reasonably neat product obtained.

Layer-wound coils *per se* cannot have the winding extending the full width of the layer insulation. Some margin is usually needed for electrical reasons, and the first and last turns of such a coil would tend to fall off unless an inordinate amount of tape were used. The wire must be spaced somewhat in from the edges of the layer insulation. This spacing is usually symmetrical. The distance from the edge of the layer insulation to the nearest conductor of the winding is called the *margin*. High voltage considerations may dictate a greater margin than mechanical considerations. For example, a margin of ⅛–¼ in. is adequate for mechanical support of 10–12 AWG wire with 0.015 in. Copaco cuffed layer insulation. A margin of ⅛ in. is adequate for 16–20 AWG with 0.010 in. Copaco or Nomex 410. A margin of 0.08 in. is adequate for 28–32 AWG with 0.010 Nomex 411. Stiffer materials or cuffed materials may only need two wire diameters of margin width for mechanical support, while softer or more flexible materials may need five to ten wire diameters.

When bringing out self-leads from layer-wound coils it may be necessary or desirable to add an insulating sleeving over the lead. A number of sleevings serve well for this purpose. For lower temperature applications (Class 105 or 130) vinyl-glass sleeving is available from most magnet wire and insulation suppliers. For Class 155 or higher braided fiberglass sleeving is necessary. All of these materials will become stiff when the impregnant soaks into the sleeving. This may be good electrically but does not leave a flexible lead. It may be necessary for this sleeving to be cut short, and an unimpregnated length of sleeving slipped over it and bonded to it after impregnation (a dab of epoxy will usually suffice) to obtain a flexible insulated self lead.

Flying leads of insulated stranded wire are often specified by systems designers. These must be anchored mechanically inside the coil so no tensile stress is transmitted to the solder joint or the wire of the winding. The lead and its solder joint must also have adequate electrical insulation to whatever is in its vicinity.

It is convenient in many cases to bring the leads of the winding out to a terminal block, solder pins or some such, and let the user bring his wires to the transformer. Many bobbins have molded-in pins or lugs for just this purpose.

13. FOIL-WOUND COILS

High frequency or space factor considerations may make it advisable to use copper foil as the winding conductor. It is important that the sheet, once slit to width, have a minimal burr and be free of oil. A supplier of custom widths of copper foil is Thinsheet [(203) 756-7414]. In large quantities copper sheet mills can actually provide foil rolled out in such a manner that the edge not only has no burr but is radiused or chamfered. If this is critical (hundreds or thousands of volts per turn) the engineer may have to sand the edges of an unrolled length of foil by hand and then solvent wipe to remove metallic dust before winding.

Starts, finishes and taps are usually accomplished by soldering a narrower strip of thicker sheet perpendicular to the foil to be wound. The foil and a strip of wider layer insulation are then wound two-in-hand (either manually fed or positioned by a fixture built in-house) onto the winding form.

It is important that the materials be fed perpendicular to the winding form, that the foil be kept centered on the layer insulation in order to preserve electrical design specifications, and that adequate winding tension be maintained.

If the materials are not held perpendicular the winding will be conical. If the foil is not centered it may creep out from under the insulation and turn-to-turn shorts can result. If inadequate tension is maintained the winding build can be excessive, the turns can slide around, and the inside can literally drop out of the coil.

14. ASSEMBLING CORE AND COIL

It is of the utmost importance that cleanliness be maintained when a split core is assembled onto a prefabricated coil. Whether the core material is tape wound, or a ferrite or powdered iron core, it is important that no dust, flakes, or shreds of insulation from the coil or whatever be allowed to end up between the core faces. A "gapless" design is not obtainable when a flake of epoxy or a shred of paper a few mils thick is trapped between the core faces. A carefully measured thickness of gap material for an inductor will not yield the correct inductance when extraneous materials add an unpredictable amount. Crud in the gap will also cause acoustic noise in units with AC excitation.

Tape-wound cores may be assumed to have an oil film on them when received. They must be degreased before adhesive bonding the core faces together or an adhesive bond will not be obtained. Ferrite or powdered iron cores have a residual "dust" on the core faces from the grinding process. They must be ultrasonically cleaned immediately before adhesive bonding. If this is done it is possible to obtain a bond strength exceeding that of the material itself. If this is not done it is impossible to obtain consistent high-strength bonds. An epoxy adhesive that has been found very useful for bonding cores is Armstrong C-7/W mixed 1 : 1 by weight and cured two hours at 160°F. Another is Hysol EA-956.

Tape-wound cores may be fastened together by banding. Tinned steel bands and solder seals are available from Electrical Specialty Products, Sharon, Pennsylvania; Westinghouse Electric Corp., Greenville, Pennsylvania; Gerrand and Co., Des Plaines, Illinois; or their distributors. Crimped or soldered steel bands are useful when time for adhesive bonding is not available, where it is desired to band the core halves together and simultaneously attach them to a mounting foot, or where other considerations make this desirable.

Adhesive bonding is an excellent method of core assembly. So is banding. I have assembled many thousands of transformers by each method. Each has its pros and cons. When adhesive bonding is used the cores must be either clamped together or held together by a current passed through a winding (the latter method was suggested to me by Dick Wood of National Magnetics). Ferrite cores are not usually banded because the sharp outer corners interfere with proper tensioning, and also because of the high frequency loss of the steel band. Gapped structures may have their properties altered somewhat because the steel band shunts the gap. Being brittle, they can easily be chipped or broken during banding.

A mechanical clamping arrangement can be contrived wherein end plates are bolted to each other, compressing the core halves between them. This method is not often used because of the cost of the hardware and the propensity for acoustic noise. It occasionally can be implemented in a cost-effective manner and has its place.

When assembling a core around a coil, insulation is occasionally required between the core and coil. If impregnation or potting is to follow, a piece of insulating paper or laminate can be inserted in the margin (the space between coil and core in the G dimension) to give some minimum thickness of insulation after processing.

15. HOW TO WIND A TOROID

In many cases a toroidal core is to be wound with a modest number of turns (say, less than 100) of a wire size which is convenient for hand winding. If the length of wire to be wound is a few feet or less a shuttle may not be necessary. For greater lengths it is convenient to use a shuttle to carry the wire through the hole. A simple shuttle may be made by cutting a piece of $\frac{1}{16}$ in. G-10 or Plexiglass to a handy length, say 4–8 in., and a width narrow enough that it can easily be passed through the hole of the toroid (when it is fully wound and the diameter of the hole is reduced from that of the bare core). Sand off all sharp corners so that wire insulation will not be abraded away from repeated passage of the shuttle through the hole. Plexiglass or birch (tongue depressors) are preferred materials in this respect, while G–10 has greater strength. Then cut a rectangular notch in each end of the shuttle. The width and depth of the notch must give sufficient area to contain the number of turns that will be wrapped on the shuttle to provide the winding length. File a smooth radius on all sides of the notches. A small hole near one end is often helpful in securing the start of the wire to the shuttle.

One way or another, we now have core and wire and are ready to wind. First, make sure that the core has adequate insulation. Some ferrite cores, for instance, have sharp corners and can easily cut through the insulation on the wire. Next, take a length of tape somewhat more than the circumference of the core for small cores, or enough to cover a 120° segment of the OD for larger cores (once you get the hang of this you can use any amount of tape any way you find it expedient). Tape the winding start against the OD, roughly centered on the tape (Fig. 9–11).

Then, assuming you (the winder) are right-handed, place the left thumb over the start and the left forefinger on the opposite side of the core. Other fingers may be used as convenient. With the right hand feed the wire up through the hole, pull toward yourself, roll the core slightly between the thumb and forefinger to expose the last turn wound, and pull the wire toward yourself and down. Roll the core back slightly so the left thumb holds the turn last wound in place. The right hand is now free to fiddle with the wire, apply tape, manipulate a coffee cup, or wind another turn.

If the winding does not cover 360° the last turn may be secured with a patch of tape, a full 360° wrap around the OD, or occasionally a wrap over the last turn, wound just as the winding turns were. If the winding happens to cover 360° it is possible to do without any more tape by threading the last turn under the tape holding the first turn in place. If the wire is sufficiently stiff tape may not be needed at all.

Tapes are sometimes used as insulation between windings on toroids.

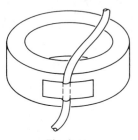

Fig. 9–11.

There is a danger here for designs which require thorough impregnation in order to meet electrical, thermal, or other requirements. Most tapes do not impregnate. The layers of adhesive-coated film or fabric may have excellent high voltage *test* properties, but air voids will be entrapped within the tape laps, preventing impregnant from flowing through to the winding beneath. A glass cloth is available, woven in various widths and thicknesses. The range includes ⅛ in. wide × 0.003 in. thick and ½ in. wide × 0.015 in. thick. It impregnates very nicely.

When it is necessary to machine-wind a toroid, either because there are too many turns of wire, especially fine wire, and/or so many units to be done that machine winding offers an economy, different toroidal winders can be evaluated to decide where the best options lie.

16. IMPREGNATION AND POTTING

Impregnation means soaking the unit thoroughly with something that displaces all the air inside the fabricated structure. Either simple soaking or vacuum-assisted impregnation qualifies, although vacuum impregnation offers far superior results. Potting generally means to cast the unit into a block of cured resin, whether the finished unit is surrounded by a metal can, a potting shell, or has an "as molded" surface. Again, the potting resin can be poured and allowed to cure, or vacuum may be used to aid the escape of trapped air.

The way vacuum impregnation aids the escape of trapped air is this: As the atmospheric pressure surrounding a part immersed in liquid is reduced, the air bubbles trapped inside the part expand. As they expand, the buoyant force on them is greater because they are displacing a greater volume of the heavier liquid. The bubbles tend to rise, leaving behind smaller bubbles of low-pressure air. When atmospheric pressure is restored the bubble contracts to a volume which leaves the remaining air at atmospheric pressure, and the bubble is of course much smaller. Assuming a good vacuum pump, the pressure on an immersed part will eventually get down to a point where the liquid is foaming or bubbling violently. It is necessary to have some extra container height to contain this foaming. At this pressure we either have a lot of air trapped in layers of paper or mat and fine wire, or the resins themselves are boiling. It is not unusual for one of the components of an impregnating or potting compound to boil at the process temperature and a pressure in the range of 5 mm Hg or less.

On the one hand this is bad because we are losing one of the components of the resin system and it will end up contaminating the vacuum pump and all the vacuum plumbing. On the other hand, when the resins boil the air in those trapped voids is being displaced with a condensible vapor. When atmospheric pressure is reestablished the vapor condenses, the liquid moves into the void space, and we have perfect impregnation with no void whatsoever. This sounds nice but in real life it does not happen.

When a part is immersed in a liquid, the liquid itself exerts a pressure on the part, depending on how much liquid is above the part. If we have a 13 mm depth of immersion, the liquid adds a pressure of 13 mm of whatever it is. For most epoxy resins, that is about 1 mm of mercury for a depth of only ½ in. As atmospheric pressure around the immersed part is reduced, the resin will boil on the surface but not down deep inside the part and its voids, because down deep the pressure is several mm Hg higher. If one observes a part immersed in a resin system in a vacuum chamber, bubbles can occasionally be seen coming to the surface from beneath. After a while bubbles begin to form and break on the surface. The latter type of "boiling" is characteristic of boiling resin.

One common method of displacing the air within a coil is by simply dipping or soaking the finished structure in a liquid resin or varnish and then (at room or elevated temperature) allowing the resin in the "saturated" coil to harden. This procedure usually provides a gross mechanical bond. With relatively large wire sizes (#10-20) it does a fair job of soaking through the coil to the inside. Many voids will, however, remain in the coil after such processing. Vacuum impregnation is far superior. In this process the coil or transformer is immersed in the liquid resin, the container is placed in a vacuum chamber, and a vacuum pump is used to reduce the pressure to 0.1-10 mm Hg. The vacuum chamber may be a vacuum oven if it is necessary to heat the resin during impregnation to reduce its viscosity.

When the pressure is reduced beyond some point most resins will boil, especially when heated. This sets a lower limit on the vacuum to which the unit may be subjected and a consequent lower limit on the extent to which air may be removed from the coil. A small amount of air may remain in the coil, and when atmospheric pressure is restored to the immersed coil small but finite air bubbles (voids) may remain. A mitigating factor is that if the pressure is reduced to the boiling point of the resin, *some* of the residual air becomes displaced with resin vapor which then condenses upon return to atmospheric pressure.

In order to obtain very good impregnation the part is placed in a container in a vacuum chamber. The pressure is reduced to less than ½ mm Hg and held there for a period of time (until the part finishes outgassing), *then* liquid resin is introduced into the vacuum chamber until the part is totally immersed. The chamber is returned to atmospheric pressure, and the part allowed to soak long enough for the resin to flow into all the interstices (a few minutes for small parts, perhaps 15–30 minutes for a 6–8 in. coil with many layers of paper and fine wire). It is then placed on a tray for draining and curing.

A variation on this process achieves a more thorough impregnation. One method is to place the coil or unit in a vacuum chamber, evacuate to 0.1 mm Hg, and then introduce the liquid resin into that chamber from a separate chamber. The coil ends up being totally immersed in the liquid. The chamber is then brought to atmospheric pressure. After this procedure, the unit (in a conformal mold or potting shell) is then placed in a pressure vessel and pressurized to 100 psi or more for sufficient time and temperature to cure the resin. Curing under pressure aids and speeds the flow of liquid resin into the coil, reduces the size of the residual air voids to a minute level, and with sufficient pressure the residual gas will dissolve in the liquid resin, the void thus vanishing.

The overall impregnation and/or potting process begins with a dry coil which has been wound and may have been assembled onto a core. The first step is to bake it in a forced air circulating oven. Material thermal limitations permitting, 24 hr at 250°F (120°C) is recommended. This baking process serves two functions: It cures the adhesive on thermoset tapes; it also drives out the residual moisture inside the coil and can volatilize the oil film present on magnet wire.

There are epoxy systems, both rigid and semiflexible, which cure at room temperature. For best impregnation only unfilled systems should be used. Some of these have less shrinkage on curing, and of course there is no shrinkage on cooling down from an elevated temperature cure because there is no elevated temperature. Less shrinkage puts less stress on strain-sensitive core materials.

Other epoxy systems, either rigid or semiflexible, require an elevated temperature cure. They offer the advantage of lower impregnating viscosity and longer working time (pot life) so that the same batch of resin may be used to impregnate successive batches of coils.

After curing the unit is ready for the next processing step, whatever that may be.

Some designs require a corona threshold test which production units, impregnated and cured according to the resin manufacturer's curing specifications, may have difficulty passing. The corona threshold may be raised by subjecting a unit to post-curing, an extended baking procedure, after the manufacturer's specified curing schedule is completed. For example, 3M

#280 epoxy is supposed to cure in 2 hr at 250°F. In one instance units so cured were found to have a corona threshold in the 6–9 kV range. The requirement was 10 kV. Post-curing in a forced air circulating oven for 24 hr at 250°F raised the corona threshold about 2 kV for all units. Post-curing for seven days (168 hr) at 250°F raised the corona threshold on all units to the 12–13 kV range. There is nothing inadequate about 3M #280. It was chosen for its excellent electrical properties. The effect of post-curing on corona threshold is common in varying degrees to all polymer systems.

Post-curing works because the chemically reactive groups on the component molecules combine during the curing reaction to form larger molecules and complex polymer matrices. The remaining reactive groups become progressively more hindered in their movements and not all the reactive groups are able to find each other and react. The manufacturer's specified curing time may be enough for, say, 99% of the reactive groups to react, but the corona threshold can be severely degraded by even a minute amount of uncured ingredients. Prolonged heating during post-curing can give sufficient time for virtually all the reactive groups to react, thus giving results which can approach the ultimate available from that resin system.

17. ENVIRONMENTAL REQUIREMENTS AND CONSEQUENCES

MIL-T-27 may be specified for a component, or the application may be a commercial or industrial grade unit which is expected to last the life of the equipment in which it is installed.

The unit may also end up being built by the lowest bidder, who will provide the minimum quality level necessary to pass incoming inspection. Even industrial or commercial grade units will be exposed to other environmental factors than a 55°C ambient and one thermal cycle per day.

The air within 50 miles or so of the ocean will contain a significant amount of salt crystals as fine as dust. These will find their way into the innards of most electronic instruments. The humidity may approach 100%, and condensation may occur on cold surfaces. Fingerprints leave a residue of oil and salt. High voltages gather the dust from the air, some of which adheres to insulating surfaces around high potential regions. Moisture will slowly permeate dielectric resins. Microscopic traces of salt and moisture will form conducting paths from pinholes in the insulation to anywhere, and it may seem that the physical universe as a whole is conspiring to cause that component to fail. It is.

We can do a great deal to prolong component life. We can eliminate weak points, and make it very difficult for pinholes to exist or moisture to degrade the insulation. Manufacturers of resins and insulating sheets and films usually have life expectancy data on their products. It is not unreasonable to expect magnetic components to have a 10,000 hour life. Knowing the degradation rate of dielectrics versus temperature, one could design components for a 100,000 hour life with a fair degree of confidence. It remains to construct the unit adequately to withstand the environment.

MIL–T–27 specifies three grades of transformers and specifies certain tests for these grades (Table 9–3).

The methodology of these tests is given in MIL–STD–202 as referenced by MIL–T–27. Electrical testing in conjunction with accelerated environmental stress is designed to assure that the unit is capable of a 10,000 hour life expectancy (about 14 months of continuous operation). Any commercial unit worth its salt should have several times this life expectancy, and most do.

Moisture and abrasion resistance of the magnet wire is crucial in achieving a long life in open-type transformers, which may be merely bobbin wound and then dipped in varnish. Moisture resistance under electrical stress is important because such units have no seal against ambient air and the internal humidity will be the external humidity. Abrasion resistance is important because the winding process unavoidably abrades the wire as it passes over other strands on the spool, the edge of the spool, pulleys, wire guides, bobbin walls, etc. Using more abrasion-resistant magnet wire insulations and giving careful attention to winding procedures and machine design can minimize abrasion. An impregnation with varnish or epoxy is much more beneficial than dipping in obtaining a long life from open type units.

Encapsulation for commercial or industrial units should be considered for

Table 9–3. MIL–T–27 Tests for Transformers.

TEST	GRADE 4 METAL ENCASED	GRADE 5 ENCAPSULATED	GRADE 6 OPEN TYPE
Seal	×	×	—
Thermal shock	×	×	×
Immersion	×	×	—
Moisture resistance	×	×	×
Vibration	×	×	×
Shock	×	×	×
Flammability	—	×	—
Salt spray (when specified)	×	×	—

operating voltages higher than ~ 500–1000 volts. This includes not only molded or potted units but those processed by immersing the impregnated and cured unit in a thixotropic resin system (usually an epoxy), subjecting the unit to a vacuum, then removing and curing. The process puts a thick conformal resin coating over the unit. It is very advantageous for transformers with tape-wound C-cores or stamped lamination cores that have high operating voltages, as it assures an essentially void-free unit even in the margins where the thin impregnating resins drain out.

Metal-encased, hermetically sealed units are desirable or mandatory in many circumstances. For liquid or gas filled units a hermetic seal is obviously necessary. The steel case can afford magnetic shielding, either to protect neighboring circuitry from the transformer or inductor, or to reduce stray field pickup by the shielded component. Metal-encased hermetically sealed units offer the best protection against moisture, contamination, reduced atmospheric pressure, and other undesirable environmental influences.

All other things being equal, metal-encased hermetically sealed units are the most expensive to manufacture. Molded encapsulated units are usually less expensive to produce. Small units potted in potting shells are even less expensive, as are conformally coated units. The cost difference between the latter three depends on the particular circumstance. The high volume, small units may be very economically encapsulated by transfer molding. The tooling for this can be quite expensive. Small units in moderate volume may be most economically potted in a potting shell, while it is often more convenient to conformally coat larger units (over perhaps a few cubic inches) than to mold or pot them. Heat transfer requirements may not permit the high thermal resistance of a potting shell but require a high thermal conductivity epoxy as the external surface.

When cost is a significant factor and the unit will not be build in-house, the unit should be properly specified and quotes obtained from magnetic component manufacturers. Life expectancy is a crucial and often invisible part of the unit cost. If a system has components with a definite life expectancy (such as a flash lamp) then the associated magnetic components (such as the trigger transformer) should be specified for that same life expectancy. Competing designs should be evaluated side by side in a full life test if at all possible.

18. MOUNTING

The military specification MIL–T–27 indicates standard sizes of drawn metal cans. While the sizes may look a bit odd, it turns out that they were chosen for a comfortable fit around standard sizes of E–I laminations in square stacks. In the packaging of magnetic components one is not necessarily restricted to those can sizes. If a metal can with threaded studs or inserts is required it is not difficult to have custom cans made to enclose any shape. Steel cans afford electrostatic and a fair degree of magnetic shielding and are often desirable or mandatory in industrial and military equipment.

If for some reason a steel can is not desired or feasible, but an enclosed or potted unit with outer dimensions as close as possible to the actual component is needed, the component may be potted in a potting shell (if small) or molded into a block with the appropriate molds or tooling. Mounting may be by molded-in threaded studs, threaded inserts, or molded or drilled through-holes.

If the unit is fairly small, whether molded or not, adhesive bonding provides a convenient and entirely adequate mounting means. Adhesives such as silicone or polysulfide would usually be preferred, as they are flexible and can readily accommodate local stress.

In commercial equipment small magnetic components such as toroids may be easily mounted on a printed circuit board with a nylon cable tie or two around the part and through a couple of holes in the board. Toroids may also be mounted with a screw through the middle, holding a fiber washer or some such against the top of the toroid and compressing it against the mounting surface.

Most sizes of pot cores have metal clips for mounting. Pot cores may also be mounted with a screw through the middle (Nylon screws would be preferred, especially if the pot core is gapped) or by adhesive bonding. Smaller pot cores may be mounted with printed-circuit-mount bobbins, as these have molded-in pins to which the windings are terminated. The bobbin pins are then soldered to the PC board. The core is secured to the bobbin either by standard pot core hardware or by adhesive bonding the pot core halves together.

Stamped and formed mounting brackets such as those offered by Hallmark Metals can be used to mount tape-wound cut cores (C-cores). The C-core may be secured to the bracket by banding or, for small units, adhesive bonding. If shock and vibration requirements permit, the core halves may be adhesive bonded to each other with an epoxy adhesive and the core then bonded (usually with a polysulfide) to a plain flat plate with mounting holes.

Larger C-core units may be mounted by two lengths of metal angle stock, held against the core by bolts between the angles. The bolt spacing is slightly more than the coil length, so that the bolts lie within the lower corner spaces of the core. Mounting holes in the metal angles secure the unit to the mounting surfaces. Insulating sheet materials must be inserted between the sides of the angle brackets and the edges of the core laminations, or a partial, shorted turn can result from the mechanical pressure of the metal angles against the lamination edges.

E–I laminations have been in use for a very long time and are an extremely economical means of obtaining a magnetic structure. It is therefore not surprising to find many ways to mount standard size laminations.

Stamped metal shells called *end bells* have bent-over sheet metal ears for fastening the unit to the chassis. The end bells are bolted to the lamination stack on each end, covering the coil structure. Holes are provided for flying leads to exit the end bell. L-shaped angle brackets are available for mounting stacks of laminations—the angle is bolted to the stack of laminations and the foot of the L is bolted to the mounting surface.

When bolts are passed through the punched holes in stacks of laminations, it is important to ensure that at least one end of the bolt is electrically isolated from that side of the stack. A shorted turn can result if the mounting hardware makes electrical contact across all or a part of the lamination stack.

Printed circuit bobbins are available for the smaller size laminations. Their strength is adequate only for commercial applications. Even so, it may be necessary to use adhesive bonding or mechanical means to secure the core to the PC board and to attach the PC board to the chassis at that place.

Stamped and formed metal brackets called channel frames are available to fit a wide range of lamination sizes. A channel frame is a sheet steel channel bent in two places such that it wraps around three of the four sides of a lamination stack. Sheet metal ears fold over and retain the lamination stack within the channel frame. Ears (or tabs on smaller sizes) are the mounting means.

19. STRESS-FREE PROCESSING

Special processing to minimize or isolate mechanical stress or coupling may be necessary for a variety of reasons. Ferrite cores should not be impregnated with anything which develops a curing stress, as it reduces the permeability. Even varnish or epoxy dipping and curing can noticeably degrade the permeability. When such processing is necessary, cores with an impermeable coating applied by the manufacturer should be used. As an alternative the impregnating and potting compounds should be restricted to silicone elastomers or a low shrinkage urethane such as 3M #221. Strain-sensitive nickel alloy cores or lamination stacks are easily degraded by impregnation with the so-called semiflexible epoxy systems. No impregnation, potting with 3M #221, or a silicone elastomer is recommended. Mounting such cores on a mounting bracket or plate with polysulfide rubber can provide stress-free mounting.

All magnetic materials are magnetostrictive. Under pulse or AC excitation they will change their dimensions very slightly. This can generate noise if the core is acoustically coupled to an acoustic radiator such as a chassis. Silicones, polysulfides, and similar soft, flexible materials can provide acoustic isolation.

A molded or impregnated coil has a coefficient of thermal expansion which differs greatly from that of core materials. It may be desirable for some or all of the above reasons to isolate the coil from the core while providing some means of mechanical attachment of coil to core so the coil does not rattle around on the core. The previously mentioned materials can perform this function. These are the primary reasons for designing a stress-free processing procedure.

Molding or impregnating the coil separately from the core is practical for most split-core designs and offers several advantages. Not only does it allow compensation for the differential coefficient of expansion, but it provides acoustic isolation between coil and core. It is not only chassis contact that will cause acoustic noise from a magnetostrictive core, but contact with anything hard. Separate processing also allows finishing operations to be done on the coil that might be impractical or inconvenient if the core were present. If the design is tight or marginal there may be a significant percentage of electrical rejects. If these can be culled out before the core is made a permanent part of the unit the cost of a reject is reduced. This can be crucial with expensive cores.

Various mounting means are available which reduce, minimize, or essentially eliminate stressing or coupling to the core. The unit may be potted in a deep drawn can such as the MIL-T-27 sizes or other, using appropriate materials. Potting shells may also be used. It is usually desirable to have something solid for a mounting plate for larger units, so molding the unit in an elastomer is rarely suitable unless the unit is to be mounted by adhesive bonding. Any significant amount of heat generated by the transformer or inductor may cause problems. Elastomers have notoriously poor thermal conductivity. Beryllia-filled silicone rubber is expensive, besides having a thermal conductivity which varies with temperature. Heat causes the rubber to expand, moving the beryllia grains apart and lowering the thermal conductivity.

For commercial designs it may be advantageous to use a bobbin with printed circuit pins molded in.

The core would usually be a stack of laminations, as bobbins are available to fit standard lamination sizes. A custom C-core with the same dimensions as a lamination stack could of course be specified if magnetic properties not available in laminations were required. The core would be assembled on the (wound and impregnated) bobbin, and could be secured with a wrap of thermosetting tape to hold the laminations or core halves together. A bit of 3M #221 dribbled in between core and coil might be desirable, or a channel frame could be slipped over the lamination stack and held in place with polysulfide rubber in lieu of mechanical crimping.

A channel frame can be mechanically crimped over a stack of nickel laminations without damage if a piece of 25 mil Copaco is used as a cushion between the lamination stack and the crimp ears of the channel frame. Do not crimp harder than necessary to bend the ears over or excessive strain can still be put on the core and lower the permeability.

While means have been shown for isolating magnetostrictive cores and reducing acoustic noise, bear in mind that all these methods offer only a relative degree of attenuation. Nothing has infinite attenuation. In some applications one may have the option of either a tape-wound or ferrite core design. Ferrites have much less magnetostriction than tape-wound cores, and will generate much less acoustic noise with little or moderate acoustic isolation. Ferrite cores are also among the least expensive.

AFTERWORD

This is the end of this book, but not the end of the subject. Many variations of these fabrication techniques exist, and many more can be created. The foregoing were the basics. From these one can create or evolve as many unique design implementations as there are specific design requirements.

New mathematical techniques are constantly being developed, the most notable among recent ones perhaps being a mathematical description of core loss by James Triner. If core manufacturers were to specify their materials with the same mathematical parameters which Mr. Triner suggests, a unified power transformer design algorithm could be set forth in general terms.

There is much material which was not covered in this volume because of the limited amount of personal time available for this work and an editor who would not tolerate a manuscript more than two years late. Wideband low level communications transformers and high frequency power transformers are worthy of a chapter each, although much pertinent data for the latter is to be found in the chapter on pulse transformers (Chapter 6).

There are far more uses or potential uses for nonlinear magnetic components than the few mentioned in the chapter on nonlinear magnetics (Chapter 4). With the current popularity of high frequency converters many circuit functions may be simplified or performance advantages obtained if the circuit designer is aware of the basic capabilities and limitations of these components, and the ease with which they may be designed and fabricated.

Take this knowledge, combine it with what you already know, and use it well. I wish you success.

REFERENCES

Ahearn, J. H., et al. 1957. "Cooling of electronic equipment by means of inert vapor," National Conference on Aeronautical Electronics, May 15, 1957.

Arnold Engineering. 1979. *Silectron Cores*. Marengo, Illinois: Arnold Engineering Bulletin SC-107B.

Aslin, Harlin. 1977. "Pulse transformer design study," contract #ECON-76-1292F, November. San Leandro, California: Physics International Co., 2700 Merced Street.

Baum, C. E., et al. 1978. "Sensors for electromagnetic pulse measurements, both inside and away from nuclear source regions," *IEEE Transactions on Antennas and Propagation*, AP-26(1), January.

Bir, D. L., et al. 1980. "Basic principles governing the design of magnetic switches." Lawrence Livermore Laboratory #UCID-18831, Livermore, California, 18 November.

Coate, Godfrey T., and Laurence R. Swain, Jr. 1966. "Winding and eddy-current losses," Section 2.1.5 in *High Power Semiconductor-Magnetic Pulse Generators*. Cambridge: MIT Press, pp. 17–21.

Frungel, F. B. A. 1965. *High Speed Pulse Technology* (4 volumes). New York: Academic Press.

General Electric Company. 1982. *Heat Transfer and Fluid Flow Data Books*. Schenectady, NY: General Electric Company, 120 Erie Blvd.

Glascoe, G. N. and J. V. Lebacqz. 1964. *Pulse Generators*. Volume 5 in MIT Radiation Laboratory Series. Boston Technical Publishing Company.

Grossman, Nathan. 1967. *Transformers for Electronic Circuits*. New York: McGraw-Hill.

Infinetics, Inc., 1965. "Magnetic core and nickel alloy tape-wound toroids." Technical Data Form 87-1. Wilmington, Delaware: Infinetics, Inc., 1601 Jessup Street.

Kays, W. and A. London. 1964. *Compact Heat Exchangers*. 2nd ed. New York: McGraw-Hill.

Lee, Reuben. 1955. *Transformers and Electronic Circuits*, 2nd ed. New York: John Wiley & Sons.

McCoy, H. E., Jr. and C. R. Brinddham. 1981. "Changes in the properties of polymer films when exposed to typical transformer environments." Conf-810913-9. Springfield, Virginia: National Technical Information Service, 5285 Port Royal Road.

McLyman, W. T. 1977a. *Spacecraft Transformer and Inductor Design*. Pasadena, California: Jet Propulsion Laboratory Publication 77-35.

McLyman, W. T. 1977b. *Transformer Design Tradeoffs*. Pasadena, California: California Institute of Technology, NASA Technical Memorandum 33-767.

(NOTE: McLyman's second book has a rather extensive tabulation of a parameter k_g which is the square of the core area times the window area divided by the mean length of turn. This parameter is introduced in Chapters 2 and 3 of this text as $(D^2E^2FG)/U$. The reader will have to divide k_g by the wound core volume to determine relative volumetric efficiency for those tabulated cores. Relative volumetric efficiency is tabulated directly in the Appendix of this book for normalized core shapes, which concept is developed in Chapter 2.)

McLyman, W. T. 1978. *Transformer and Inductor Design Handbook*. New York: Marcel Dekker Inc.

Magnetic Metals Company, Hayes Avenue at 21st Street, Camden, NJ 08101.

Minnesota Mining and Manufacturing Company (3M), n.d. *Fluorinert Data Book,* Y-ILBG(R) (87-1)RC.

Nunnally, W. C. 1981. "Magnetic switches and circuits." Los Alamos, New Mexico: Los Alamos National Laboratory, LA-8862MS, September.

Nordenberg, Harold. 1967. *Electronic Transformers*. New York: Reinhold Publishing Co.

Oxner, Ed. 1982. "Correlating the charge transfer characteristics of power mosfets with switching speed," *Powercon 9*. Ventura, California: Power Concepts, Inc.

Reference Data for Radio Engineers. 1970. 5th ed. Cleveland, OH: Howard W. Sams & Company.

Rippel, W., and W. McLyman. 1982. "Design techniques for minimizing the parasitic capacitance and leakage inductance of switched-mode power transformers," *Powercon 9*. Ventura, California: Power Concepts, Inc.

Robenein, G. J. 1980. "High voltage air core pulse transformers," SAND 80-0451, Sandia National Laboratories.

Rohensow, W. M. and J. P. Hartnett. 1963. *Handbook of Heat Transfer*. New York: McGraw-Hill.

Rottwein, G. J., 1979a. "Design of pulse transformer for charging," SAND-79-0698c. Sandia National Laboratories, Albuquerque, New Mexico.

Rottwein, G. J. 1979b. "Development of a 3 MV pulse transformer," SAND-79-0813.. Sandia National Laboratories, Albuquerque, New Mexico.

Schade, O. H. "Schade's Curves," *Proceedings of IRE* (now IEEE), July 1943, p. 346.

Sower, G. 1981. "I-dot probes for pulsed power monitors," *Proceedings of Third International Pulsed Power Conference*. Albuquerque, New Mexico: IEEE.

Storm, H. F. 1950a. "Series connected saturable reactor with control source of comparatively low impedance," *AIEE Transactions,* **69**(Part II), pp. 756–765.

Storm, H. F. 1950b. "Series connected saturable reactor with control source of comparatively high impedance," *AIEE Transactions,* **69**(Part II), pp. 1299–1304.

Triner, J. E. 1982. "Advances in core loss calculations for magnetic materials," NASA Technical Memorandum 82947. Cleveland, OH: Nasa-Lewis Research Center.

U. S. Department of the Army, 1979. *Dielectric Embedding of Electrical or Electronic Components,* Engineering Design Handbook DARCOM P-P-706-315. HQ U.S. Army Materiel Development and Readiness Command, 5001 Eisenhower Avenue, Alexandria, Virginia 22333.

van Sant, J. H. 1980. "Conduction heat transfer solutions." Livermore, California: Lawrence Livermore National Laboratory, UCFL #52863.

INTRODUCTION TO
APPENDICES A AND B

In the analysis for minimum copper loss (chokes and resonant charging reactors), we differentiated with respect to a core dimension in order to find the maximum value of D^2E^2FG/U (the smallest denormalization constant) for each of many overall form factors (P,Q sets).

In the analysis for minimum total transformer loss we differentiated with respect to a core dimension to find the minimum of [the geometric portion of Eq. (2.25)], which again gives the smallest denormalization constant for each of many P,Q sets.

In both analyses the plots of the magnitude of this geometric parameter versus P and Q show that there is a ridge of relative maxima running along certain P,Q values. This ridge is denoted by a dotted line in each of the P,Q plots. Overall form factors or relative shapes which would be preferred are to be found on that ridge.

The question of how far out on that ridge one should go to pick a relatively more efficient design is answered by a second set of P,Q plots. For each of the two analyses, the magnitude of the geometric parameter *per unit volume* is plotted for the P,Q range. One sees, then, that as one moves along the ridge of optima this *volumetric* efficiency increases faster at first and then more slowly. Where it increases more slowly one gains less (although one still gains) by moving further in that direction.

In the following pages the reader will find the copper-loss-only analysis (entitled "chokes and resonant charging reactors") and the combined-core-and-copper-loss analysis (entitled "transformers"). Preceeding each case analysis are the two P-Q plots. They are Case I, geometric parameter; Case I, geometric parameter per unit volume; Case II, geometric parameter; Case II, geometric parameter per unit volume; and so forth. These plots have been previously presented in the text (Chapters 2 and 3), but are shown here in full for the convenience of reference.

In general, the reader will note that for chokes and resonant charging reactors the preferred form factors tend to be somewhat long, and higher than the width. For transformers the preferred form factors tend toward square shapes, with the length comparable to the height and both large compared to the width. In this regard it might be noted that square stacks of E-I laminations are not too far from an optimum, while double stacks are somewhat further removed.

Ferrite cores can be pressed in any shape, and it takes only the manufacturer's recognition that he could offer a better product for such optimized shapes in ferrite to become available.

Tape-wound C-cores may today be made in almost any shape, and the designer will find that optimized designs can be readily fabricated using tape-wound cores. The increasing costs of assembly labor and the higher performance offered by special alloys or thin gauges also serve to make tape-wound cores a preferred means of realizing these designs today.

APPENDIX A
CHOKE/RESONANT
CHARGING REACTOR
DESIGN TABLES

OPTMIZATION PARAMETER: POWER LOSS = $U/(D^2 * E^2 * F * G)$

CASE # 1

$E=(P-1+D)/2$ $F=(1-D)/2$ $G=Q-P+1-D$
$U=3*D+P-1+1.5708*(1-D)$
$Um=3*(1-D)+2*(Q-P)+1.5708*(P-1+D)$

CASE # 2

$E=(P-1+D)/2$ $F=(1-D)/2$ $G=(2*Q-P+1-D)/2$
$U=3*D+P-1+1.5708*(1-D)$
$Um=2*Q-P+2-2*D+0.7854*(P-1+D)$

CASE # 3

$E=(P-2+2*D)/2$ $F=1-D$ $G=Q-P+2-2*D$
$U=4*D+P-2+1.5708*(1-D)$
$Um=6-6*D+2*Q-2*P+1.5708*(P-2+2*D)$

CASE # 4

$E=(P-3+3*D)/3$ $F=1-D$ $G=(3*Q-2*P+6-6*D)/3$
$U=2*(P-3+6*D)/3+1.5708*(1-D)$
$Um=3*Q-2*P+10-10*D+1.713864*(P-3+3*D)$

CASE # 1   COPYRIGHT 1982 STEVE SMITH   DATA DATE 1/ 3/82   CHOKE/RESONANT CHARGE REACTOR DESIGN TABLE

SCALE BY 1.000E-04   FUNCTION 1/f(D) $f(D)=U/(F*G*(D*E)^2)$

Q/P	0.33	0.39	0.47	0.56	0.68	0.82	1.00	1.20	1.50	1.80	2.20	2.70	3.30	3.90	4.70	5.60	6.80	8.20	10.00
0.33	0	0																	
0.39	0	0																	
0.47	0	0																	
0.56	1	1																	
0.68	1	2																	
0.82	2	3																	
1.00	2	3																	
1.20	3	5																	
1.50	4	6																	
1.80	6	9																	
2.20	7	12	18	27	38	54	70	88	66	19									
2.70	9	14	22	34	51	73	98	129	146	146	96	17							
3.30	11	17	28	44	65	96	131	181	221	248	232	134	14						
3.90	13	20	34	53	80	118	165	233	294	353	377	321	175	18					
4.70	16	25	42	66	99	149	209	303	393	493	573	587	498	231	3				
5.60	20	33	50	80	121	183	260	382	503	651	795	890	884	708	295				
6.80	25	40	62	98	150	228	327	487	651	862	1092	1296	1404	1382	1115	384			
8.20	30	49	76	120	184	281	405	609	824	1108	1438	1771	2013	2175	2117	1624	489		
10.00	37	60	93	148	228	349	505	766	1046	1425	1884	2381	2796	3196	3413	3290	2527	625	
12.00	45	73	113	180	276	425	617	941	1292	1777	2379	3060	3667	4332	4855	5148	4875	3520	
15.00	56	91	142	227	349	539	785	1204	1662	2305	3123	4079	4974	6037	7020	7939	8408	8012	
18.00	68	110	171	273	422	653	952	1466	2032	2834	3866	5097	6281	7742	9185	10731	11943	12514	
22.00	83	135	210	336	519	804	1176	1816	2525	3538	4857	6456	8024	10015	12072	14455	16658	18520	
27.00	102	166	258	414	640	994	1456	2253	3142	4418	6096	8153	10202	12857	15681	19109	22553	26029	
33.00	125	203	316	508	786	1221	1791	2778	3882	5475	7583	10191	12816	16268	20013	24695	29627	35041	

CHOKE/RESONANT CHARGE REACTOR DESIGN TABLE

CASE # 1   COPYRIGHT 1982 STEVE SMITH   DATA DATE 1/ 3/82

$f(D)=U/(F*G*(D*E)^2)$

SCALE BY 1.000E-04   FUNCTION 1/[P*Q*f(D)]

Q/P	0.33	0.39	0.47	0.56	0.68	0.82	1.00	1.20	1.50	1.80	2.20	2.70	3.30	3.90	4.70	5.60	6.80	8.20	10.00
0.39	3	3	3	3	2	1	0												
0.47	3	4	4	3	3	2	0	0											
0.56	4	4	5	5	4	3	2	0	0										
0.68	4	5	6	6	6	4	3	2	0	0									
0.82	5	5	7	8	8	7	5	4	3	2									
1.00	5	6	8	9	10	9	7	8	6	7									
1.20	6	7	9	11	12	12	11	12	13	14									
1.50	6	7	10	12	14	15	15	18	20	22	5								
1.80	6	8	10	13	15	18	19	22	26	30	9	3							
2.20	6	8	11	13	17	20	22	27	32	37	14	8							
2.70	7	8	11	14	18	21	25	30	37	42	25	13	2						
3.30	7	9	12	15	19	23	27	33	43	46	34	26	12	1					
3.90	7	9	12	15	19	24	29	35	45	50	41	36	25	11	0				
4.70	7	9	12	16	20	25	32	37	48	53	48	45	38	27	10	0			
5.60	7	9	13	16	21	26	33	39	50	56	53	53	48	40	27	9			
6.80	7	9	13	16	21	26	34	42	51	58	58	59	58	53	43	29	8		
8.20	7	9	13	17	22	27	34	43	52	60	61	65	65	63	56	46	29	7	
10.00	7	10	13	17	22	28	35	44	53	62	65	70	72	72	68	61	48	31	6
12.00	7	10	13	17	22	28	36	44	54	63	67	73	77	78	77	72	63	50	29
15.00	7	10	13	17	22	29	36	44	55	64	70	77	82	85	86	84	78	68	53
18.00	7	10	13	17	23	29	37	45	56	64	72	80	86	89	92	91	88	81	70
22.00	7	10	13	17	23	29	37	45	56	65	73	82	89	94	97	98	97	92	84
27.00	7	10	13	17	23	29	37	45	56	65	74	84	92	97	101	104	104	102	96
33.00	7	10	13	17	23	29	37	45	56	65	75	85	94	100	105	108	110	109	106

CASE # 1   COPYRIGHT 1982 STEVE SMITH   DATA DATE 1/3/82   CHOKE/RESONANT CHARGE REACTOR DESIGN TABLE

$$f(D)=U/(F*G*(D*E)^2)$$

P	Q	D	E	F	G	U	Um	DE	FG	f(D)	1/f(D)	1/(PQf(D))	DEUm
0.33	0.33	.846	.088	0.077	0.154	2.110	0.738	0.0744	0.012	.3210E+5	.3115E-4	.2860E-3	.5498E-1
0.33	0.39	.860	.095	0.070	0.200	2.130	0.838	0.0817	0.014	.2279E+5	.4387E-4	.3409E-3	.6850E-1
0.33	0.47	.872	.101	0.064	0.268	2.147	0.981	0.0881	0.017	.1614E+5	.6196E-4	.3995E-3	.8643E-1
0.33	0.56	.880	.105	0.060	0.350	2.158	1.150	0.0924	0.021	.1204E+5	.8306E-4	.4495E-3	.1062E+0
0.33	0.68	.887	.107	0.058	0.465	2.166	1.383	0.0951	0.027	.8944E+4	.1117E-3	.4980E-3	.1315E+0
0.33	0.82	.889	.109	0.056	0.601	2.171	1.657	0.0973	0.033	.6870E+4	.1456E-3	.5379E-3	.1613E+0
0.33	1.00	.892	.111	0.054	0.778	2.176	2.013	0.0990	0.042	.5282E+4	.1893E-3	.5736E-3	.1993E+0
0.33	1.20	.894	.112	0.053	0.976	2.178	2.410	0.1001	0.052	.4201E+4	.2381E-3	.6012E-3	.2413E+0
0.33	1.50	.896	.113	0.052	1.274	2.181	3.007	0.1012	0.066	.3212E+4	.3113E-3	.6289E-3	.3045E+0
0.33	1.80	.897	.114	0.051	1.573	2.183	3.606	0.1018	0.081	.2600E+4	.3847E-3	.6476E-3	.3671E+0
0.33	2.20	.898	.114	0.051	1.972	2.184	4.404	0.1024	0.101	.2072E+4	.4826E-3	.6647E-3	.4509E+0
0.33	2.70	.899	.114	0.051	2.471	2.186	5.403	0.1029	0.125	.1653E+4	.6049E-3	.6789E-3	.5561E+0
0.33	3.30	.900	.115	0.050	3.070	2.187	6.601	0.1035	0.154	.1330E+4	.7518E-3	.6904E-3	.6832E+0
0.33	3.90	.900	.115	0.050	3.670	2.187	7.801	0.1035	0.184	.1113E+4	.8988E-3	.6984E-3	.8074E+0
0.33	4.70	.900	.115	0.050	4.470	2.187	9.401	0.1035	0.224	.9135E+3	.1095E-2	.7058E-3	.9730E+0
0.33	5.60	.901	.116	0.049	5.369	2.189	11.200	0.1041	0.266	.7604E+3	.1315E-2	.7116E-3	.1166E+1
0.33	6.80	.901	.116	0.049	6.569	2.189	13.600	0.1041	0.325	.6215E+3	.1609E-2	.7174E-3	.1415E+1
0.33	8.20	.901	.116	0.049	7.969	2.189	16.400	0.1041	0.394	.5123E+3	.1952E-2	.7214E-3	.1707E+1
0.33	10.00	.901	.116	0.049	9.769	2.189	20.000	0.1041	0.484	.4179E+3	.2393E-2	.7251E-3	.2081E+1
0.33	12.00	.901	.116	0.049	11.769	2.189	24.000	0.1041	0.583	.3469E+3	.2883E-2	.7280E-3	.2498E+1
0.33	15.00	.901	.116	0.049	14.769	2.189	30.000	0.1041	0.731	.2764E+3	.3618E-2	.7308E-3	.3122E+1
0.33	18.00	.901	.116	0.049	17.769	2.189	36.000	0.1041	0.880	.2298E+3	.4352E-2	.7327E-3	.3746E+1
0.33	22.00	.902	.116	0.049	21.769	2.190	44.000	0.1041	1.078	.1875E+3	.5332E-2	.7355E-3	.4579E+1
0.33	27.00	.902	.116	0.049	26.768	2.190	53.998	0.1046	1.312	.1525E+3	.6557E-2	.7355E-3	.5650E+1
0.33	33.00	.902	.116	0.049	32.768	2.190	65.998	0.1046	1.606	.1246E+3	.8027E-2	.7371E-3	.6906E+1
0.39	0.33	.802	.096	0.099	0.138	2.107	0.776	0.0770	0.014	.2602E+5	.3844E-4	.2986E-3	.5971E-1
0.39	0.39	.822	.106	0.089	0.178	2.136	0.867	0.0871	0.016	.1776E+5	.5632E-4	.3703E-3	.7554E-1
0.39	0.47	.840	.115	0.080	0.240	2.161	1.001	0.0966	0.019	.1206E+5	.8290E-4	.4522E-3	.9672E-1
0.39	0.56	.853	.122	0.074	0.317	2.180	1.163	0.1036	0.023	.8710E+4	.1148E-3	.5257E-3	.1205E+0
0.39	0.68	.862	.126	0.069	0.428	2.193	1.390	0.1086	0.030	.6294E+4	.1589E-3	.5991E-3	.1510E+0
0.39	0.82	.869	.130	0.065	0.561	2.203	1.660	0.1125	0.037	.4734E+4	.2113E-3	.6606E-3	.1868E+0
0.39	1.00	.873	.132	0.063	0.737	2.208	2.014	0.1148	0.047	.3581E+4	.2793E-3	.7161E-3	.2312E+0
0.39	1.20	.876	.133	0.062	0.934	2.213	2.410	0.1165	0.058	.2815E+4	.3552E-3	.7590E-3	.2808E+0
0.39	1.50	.879	.135	0.060	1.231	2.217	3.006	0.1182	0.074	.2130E+4	.4695E-3	.8026E-3	.3553E+0
0.39	1.80	.880	.135	0.060	1.530	2.218	3.604	0.1188	0.092	.1712E+4	.5840E-3	.8319E-3	.4282E+0
0.39	2.20	.882	.136	0.059	1.928	2.221	4.401	0.1200	0.114	.1357E+4	.7368E-3	.8588E-3	.5279E+0

CASE # 1 COPYRIGHT 1982 STEVE SMITH DATA DATE 1/ 3/82 CHOKE/RESONANT CHARGE REACTOR DESIGN TABLE

$$f(D)=U/(F*G*(D*E)^2)$$

P	Q	D	E	F	G	U	Um	DE	FG	f(D)	1/f(D)	1/(PQf(D))	DEUm
0.39	2.70	.883	0.137	0.058	2.427	2.223	5.400	0.1205	0.142	.1078E+4	.9279E-3	.8812E-3	.6508E+0
0.39	3.30	.884	0.137	0.058	3.026	2.224	6.598	0.1211	0.176	.8640E+3	.1157E-2	.8993E-3	.7991E+0
0.39	3.90	.884	0.137	0.058	3.626	2.224	7.798	0.1211	0.210	.7211E+3	.1387E-2	.9118E-3	.9444E+0
0.39	4.70	.885	0.137	0.058	4.425	2.226	9.397	0.1217	0.254	.5907E+3	.1693E-2	.9235E-3	.1143E+1
0.39	5.60	.885	0.137	0.058	5.325	2.226	11.197	0.1217	0.306	.4909E+3	.2037E-2	.9328E-3	.1363E+1
0.39	6.80	.885	0.137	0.058	6.525	2.226	13.597	0.1217	0.375	.4006E+3	.2496E-2	.9413E-3	.1655E+1
0.39	8.20	.886	0.138	0.057	7.924	2.227	16.396	0.1223	0.452	.3298E+3	.3032E-2	.9481E-3	.2005E+1
0.39	10.00	.886	0.138	0.057	9.724	2.227	19.996	0.1223	0.554	.2688E+3	.3721E-2	.9540E-3	.2445E+1
0.39	12.00	.886	0.138	0.057	11.724	2.227	23.996	0.1223	0.668	.2229E+3	.4486E-2	.9585E-3	.2934E+1
0.39	15.00	.886	0.138	0.057	14.724	2.227	29.996	0.1223	0.839	.1775E+3	.5634E-2	.9630E-3	.3667E+1
0.39	18.00	.886	0.138	0.057	17.724	2.227	35.996	0.1223	1.010	.1475E+3	.6782E-2	.9660E-3	.4401E+1
0.39	22.00	.886	0.138	0.057	21.724	2.227	43.996	0.1223	1.238	.1203E+3	.8312E-2	.9688E-3	.5379E+1
0.39	27.00	.887	0.139	0.056	26.723	2.229	53.994	0.1228	1.510	.9780E+2	.1023E-1	.9710E-3	.6633E+1
0.39	33.00	.887	0.139	0.056	32.723	2.229	65.994	0.1228	1.849	.7987E+2	.1252E-1	.9729E-3	.8107E+1
0.47	0.33	.735	0.103	0.132	0.125	2.091	0.837	0.0753	0.017	.2225E+5	.4495E-4	.2898E-3	.6306E-1
0.47	0.39	.762	0.116	0.119	0.158	2.130	0.918	0.0884	0.019	.1450E+5	.6897E-4	.3763E-3	.8118E-1
0.47	0.47	.791	0.131	0.105	0.209	2.171	1.037	0.1032	0.022	.9338E+4	.1072E-3	.4852E-3	.1070E+0
0.47	0.56	.812	0.141	0.094	0.278	2.201	1.187	0.1145	0.026	.6626E+4	.1556E-3	.5912E-3	.1359E+0
0.47	0.68	.829	0.149	0.086	0.381	2.226	1.403	0.1239	0.033	.4448E+4	.2248E-3	.7034E-3	.1738E+0
0.47	0.82	.840	0.155	0.080	0.510	2.241	1.667	0.1302	0.041	.3241E+4	.3086E-3	.8007E-3	.2170E+0
0.47	1.00	.847	0.159	0.077	0.683	2.251	2.017	0.1342	0.052	.2391E+4	.4183E-3	.8900E-3	.2708E+0
0.47	1.20	.852	0.161	0.074	0.878	2.258	2.410	0.1372	0.065	.1847E+4	.5413E-3	.9598E-3	.3306E+0
0.47	1.50	.857	0.163	0.072	1.173	2.266	3.003	0.1401	0.084	.1376E+4	.7268E-3	.1031E-2	.4207E+0
0.47	1.80	.859	0.165	0.071	1.471	2.268	3.600	0.1413	0.104	.1096E+4	.9128E-3	.1079E-2	.5087E+0
0.47	2.20	.861	0.166	0.070	1.869	2.271	4.397	0.1425	0.130	.8612E+3	.1161E-2	.1123E-2	.6265E+0
0.47	2.70	.863	0.167	0.069	2.367	2.274	5.394	0.1437	0.162	.6793E+3	.1472E-2	.1160E-2	.7751E+0
0.47	3.30	.864	0.167	0.068	2.966	2.276	6.593	0.1443	0.202	.5420E+3	.1845E-2	.1190E-2	.9512E+0
0.47	3.90	.865	0.168	0.067	3.565	2.277	7.791	0.1449	0.241	.4508E+3	.2218E-2	.1210E-2	.1129E+1
0.47	4.70	.865	0.168	0.067	4.365	2.277	9.391	0.1449	0.295	.3681E+3	.2716E-2	.1230E-2	.1361E+1
0.47	5.60	.866	0.168	0.067	5.264	2.278	11.190	0.1455	0.353	.3052E+3	.3276E-2	.1245E-2	.1628E+1
0.47	6.80	.866	0.168	0.067	6.464	2.278	13.590	0.1455	0.433	.2486E+3	.4023E-2	.1259E-2	.1977E+1
0.47	8.20	.867	0.168	0.067	7.863	2.280	16.388	0.1461	0.523	.2043E+3	.4895E-2	.1270E-2	.2394E+1
0.47	10.00	.867	0.168	0.067	9.663	2.280	19.988	0.1461	0.643	.1662E+3	.6015E-2	.1280E-2	.2920E+1
0.47	12.00	.867	0.168	0.067	11.663	2.280	23.988	0.1461	0.776	.1377E+3	.7260E-2	.1287E-2	.3504E+1
0.47	15.00	.867	0.168	0.067	14.663	2.280	29.988	0.1461	0.975	.1096E+3	.9128E-2	.1295E-2	.4381E+1
0.47	18.00	.867	0.168	0.067	17.663	2.280	35.988	0.1461	1.175	.9095E+2	.1100E-1	.1300E-2	.5258E+1

CASE # 1   COPYRIGHT 1982 STEVE SMITH   DATA DATE 1/ 3/82   CHOKE/RESONANT CHARGE REACTOR DESIGN TABLE

$$f(D) = U/(F \cdot G \cdot (D \cdot E)^2)$$

P	Q	D	E	F	G	U	Um	DE	FG	f(D)	1/f(D)	1/(PQf(D))	DEUm
0.47	22.00	.868	0.169	0.066	21.662	2.281	43.987	0.1467	1.430	.7415E+2	.1349E-1	.1304E-2	.6453E+1
0.47	27.00	.868	0.169	0.066	26.662	2.281	53.987	0.1467	1.760	.6025E+2	.1660E-1	.1388E-2	.7919E+1
0.47	33.00	.868	0.169	0.066	32.662	2.281	65.987	0.1467	2.156	.4918E+2	.2033E-1	.1311E-2	.9680E+1
0.56	0.33	.656	0.108	0.172	0.114	2.068	0.911	0.0708	0.020	.1102E+5	.4758E-4	.2575E-3	.6456E-1
0.56	0.39	.688	0.124	0.156	0.142	2.114	0.986	0.0853	0.022	.1311E+5	.7626E-4	.3492E-3	.8408E-1
0.56	0.47	.726	0.143	0.137	0.184	2.168	1.091	0.1038	0.025	.7981E+4	.1253E-3	.4761E-3	.1133E+0
0.56	0.56	.758	0.159	0.121	0.242	2.214	1.226	0.1205	0.029	.5206E+4	.1921E-3	.6126E-3	.1477E+0
0.56	0.68	.786	0.173	0.107	0.334	2.254	1.425	0.1360	0.036	.3411E+4	.2931E-3	.7698E-3	.1938E+0
0.56	0.82	.805	0.183	0.098	0.455	2.281	1.678	0.1469	0.044	.2383E+4	.4197E-3	.9140E-3	.2466E+0
0.56	1.00	.818	0.189	0.091	0.622	2.300	2.020	0.1546	0.057	.1700E+4	.5882E-3	.1050E-2	.3123E+0
0.56	1.20	.826	0.193	0.087	0.814	2.311	2.408	0.1594	0.071	.1284E+4	.7787E-3	.1159E-2	.3339E+0
0.56	1.50	.833	0.197	0.083	1.107	2.321	2.998	0.1637	0.092	.9373E+3	.1067E-2	.1270E-2	.4908E+0
0.56	1.80	.836	0.198	0.082	1.404	2.326	3.594	0.1655	0.115	.7372E+3	.1356E-2	.1466E-2	.5949E+0
0.56	2.20	.839	0.200	0.081	1.801	2.330	4.390	0.1674	0.145	.5736E+3	.1743E-2	.1415E-2	.7348E+0
0.56	2.70	.841	0.201	0.079	2.299	2.333	5.387	0.1686	0.183	.4489E+3	.2228E-2	.1473E-2	.9083E+0
0.56	3.30	.843	0.202	0.079	2.897	2.336	6.584	0.1699	0.227	.3559E+3	.2809E-2	.1520E-2	.1118E+1
0.56	3.90	.844	0.202	0.078	3.496	2.338	7.783	0.1705	0.273	.2949E+3	.3391E-2	.1553E-2	.1327E+1
0.56	4.70	.845	0.203	0.077	4.295	2.338	9.381	0.1711	0.333	.2399E+3	.4168E-2	.1583E-2	.1605E+1
0.56	5.60	.846	0.203	0.077	5.194	2.340	11.180	0.1717	0.400	.1984E+3	.5041E-2	.1608E-2	.1920E+1
0.56	6.80	.846	0.203	0.077	6.394	2.340	13.580	0.1717	0.492	.1611E+3	.6206E-2	.1630E-2	.2332E+1
0.56	8.20	.847	0.204	0.077	7.793	2.341	16.378	0.1724	0.596	.1322E+3	.7565E-2	.1647E-2	.2823E+1
0.56	10.00	.847	0.204	0.077	9.593	2.341	19.978	0.1724	0.734	.1074E+3	.9312E-2	.1663E-2	.3444E+1
0.56	12.00	.848	0.204	0.076	11.592	2.343	23.977	0.1730	0.881	.8886E+2	.1125E-1	.1675E-2	.4148E+1
0.56	15.00	.848	0.204	0.076	14.592	2.343	29.977	0.1730	1.109	.7059E+2	.1417E-1	.1686E-2	.5186E+1
0.56	18.00	.848	0.204	0.076	17.592	2.343	35.977	0.1730	1.337	.5855E+2	.1708E-1	.1694E-2	.6224E+1
0.56	22.00	.848	0.204	0.076	21.592	2.343	43.977	0.1730	1.641	.4771E+2	.2096E-1	.1701E-2	.7608E+1
0.56	27.00	.849	0.204	0.076	26.591	2.344	53.975	0.1736	2.008	.3874E+2	.2582E-1	.1707E-2	.9371E+1
0.56	33.00	.849	0.204	0.076	32.591	2.344	65.975	0.1736	2.461	.3160E+2	.3164E-1	.1712E-2	.1145E+2
0.68	0.33	.546	0.113	0.227	0.106	2.031	1.017	0.0617	0.024	.2260E+5	.4424E-4	.1972E-3	.6275E-1
0.68	0.39	.584	0.132	0.208	0.126	2.085	1.083	0.0771	0.026	.1339E+5	.7468E-4	.2816E-3	.8346E-1
0.68	0.47	.630	0.155	0.185	0.160	2.151	1.177	0.0976	0.030	.7622E+4	.1312E-3	.4105E-3	.1149E+0
0.68	0.56	.675	0.178	0.163	0.205	2.216	1.293	0.1198	0.033	.4633E+4	.2158E-3	.5668E-3	.1549E+0
0.68	0.68	.720	0.200	0.140	0.280	2.280	1.468	0.1440	0.039	.2885E+4	.3565E-3	.7711E-3	.2114E+0
0.68	0.82	.754	0.217	0.123	0.386	2.328	1.700	0.1636	0.047	.1832E+4	.5459E-3	.9790E-3	.2781E+0
0.68	1.00	.778	0.229	0.111	0.542	2.363	2.025	0.1782	0.060	.1237E+4	.8082E-3	.1189E-2	.3609E+0
0.68	1.20	.792	0.236	0.104	0.728	2.383	2.405	0.1869	0.076	.9008E+3	.1110E-2	.1360E-2	.4496E+0

CASE # 1   COPYRIGHT 1982 STEVE SMITH   DATA DATE 1/3/82   CHOKE/RESONANT CHARGE REACTOR DESIGN TABLE

$$f(D)=U/(F*G*(D*E)^2)$$

P	Q	D	E	F	G	U	Um	DE	FG	f(D)	1/f(D)	1/(PQf(D))	DEUm
0.68	1.50	.802	.241	.099	1.018	2.397	2.991	0.1933	0.101	.6367E+3	.1571E-2	.1540E-2	.5781E+0
0.68	1.80	.808	.244	.096	1.312	2.406	3.583	0.1972	0.126	.4914E+3	.2035E-2	.1663E-2	.7063E+0
0.68	2.20	.813	.247	.093	1.707	2.413	4.375	0.2004	0.160	.3764E+3	.2657E-2	.1776E-2	.8769E+0
0.68	2.70	.816	.248	.092	2.204	2.417	5.371	0.2024	0.203	.2911E+3	.3436E-2	.1871E-2	.1087E+1
0.68	3.30	.818	.249	.091	2.802	2.420	6.568	0.2037	0.255	.2288E+3	.4371E-2	.1948E-2	.1338E+1
0.68	3.90	.820	.250	.090	3.400	2.423	7.765	0.2050	0.306	.1884E+3	.5308E-2	.2001E-2	.1592E+1
0.68	4.70	.821	.251	.090	4.199	2.424	9.364	0.2057	0.376	.1525E+3	.6557E-2	.2052E-2	.1926E+1
0.68	5.60	.822	.251	.089	5.098	2.426	11.163	0.2063	0.454	.1256E+3	.7963E-2	.2091E-2	.2303E+1
0.68	6.80	.823	.252	.088	6.297	2.427	13.561	0.2070	0.557	.1017E+3	.9837E-2	.2127E-2	.2807E+1
0.68	8.20	.824	.252	.088	7.696	2.428	16.360	0.2076	0.677	.8316E+2	.1202E-1	.2157E-2	.3397E+1
0.68	10.00	.824	.252	.088	9.496	2.428	19.960	0.2076	0.836	.6740E+2	.1484E-1	.2182E-2	.4145E+1
0.68	12.00	.825	.253	.088	11.495	2.430	23.958	0.2083	1.006	.5567E+2	.1796E-1	.2201E-2	.4991E+1
0.68	15.00	.825	.253	.088	14.495	2.430	29.958	0.2083	1.268	.4415E+2	.2265E-1	.2221E-2	.6241E+1
0.68	18.00	.825	.253	.088	17.495	2.430	35.958	0.2083	1.531	.3658E+2	.2734E-1	.2233E-2	.7491E+1
0.68	22.00	.826	.253	.087	21.494	2.431	43.957	0.2090	1.870	.2977E+2	.3359E-1	.2245E-2	.9186E+1
0.68	27.00	.826	.253	.087	26.494	2.431	53.957	0.2090	2.305	.2415E+2	.4140E-1	.2255E-2	.1128E+2
0.68	33.00	.826	.253	.087	32.494	2.431	65.957	0.2090	2.827	.1969E+2	.5078E-1	.2263E-2	.1378E+2
0.68	0.33	.417	.1118	.292	.093	1.987	1.141	0.0494	0.027	.3001E+5	.3332E-4	.1231E-3	.5640E-1
0.82	0.39	.459	.140	.271	.111	2.047	1.201	0.0640	0.030	.1663E+5	.6014E-4	.1881E-3	.7692E-1
0.82	0.47	.512	.166	.244	.138	2.123	1.286	0.0850	0.034	.8726E+4	.1146E-3	.2973E-3	.1093E+0
0.82	0.56	.566	.193	.217	.174	2.200	1.388	0.1092	0.038	.4882E+4	.2048E-3	.4461E-3	.1517E+0
0.82	0.68	.629	.225	.186	.231	2.290	1.538	0.1412	0.043	.2680E+4	.3732E-3	.6692E-3	.2172E+0
0.82	0.82	.683	.252	.159	.317	2.367	1.741	0.1718	0.050	.1597E+4	.6264E-3	.9315E-3	.2991E+0
0.82	1.00	.725	.273	.137	.455	2.427	2.041	0.1976	0.063	.9939E+3	.1006E-2	.1227E-2	.4032E+0
0.82	1.20	.750	.285	.125	.630	2.463	2.405	0.2138	0.079	.6845E+3	.1461E-2	.1485E-2	.5141E+0
0.82	1.50	.769	.295	.116	.911	2.490	2.978	0.2265	0.105	.4614E+3	.2167E-2	.1762E-2	.6745E+0
0.82	1.80	.778	.299	.111	1.202	2.503	3.565	0.2326	0.133	.3466E+3	.2885E-2	.1954E-2	.8294E+0
0.82	2.20	.785	.303	.107	1.595	2.513	4.355	0.2375	0.171	.2599E+3	.3848E-2	.2133E-2	.1034E+1
0.82	2.70	.790	.305	.105	2.090	2.520	5.348	0.2410	0.219	.1978E+3	.5056E-2	.2284E-2	.1289E+1
0.82	3.30	.793	.306	.104	2.687	2.524	6.544	0.2431	0.278	.1536E+3	.6509E-2	.2405E-2	.1591E+1
0.82	3.90	.796	.308	.102	3.284	2.528	7.740	0.2452	0.335	.1256E+3	.7963E-2	.2490E-2	.1898E+1
0.82	4.70	.798	.309	.102	4.083	2.530	9.338	0.2459	0.414	.1010E+3	.9903E-2	.2570E-2	.2296E+1
0.82	5.60	.799	.310	.100	4.981	2.533	11.135	0.2473	0.501	.8274E+2	.1209E-1	.2632E-2	.2754E+1
0.82	6.80	.800	.310	.100	6.180	2.534	13.534	0.2480	0.618	.6667E+2	.1500E-1	.2690E-2	.3356E+1
0.82	8.20	.801	.311	.100	7.579	2.536	16.332	0.2487	0.754	.5436E+2	.1840E-1	.2736E-2	.4062E+1
0.82	10.00	.802	.311	.099	9.378	2.537	19.931	0.2494	0.928	.4392E+2	.2277E-1	.2776E-2	.4971E+1

CASE # 1   COPYRIGHT 1982 STEVE SMITH   DATA DATE 1/ 3/82   CHOKE/RESONANT CHARGE REACTOR DESIGN TABLE

$$f(D)=U/(F*G*(D*E)^2)$$

P	Q	D	E	F	G	U	Um	DE	FG	f(D)	1/f(D)	1/(PQf(D))	DEUm
0.82	12.00	.802	0.311	0.099	11.378	2.537	23.931	0.2494	1.126	.3620E+2	.2762E-1	.2807E-2	.5969E+1
0.82	15.00	.803	0.311	0.099	14.377	2.538	29.930	0.2501	1.416	.2865E+2	.3490E-1	.2838E-2	.7486E+1
0.82	18.00	.803	0.311	0.099	17.377	2.538	35.930	0.2501	1.712	.2370E+2	.4219E-1	.2858E-2	.8987E+1
0.82	22.00	.803	0.311	0.099	21.377	2.538	43.930	0.2501	2.106	.1927E+2	.5190E-1	.2877E-2	.1099E+2
0.82	27.00	.804	0.312	0.098	26.376	2.540	53.928	0.2508	2.585	.1562E+2	.6404E-1	.2892E-2	.1353E+2
0.82	33.00	.804	0.312	0.098	32.376	2.540	65.928	0.2508	3.173	.1272E+2	.7861E-1	.2905E-2	.1654E+2
1.00	0.33	.256	0.128	0.372	0.074	1.937	1.294	0.0328	0.028	.6552E+5	.1526E-4	.4625E-4	.4241E-1
1.00	0.39	.300	0.150	0.350	0.090	2.012	1.351	0.0450	0.031	.3135E+5	.3190E-4	.8100E-4	.6081E-1
1.00	0.47	.358	0.179	0.321	0.112	2.082	1.428	0.0641	0.036	.1411E+5	.7090E-4	.1508E-3	.9153E-1
1.00	0.56	.420	0.210	0.290	0.140	2.171	1.520	0.0882	0.041	.6874E+4	.1455E-3	.2598E-3	.1340E+0
1.00	0.68	.496	0.248	0.252	0.184	2.280	1.651	0.1230	0.046	.3249E+4	.3078E-3	.4526E-3	.2031E+0
1.00	0.82	.573	0.287	0.214	0.247	2.390	1.821	0.1642	0.053	.1681E+4	.5947E-3	.7223E-3	.2990E+0
1.00	1.00	.645	0.323	0.178	0.355	2.493	2.078	0.2080	0.063	.9142E+3	.1094E-2	.1094E-2	.4323E+0
1.00	1.20	.692	0.346	0.154	0.508	2.560	2.411	0.2394	0.078	.5708E+3	.1752E-2	.1460E-2	.5773E+0
1.00	1.50	.727	0.364	0.137	0.717	2.610	2.961	0.2643	0.106	.3542E+3	.2823E-2	.1882E-2	.7825E+0
1.00	1.80	.743	0.371	0.129	1.057	2.633	3.538	0.2760	0.136	.2544E+3	.3931E-2	.2184E-2	.9766E+0
1.00	2.00	.754	0.377	0.123	1.446	2.648	4.322	0.2843	0.178	.1843E+3	.5426E-2	.2467E-2	.1229E+1
1.00	2.70	.762	0.381	0.119	1.938	2.660	5.311	0.2903	0.231	.1368E+3	.7308E-2	.2707E-2	.1542E+1
1.00	3.30	.767	0.384	0.117	2.533	2.667	6.504	0.2941	0.295	.1045E+3	.9573E-2	.2901E-2	.1913E+1
1.00	3.90	.770	0.385	0.114	3.130	2.671	7.700	0.2965	0.360	.8445E+2	.1184E-1	.3036E-2	.2283E+1
1.00	4.70	.772	0.386	0.114	3.928	2.674	9.297	0.2980	0.448	.6725E+2	.1487E-1	.3164E-2	.2770E+1
1.00	5.60	.774	0.387	0.113	4.826	2.677	11.094	0.2995	0.545	.5471E+2	.1828E-1	.3264E-2	.3323E+1
1.00	6.80	.776	0.388	0.112	6.024	2.680	13.491	0.3011	0.675	.4381E+2	.2282E-1	.3356E-2	.4062E+1
1.00	8.00	.777	0.389	0.112	7.423	2.681	16.290	0.3019	0.828	.3555E+2	.2813E-1	.3430E-2	.4917E+1
1.00	10.00	.778	0.389	0.111	9.222	2.683	19.888	0.3026	1.024	.2861E+2	.3495E-1	.3495E-2	.6019E+1
1.00	12.00	.779	0.390	0.111	11.221	2.684	23.887	0.3034	1.240	.2351E+2	.4253E-1	.3544E-2	.7248E+1
1.00	15.00	.779	0.390	0.111	14.221	2.684	29.887	0.3034	1.571	.1855E+2	.5390E-1	.3593E-2	.9068E+1
1.00	18.00	.780	0.390	0.110	17.220	2.686	35.885	0.3042	1.894	.1532E+2	.6527E-1	.3626E-2	.1092E+2
1.00	22.00	.780	0.390	0.110	21.220	2.686	43.885	0.3042	2.334	.1243E+2	.8043E-1	.3656E-2	.1335E+2
1.00	27.00	.780	0.390	0.110	26.220	2.686	53.885	0.3042	2.884	.1006E+2	.9938E-1	.3681E-2	.1639E+2
1.00	33.00	.781	0.391	0.109	32.219	2.687	65.884	0.3050	3.528	.8188E+1	.1221E+0	.3701E-2	.2009E+2
1.20	0.33	.093	0.147	0.454	0.037	1.904	1.441	0.0136	0.017	.6112E+6	.1636E-5	.4112E-5	.1964E-1
1.20	0.39	.137	0.169	0.432	0.053	1.967	1.498	0.0231	0.023	.1614E+6	.6197E-5	.1324E-4	.3459E-1
1.20	0.47	.195	0.198	0.375	0.075	2.049	1.575	0.0385	0.030	.4577E+5	.2185E-4	.3874E-4	.6068E-1
1.20	0.56	.260	0.230	0.370	0.100	2.142	1.663	0.0598	0.037	.1619E+5	.6176E-4	.9190E-4	.9942E-1
1.20	0.68	.344	0.272	0.328	0.136	2.262	1.783	0.0936	0.045	.5793E+4	.1726E-3	.2115E-3	.1668E+0

CASE # 1   COPYRIGHT 1982 STEVE SMITH   DATA DATE 1/ 3/82   CHOKE/RESONANT CHARGE REACTOR DESIGN TABLE

$f(D) = U/(f*G*(D*E)^2)$

P	Q	D	E	F	G	U	Um	DE	FG	f(D)	1/f(D)	1/(PQf(D))	DEUm
1.20	0.82	.434	0.317	0.283	0.186	2.391	1.934	.1376	0.053	.2400E+4	.4167E-3	.4235E-3	.2661E+0
1.20	1.00	.534	0.367	0.233	0.266	2.534	2.151	.1960	0.062	.1065E+4	.9394E-3	.7828E-3	.4215E+0
1.20	1.20	.614	0.407	0.193	0.386	2.648	2.437	.2499	0.074	.5692E+3	.1757E-2	.1220E-2	.6089E+0
1.20	1.50	.678	0.439	0.161	0.622	2.740	2.945	.2976	0.100	.3088E+3	.3238E-2	.1799E-2	.8766E+0
1.20	1.80	.706	0.453	0.147	0.894	2.780	3.505	.3198	0.131	.2068E+3	.4836E-2	.2239E-2	.1121E+1
1.20	2.20	.725	0.463	0.137	1.275	2.807	4.278	.3353	0.175	.1424E+3	.7022E-2	.2660E-2	.1434E+1
1.20	2.70	.736	0.468	0.132	1.764	2.823	5.262	.3444	0.233	.1022E+3	.9787E-2	.3021E-2	.1813E+1
1.20	3.30	.743	0.472	0.129	2.357	2.833	6.452	.3503	0.303	.7621E+2	.1312E-1	.3314E-2	.2260E+1
1.20	3.90	.747	0.474	0.127	2.953	2.838	7.647	.3537	0.374	.6074E+2	.1646E-1	.3518E-2	.2705E+1
1.20	4.70	.750	0.475	0.125	3.750	2.843	9.242	.3563	0.469	.4778E+2	.2093E-1	.3711E-2	.3293E+1
1.20	5.60	.752	0.476	0.124	4.648	2.846	11.039	.3580	0.576	.3853E+2	.2595E-1	.3862E-2	.3952E+1
1.20	6.80	.754	0.477	0.123	5.846	2.848	13.437	.3597	0.719	.3062E+2	.3265E-1	.4002E-2	.4833E+1
1.20	8.20	.756	0.478	0.122	7.244	2.851	16.234	.3614	0.884	.2471E+2	.4048E-1	.4113E-2	.5866E+1
1.20	10.00	.757	0.479	0.122	9.043	2.853	19.832	.3622	1.099	.1979E+2	.5053E-1	.4221E-2	.7184E+1
1.20	12.00	.758	0.479	0.122	11.042	2.854	23.831	.3631	1.336	.1648E+2	.6171E-1	.4286E-2	.8653E+1
1.20	15.00	.759	0.480	0.121	14.041	2.856	29.829	.3639	1.692	.1274E+2	.7848E-1	.4360E-2	.1086E+2
1.20	18.00	.759	0.480	0.120	17.040	2.856	35.828	.3648	2.045	.1050E+2	.9525E-1	.4410E-2	.1307E+2
1.20	22.00	.760	0.480	0.120	21.040	2.857	43.828	.3648	2.525	.8503E+1	.1176E+0	.4455E-2	.1599E+2
1.20	27.00	.760	0.480	0.120	26.040	2.857	53.827	.3648	3.125	.6870E+1	.1456E+0	.4492E-2	.1964E+2
1.20	33.00	.760	0.480	0.120	32.039	2.858	65.827	.3657	3.829	.5584E+1	.1791E+0	.4523E-2	.2447E+2
1.50	0.68	.123	0.312	0.439	0.057	2.247	1.970	.0383	0.025	.6123E+5	.1633E-4	.1601E-4	.7546E-1
1.50	0.82	.220	0.360	0.390	0.100	2.385	2.111	.0792	0.039	.9750E+4	.1026E-3	.8338E-4	.1672E+0
1.50	1.00	.339	0.420	0.331	0.161	2.555	2.301	.1422	0.053	.2375E+4	.4211E-3	.2808E-3	.3272E+0
1.50	1.20	.457	0.479	0.272	0.243	2.724	2.532	.2187	0.066	.8634E+3	.1158E-2	.6434E-3	.5537E+0
1.50	1.50	.582	0.541	0.209	0.418	2.903	2.954	.3149	0.087	.3351E+3	.2984E-2	.1326E-2	.9300E+0
1.50	1.80	.647	0.574	0.177	0.653	2.995	3.461	.3711	0.115	.1888E+3	.5297E-2	.1962E-2	.1284E+1
1.50	2.20	.684	0.592	0.158	1.016	3.048	4.209	.4049	0.161	.1158E+3	.8635E-2	.2617E-2	.1704E+1
1.50	2.70	.703	0.602	0.149	1.497	3.076	5.181	.4229	0.222	.7737E+2	.1292E-1	.3191E-2	.2191E+1
1.50	3.30	.715	0.608	0.143	2.085	3.093	6.364	.4344	0.297	.5517E+2	.1813E-1	.3662E-2	.2764E+1
1.50	3.90	.721	0.611	0.140	2.679	3.101	7.555	.4402	0.374	.4283E+2	.2335E-1	.3991E-2	.3325E+1
1.50	4.70	.725	0.613	0.137	3.475	3.107	9.149	.4441	0.478	.3298E+2	.3033E-1	.4301E-2	.4063E+1
1.50	5.60	.729	0.615	0.136	4.371	3.113	10.944	.4480	0.592	.2619E+2	.3818E-1	.4546E-2	.4902E+1
1.50	6.80	.731	0.616	0.135	5.569	3.116	13.341	.4499	0.749	.2055E+2	.4867E-1	.4772E-2	.6002E+1
1.50	8.20	.733	0.617	0.134	6.967	3.118	16.138	.4519	0.930	.1642E+2	.6091E-1	.4952E-2	.7293E+1
1.50	10.00	.735	0.617	0.132	8.765	3.121	19.735	.4539	1.161	.1305E+2	.7665E-1	.5110E-2	.8957E+1
1.50	12.00	.736	0.618	0.132	10.764	3.123	23.734	.4548	1.421	.1062E+2	.9414E-1	.5230E-2	.1080E+2

CASE # 1   COPYRIGHT 1982 STEVE SMITH   DATA DATE 1/ 3/82   CHOKE/RESONANT CHARGE REACTOR DESIGN TABLE

$$f(D)=U/(F*G*(D*E)^2)$$

P	Q	D	E	F	G	U	Um	DE	FG	f(D)	1/f(D)	1/(PQf(D))	DEUm
1.50	15.00	.737	0.619	0.132	13.763	3.124	29.732	0.4558	1.810	.8308E+1	.1204E+0	.5350E-2	.1355E+2
1.50	18.00	.738	0.619	0.131	16.762	3.126	35.731	0.4568	2.196	.6821E+1	.1466E+0	.5430E-2	.1632E+2
1.50	22.00	.738	0.619	0.131	20.762	3.126	43.731	0.4568	2.720	.5507E+1	.1816E+0	.5503E-2	.1998E+2
1.50	27.00	.739	0.620	0.131	25.761	3.127	53.729	0.4578	3.362	.4438E+1	.2253E+0	.5564E-2	.2460E+2
1.50	33.00	.739	0.620	0.131	31.761	3.127	65.729	0.4578	4.145	.3600E+1	.2778E+0	.5612E-2	.3009E+2
1.80	1.20	.135	0.467	0.433	0.065	2.564	2.464	0.0631	0.028	.2290E+5	.4368E-4	.2427E-4	.1555E+0
1.80	1.20	.267	0.533	0.367	0.133	2.752	2.675	0.1424	0.049	.2783E+4	.3593E-3	.1664E-3	.3810E+0
1.80	1.50	.442	0.621	0.279	0.258	3.003	3.025	0.2745	0.072	.5536E+3	.1806E-2	.6690E-3	.8303E+0
1.80	1.80	.563	0.682	0.219	0.437	3.175	3.452	0.3837	0.095	.2259E+3	.4427E-2	.1366E-2	.1324E+1
1.80	2.20	.639	0.720	0.181	0.761	3.284	4.143	0.4598	0.137	.1131E+3	.8841E-2	.2233E-2	.1905E+1
1.80	2.70	.675	0.738	0.163	1.225	3.336	5.092	0.4978	0.199	.6761E+2	.1479E-1	.3043E-2	.2535E+1
1.80	3.30	.692	0.746	0.154	1.808	3.360	6.268	0.5162	0.278	.4528E+2	.2208E-1	.3718E-2	.3236E+1
1.80	3.90	.701	0.750	0.150	2.399	3.373	7.455	0.5261	0.359	.3398E+2	.2943E-1	.4193E-2	.3922E+1
1.80	4.70	.707	0.754	0.147	3.193	3.381	9.046	0.5327	0.468	.2547E+2	.3926E-1	.4641E-2	.4819E+1
1.80	5.60	.712	0.756	0.144	4.088	3.388	10.839	0.5383	0.589	.1987E+2	.5034E-1	.4994E-2	.5834E+1
1.80	6.80	.715	0.758	0.143	5.285	3.393	13.235	0.5416	0.753	.1536E+2	.6512E-1	.5320E-2	.7168E+1
1.80	8.20	.717	0.759	0.142	6.683	3.396	16.032	0.5438	0.946	.1214E+2	.8237E-1	.5581E-2	.8719E+1
1.80	10.00	.719	0.760	0.141	8.481	3.398	19.629	0.5461	1.192	.9564E+1	.1046E+0	.5809E-2	.1072E+2
1.80	12.00	.720	0.760	0.140	10.480	3.400	23.628	0.5472	1.467	.7739E+1	.1292E+0	.5982E-2	.1293E+2
1.80	15.00	.722	0.761	0.139	13.478	3.403	29.625	0.5494	1.673	.6016E+1	.1662E+0	.6156E-2	.1628E+2
1.80	18.00	.722	0.761	0.139	16.478	3.403	35.625	0.5494	2.290	.4921E+1	.2032E+0	.6272E-2	.1957E+2
1.80	22.00	.723	0.762	0.139	20.477	3.404	43.623	0.5506	2.836	.3960E+1	.2525E+0	.6377E-2	.2402E+2
1.80	27.00	.724	0.762	0.138	25.476	3.406	53.622	0.5517	3.516	.3183E+1	.3142E+0	.6465E-2	.2958E+2
1.80	33.00	.724	0.762	0.138	31.476	3.406	65.622	0.5517	4.344	.2576E+1	.3882E+0	.6535E-2	.3620E+2
2.20	1.50	.198	0.699	0.401	0.102	3.054	3.202	0.1384	0.041	.3898E+4	.2566E-3	.7775E-4	.4432E+2
2.20	1.80	.378	0.789	0.311	0.222	3.311	3.545	0.2982	0.069	.5392E+3	.1855E-2	.4684E-3	.1057E+1
2.20	2.20	.546	0.873	0.227	0.454	3.551	4.105	0.4767	0.103	.1517E+3	.6594E-2	.1362E-2	.1956E+1
2.20	2.70	.632	0.916	0.184	0.868	3.674	4.982	0.5789	0.160	.6864E+4	.1457E-3	.2453E-2	.2884E+1
2.20	3.30	.666	0.933	0.167	1.434	3.723	6.133	0.6214	0.239	.4026E+2	.2484E-1	.3421E-2	.3811E+1
2.20	3.90	.680	0.940	0.160	2.020	3.743	7.313	0.6392	0.323	.2834E+2	.3528E-1	.4112E-2	.4675E+1
2.20	4.70	.690	0.945	0.155	2.810	3.757	8.899	0.6521	0.416	.2029E+2	.4929E-1	.4767E-2	.5802E+1
2.20	5.60	.695	0.948	0.153	3.705	3.764	10.692	0.6585	0.565	.1536E+2	.6509E-1	.5283E-2	.7041E+1
2.20	6.80	.698	0.950	0.150	4.900	3.771	13.085	0.6650	0.735	.1160E+2	.8619E-1	.5761E-2	.8701E+1
2.20	8.20	.703	0.952	0.149	6.297	3.776	15.880	0.6689	0.935	.9024E+1	.1108E+0	.6143E-2	.1062E+2
2.20	10.00	.705	0.953	0.148	8.095	3.778	19.477	0.6715	1.194	.7018E+1	.1425E+0	.6477E-2	.1308E+2
2.20	12.00	.706	0.953	0.147	10.094	3.780	23.476	0.6728	1.484	.5627E+1	.1777E+0	.6731E-2	.1580E+2

CHOKE/RESONANT CHARGE REACTOR DESIGN TABLE

CASE # 1   COPYRIGHT 1982 STEVE SMITH   DATA DATE 1/ 3/82

$f(D)=U/(F*G*(D*E)^2)$

P	Q	D	E	F	G	U	Um	DE	FG	f(D)	1/f(D)	1/(PQf(D))	DEUm
2.20	15.00	.708	0.954	0.146	13.092	3.783	29.473	0.6754	1.911	.4338E+1	.2305E+0	.6986E-2	.1991E+2
2.20	18.00	.709	0.955	0.146	16.091	3.784	35.472	0.6767	2.341	.3529E+1	.2834E+0	.7155E-2	.2401E+2
2.20	22.00	.709	0.955	0.146	20.091	3.784	43.472	0.6767	2.923	.2827E+1	.3538E+0	.7310E-2	.2942E+2
2.20	27.00	.710	0.955	0.145	25.090	3.786	53.470	0.6781	3.638	.2263E+1	.4418E+0	.7438E-2	.3626E+2
2.20	33.00	.710	0.955	0.145	31.090	3.786	65.470	0.6781	4.508	.1826E+1	.5475E+0	.7541E-2	.4439E+2
2.70	1.80	.066	0.883	0.467	0.034	3.365	3.776	0.0583	0.016	.6240E+5	.1603E-4	.3297E-5	.2201E+0
2.70	2.20	.317	1.009	0.342	0.183	3.724	4.217	0.3197	0.062	.5830E+3	.1715E-2	.2888E-3	.1348E+1
2.70	2.70	.534	1.117	0.233	0.466	4.034	4.907	0.5965	0.109	.1044E+3	.9576E-2	.1344E-2	.2927E+1
2.70	3.30	.627	1.164	0.187	0.973	4.167	5.974	0.7295	0.181	.4315E+2	.2318E-1	.2601E-2	.4358E+1
2.70	3.90	.656	1.178	0.172	1.544	4.208	7.133	0.7728	0.266	.2654E+2	.3768E-1	.3579E-2	.5512E+1
2.70	4.70	.672	1.186	0.164	2.328	4.231	8.710	0.7970	0.382	.1745E+2	.5732E-1	.4517E-2	.6942E+1
2.70	5.60	.681	1.191	0.160	3.219	4.244	10.497	0.8107	0.513	.1258E+2	.7952E-1	.5259E-2	.8510E+1
2.70	6.80	.686	1.193	0.157	4.414	4.251	12.890	0.8184	0.693	.9159E+1	.1092E+0	.5947E-2	.1055E+2
2.70	8.20	.690	1.195	0.155	5.810	4.257	15.684	0.8246	0.901	.6953E+1	.1438E+0	.6496E-2	.1293E+2
2.70	10.00	.693	1.197	0.154	7.607	4.261	19.280	0.8292	1.168	.5308E+1	.1884E+0	.6978E-2	.1599E+2
2.70	12.00	.695	1.198	0.153	9.605	4.264	23.277	0.8323	1.465	.4203E+1	.2379E+0	.7344E-2	.1937E+2
2.70	15.00	.696	1.198	0.152	12.604	4.266	29.276	0.8338	1.916	.3202E+1	.3123E+0	.7716E-2	.2441E+2
2.70	18.00	.697	1.199	0.151	15.603	4.267	35.274	0.8354	2.364	.2587E+1	.3866E+0	.7954E-2	.2947E+2
2.70	22.00	.698	1.199	0.151	19.602	4.268	43.273	0.8369	2.960	.2059E+1	.4857E+0	.8177E-2	.3622E+2
2.70	27.00	.699	1.200	0.151	24.601	4.270	53.271	0.8385	3.702	.1644E+1	.6096E+0	.8362E-2	.4467E+2
2.70	33.00	.699	1.200	0.151	30.601	4.270	65.271	0.8385	4.605	.1319E+1	.7583E+0	.8510E-2	.5473E+2
3.30	2.70	.256	1.278	0.372	0.144	4.237	5.047	0.3272	0.054	.7389E+3	.1353E-2	.1519E-3	.1651E+1
3.30	3.30	.525	1.413	0.238	0.475	4.621	5.863	0.7416	0.113	.7449E+2	.1342E-1	.1233E-2	.4347E+1
3.30	3.90	.617	1.459	0.192	0.983	4.753	6.931	0.8990	0.188	.3118E+2	.3208E-1	.2492E-2	.6237E+1
3.30	4.70	.652	1.476	0.174	1.748	4.803	8.481	0.9624	0.304	.1705E+2	.5865E-1	.3782E-2	.8162E+1
3.30	5.60	.666	1.483	0.167	2.634	4.823	10.261	0.9877	0.440	.1124E+2	.8898E-1	.4815E-2	.1013E+2
3.30	6.80	.675	1.488	0.163	3.825	4.836	12.648	1.0041	0.622	.7717E+1	.1296E+0	.5775E-2	.1270E+2
3.30	8.20	.683	1.490	0.160	5.220	4.843	15.441	1.0132	0.835	.5648E+1	.1771E+0	.6543E-2	.1564E+2
3.30	10.00	.683	1.492	0.159	7.017	4.847	19.037	1.0187	1.112	.4200E+1	.2381E+0	.7216E-2	.1939E+2
3.30	12.00	.686	1.493	0.157	9.014	4.851	23.032	1.0242	1.415	.3268E+1	.3060E+0	.7727E-2	.2355E+2
3.30	15.00	.688	1.494	0.156	12.012	4.854	29.030	1.0279	1.874	.2452E+1	.4079E+0	.8240E-2	.2984E+2
3.30	18.00	.688	1.495	0.155	15.011	4.856	35.028	1.0297	2.334	.1962E+1	.5097E+0	.8581E-2	.3607E+2
3.30	22.00	.690	1.495	0.155	19.010	4.857	43.027	1.0316	2.947	.1549E+1	.6456E+0	.8892E-2	.4438E+2
3.30	27.00	.690	1.495	0.155	24.010	4.857	53.027	1.0316	3.722	.1226E+1	.8153E+0	.9151E-2	.5470E+2
3.30	33.00	.691	1.496	0.155	30.009	4.858	65.025	1.0334	4.636	.9813E+0	.1019E+1	.9358E-2	.6720E+2
3.90	3.30	.254	1.577	0.373	0.146	4.834	5.992	0.4006	0.054	.5533E+3	.1808E-2	.1405E-3	.2400E+1

CHOKE/RESONANT CHARGE REACTOR DESIGN TABLE

CASE # 1   COPYRIGHT 1982 STEVE SMITH   DATA DATE 1/ 3/82

$f(D)=U/(F*G*(D*E)^2)$

P	Q	D	E	F	G	U	Um	DE	FG	f(D)	1/f(D)	1/(PQf(D))	DEUm
3.90	3.90	.519	1.710	0.241	0.481	5.213	6.814	0.8872	0.116	.5724E+2	.1747E-1	.1149E-2	.6045E+1
3.90	4.70	.624	1.762	0.188	1.176	5.363	8.264	1.0995	0.221	.2006E+2	.4984E-1	.2719E-2	.9086E+1
3.90	5.60	.652	1.776	0.174	2.048	5.403	12.023	1.1580	0.356	.1131E+2	.8844E-1	.4050E-2	.1161E+2
3.90	6.80	.666	1.783	0.167	3.234	5.423	12.403	1.1875	0.540	.7120E+1	.1404E+0	.5296E-2	.1473E+2
3.90	8.20	.672	1.786	0.164	4.628	5.431	15.195	1.2002	0.759	.4968E+1	.2013E+0	.6295E-2	.1824E+2
3.90	10.00	.677	1.789	0.162	6.423	5.438	18.788	1.2108	1.037	.357E+1	.2796E+0	.7170E-2	.2275E+2
3.90	12.00	.680	1.790	0.160	8.420	5.443	22.783	1.2172	1.347	.2727E+1	.3667E+0	.7836E-2	.2773E+2
3.90	15.00	.682	1.791	0.159	11.418	5.446	28.781	1.2215	1.815	.2010E+1	.4974E+0	.8503E-2	.3515E+2
3.90	18.00	.683	1.792	0.159	14.417	5.447	34.779	1.2236	2.285	.1592E+1	.6281E+0	.8947E-2	.4256E+2
3.90	22.00	.684	1.792	0.158	18.416	5.448	42.778	1.2257	2.910	.1246E+1	.8024E+0	.9352E-2	.5243E+2
3.90	27.00	.685	1.793	0.158	23.415	5.450	52.776	1.2279	3.688	.9802E+0	.1020E+1	.9689E-2	.6480E+2
3.90	33.00	.685	1.793	0.158	29.415	5.450	64.776	1.2279	4.633	.7802E+0	.1282E+1	.9958E-2	.7954E+2
4.70	3.90	.131	1.915	0.435	0.069	5.458	7.025	0.2509	0.030	.2891E+4	.3459E-3	.1887E-4	.1763E+1
4.70	4.70	.515	2.107	0.243	0.485	6.007	8.076	1.0854	0.118	.4335E+2	.2307E-1	.1044E-2	.8765E+1
4.70	5.60	.623	2.162	0.188	1.277	6.161	9.722	1.3466	0.241	.1411E+2	.7085E-1	.2692E-2	.1399E+2
4.70	6.80	.653	2.177	0.174	2.447	6.204	12.079	1.4213	0.425	.7234E+1	.1382E+0	.4325E-2	.1717E+2
4.70	8.20	.664	2.182	0.168	3.836	6.220	14.863	1.4488	0.644	.4598E+1	.2175E+0	.5643E-2	.2153E+2
4.70	10.00	.670	2.185	0.165	5.630	6.228	18.454	1.4640	0.929	.3128E+1	.3196E+0	.6801E-2	.2702E+2
4.70	12.00	.674	2.187	0.163	7.626	6.234	24.449	1.4740	1.243	.2300E+1	.4332E+0	.7682E-2	.3309E+2
4.70	15.00	.676	2.188	0.162	10.624	6.237	28.446	1.4791	1.721	.1656E+1	.6037E+0	.8563E-2	.4207E+2
4.70	18.00	.678	2.189	0.161	13.622	6.240	34.443	1.4841	2.193	.1292E+1	.7742E+0	.9151E-2	.5112E+2
4.70	22.00	.679	2.189	0.161	17.621	6.241	42.442	1.4867	2.828	.9985E+0	.1002E+1	.9686E-2	.6110E+2
4.70	27.00	.680	2.190	0.160	22.620	6.243	52.440	1.4892	3.619	.7778E+0	.1286E+1	.1015E-1	.7809E+2
4.70	33.00	.681	2.191	0.160	28.619	6.244	64.439	1.4917	4.565	.6147E+0	.1627E+1	.1049E-1	.9613E+2
5.60	4.70	.066	2.333	0.467	0.034	6.265	8.331	0.1540	0.016	.1664E+5	.6009E-4	.2283E-5	.1283E+1
5.60	5.60	.512	2.556	0.244	0.488	6.903	9.494	1.3087	0.119	.3388E+2	.2954E-1	.9421E-3	.1242E+2
5.60	6.80	.631	2.616	0.185	1.569	7.073	11.724	1.6504	0.289	.8970E+1	.1115E+0	.2928E-2	.1935E+2
5.60	8.20	.655	2.628	0.173	2.945	7.107	14.490	1.7210	0.508	.4723E+1	.2117E+0	.4611E-2	.2494E+2
5.60	10.00	.664	2.632	0.168	4.736	7.120	18.077	1.7476	0.796	.2930E+1	.3413E+0	.6095E-2	.3159E+2
5.60	12.00	.669	2.635	0.166	6.731	7.127	22.070	1.7625	1.114	.2060E+1	.4855E+0	.7225E-2	.3890E+2
5.60	15.00	.672	2.636	0.164	9.728	7.131	28.065	1.7714	1.595	.1425E+1	.7020E+0	.8357E-2	.4971E+2
5.60	18.00	.674	2.637	0.163	12.726	7.136	34.062	1.7773	2.074	.1089E+1	.9185E+0	.9112E-2	.6054E+2
5.60	22.00	.675	2.638	0.163	16.725	7.136	42.061	1.7803	2.718	.8283E+0	.1207E+1	.9799E-2	.7488E+2
5.60	27.00	.676	2.638	0.162	21.724	7.137	52.058	1.7833	3.519	.6377E+0	.1568E+1	.1037E-1	.9284E+2
5.60	33.00	.677	2.639	0.162	27.723	7.138	64.058	1.7863	4.477	.4997E+0	.2001E+1	.1083E-1	.1144E+3
6.80	6.80	.509	3.155	0.246	0.491	8.098	11.383	1.6056	0.121	.2606E+2	.3837E-1	.8299E-3	.1828E+2

CHOKE/RESONANT CHARGE REACTOR DESIGN TABLE

CASE # 1   COPYRIGHT 1982 STEVE SMITH   DATA DATE 1/ 3/82

$f(D)=U/(F*G*(D*E)^2)$

P	Q	D	E	F	G	U	Um	DE	FG	f(D)	1/f(D)	1/(PQf(D))	DEUm
6.80	8.20	.634	3.217	0.183	1.766	8.277	14.005	2.0396	0.323	.6157E+1	.1624E+0	.2913E-2	.2856E+2
6.80	10.00	.656	3.228	0.172	3.544	8.308	17.573	2.1176	0.610	.3044E+1	.3290E+0	.4838E-2	.3721E+2
6.80	12.00	.663	3.232	0.169	5.537	8.318	21.563	2.1425	0.933	.1942E+1	.5148E+0	.6309E-2	.4620E+2
6.80	15.00	.668	3.234	0.166	8.532	8.326	27.556	2.1603	1.416	.1260E+1	.7939E+0	.7784E-2	.5953E+2
6.80	18.00	.670	3.235	0.165	11.530	8.328	33.553	2.1675	1.902	.9319E+0	.1073E+1	.8767E-2	.7272E+2
6.80	22.00	.672	3.236	0.164	15.528	8.331	41.550	2.1746	2.547	.6918E+0	.1445E+1	.9662E-2	.9035E+2
6.80	27.00	.673	3.237	0.164	20.527	8.333	51.549	2.1782	3.356	.5233E+0	.1911E+1	.1041E-1	.1123E+3
6.80	33.00	.674	3.237	0.163	26.526	8.334	63.547	2.1817	4.324	.4049E+0	.2469E+1	.1100E-1	.1388E+3
8.20	8.20	.507	3.854	0.247	0.493	8.495	13.585	1.9537	0.122	.2047E+2	.4885E-1	.7265E-3	.2654E+2
8.20	10.00	.639	3.920	0.181	2.161	9.684	16.997	2.5046	0.390	.3958E+1	.2527E+0	.3081E-2	.4257E+2
8.20	12.00	.657	3.929	0.172	4.143	9.710	20.971	2.5810	0.711	.2051E+1	.4875E+0	.4954E-2	.5413E+2
8.20	15.00	.664	3.932	0.168	7.136	9.720	26.961	2.6108	1.199	.1189E+1	.8408E+0	.6835E-2	.7039E+2
8.20	18.00	.667	3.934	0.167	10.133	9.724	32.956	2.6236	1.687	.8373E+0	.1194E+1	.8091E-2	.8647E+2
8.20	22.00	.669	3.935	0.166	14.131	9.727	40.954	2.6322	2.339	.6003E+0	.1666E+1	.9234E-2	.1078E+3
8.20	27.00	.671	3.936	0.165	19.129	9.730	50.951	2.6407	3.147	.4434E+0	.2255E+1	.1019E-1	.1345E+3
8.20	33.00	.672	3.936	0.164	25.128	9.731	62.949	2.6450	4.121	.3375E+0	.2963E+1	.1095E-1	.1665E+3
10.00	8.20	.505	4.753	0.248	0.494	11.293	16.415	2.4000	0.122	.1600E+2	.6249E-1	.7622E-3	.3940E+2
10.00	10.00	.641	4.821	0.180	2.359	11.487	20.221	3.0899	0.423	.2841E+1	.3520E+0	.3520E-2	.6248E+2
10.00	12.00	.659	4.830	0.171	5.341	11.513	26.195	3.1826	0.911	.1248E+1	.8012E+0	.6677E-2	.8337E+2
10.00	15.00	.664	4.832	0.168	8.336	11.524	32.188	3.2084	1.400	.7991E+0	.1251E+1	.8340E-2	.1033E+3
10.00	18.00	.667	4.834	0.167	12.333	11.527	40.184	3.2239	2.053	.5399E+0	.1852E+1	.1029E-1	.1296E+3
10.00	22.00	.669	4.835	0.166	17.331	11.528	50.181	3.2343	2.868	.3842E+0	.2603E+1	.1183E-1	.1623E+3
10.00	27.00	.670	4.835	0.165	23.330	11.528	62.180	3.2395	3.849	.2854E+0	.3504E+1	.1298E-1	.2014E+3

CASE # 2   COPYRIGHT 1982 STEVE SMITH   DATA DATE 1/ 3/82   CHOKE/RESONANT CHARGE REACTOR DESIGN TABLE

SCALE BY 1.000E-04   FUNCTION 1/f(D)

$$f(D)=U/(F*G*(D*E)^2)$$

Q/P	0.33	0.39	0.47	0.56	0.68	0.82	1.00	1.20	1.50	1.80	2.20	2.70	3.30	3.90	4.70	5.60	6.80	8.20	10.00
0.33	0	0	1	1	2	2													
0.39	0	0	1	2	3	3	3												
0.47	0	1	2	3	4	4	5	4											
0.56	1	1	2	4	5	6	7	7	4										
0.68	1	2	3	5	7	9	12	13	10	5									
0.82	1	2	4	6	9	13	17	20	20	15	4								
1.00	2	3	5	8	12	17	23	30	35	33	20	2							
1.20	2	3	6	10	15	22	31	41	52	56	48	21	0						
1.50	3	5	8	13	20	28	41	57	78	92	98	81	30	0					
1.80	4	6	10	16	24	36	54	74	104	129	150	152	112	39	0				
2.20	5	8	13	20	30	46	69	96	139	178	220	243	243	186	52	0			
2.70	6	10	16	24	38	58	88	124	182	239	308	373	411	399	299	91	0		
3.30	8	12	19	29	48	73	110	158	235	313	413	521	614	659	635	487	119	0	
3.90	9	14	23	36	57	87	133	191	287	387	521	670	817	920	974	915	629	77	0
4.70	11	17	28	44	69	107	163	236	357	486	660	868	1089	1268	1428	1490	1367	914	34
5.60	13	21	34	52	84	128	197	286	436	597	818	1091	1388	1659	1939	2139	2202	1967	1165
6.80	16	25	41	64	102	158	243	353	541	745	1091	1394	1735	2182	2621	3005	3318	3379	2957
8.20	20	31	50	78	124	192	296	432	663	917	1276	1802	2181	2792	3416	4015	4621	5029	5056
10.00	24	38	61	95	152	235	364	532	821	1139	1593	2277	2888	3567	4439	5314	6296	7150	7759
12.00	29	45	74	115	184	284	440	644	995	1386	1945	2676	3576	4448	5576	6758	8158	9508	10762
15.00	36	57	92	144	230	357	554	812	1258	1756	2473	3567	4586	5755	7281	8924	10951	13045	15268
18.00	44	68	111	173	277	429	667	979	1520	2126	3002	4163	5605	7062	8987	10951	13744	16582	21682
22.00	54	84	136	212	340	527	819	1203	1870	2619	3706	5154	6963	8805	11260	13977	17468	21298	25781
27.00	66	103	167	260	418	648	1089	1482	2307	3236	4587	6393	8661	10983	14102	17586	22123	27193	33291
33.00	81	126	204	318	512	794	1236	1818	2832	3976	5643	7880	10699	13598	17513	21918	27709	34268	42303

CASE # 2   COPYRIGHT 1982 STEVE SMITH   DATA DATE 1/ 3/82   CHOKE/RESONANT CHARGE REACTOR DESIGN TABLE

SCALE BY 1.000E-04   FUNCTION 1/[P*Q*f(D)]

$f(D)=U/(F*G*(D*E)^2)$

Q/P	0.33	0.39	0.47	0.56	0.68	0.82	1.00	1.20	1.50	1.80	2.20	2.70	3.30	3.90	4.70	5.60	6.80	8.20	10.00
0.33	5	6	7	7	7	6	4	2	.	.	.	.	.	.	.	.	.	.	.
0.39	5	6	8	9	9	9	6	4	.	.	.	.	.	.	.	.	.	.	.
0.47	6	7	9	10	11	11	8	7	0	0	.	.	.	.	.	.	.	.	.
0.56	6	7	9	11	13	14	13	10	0	0	.	.	.	.	.	.	.	.	.
0.68	6	8	10	12	15	16	17	15	2	4	0	.	.	.	.	.	.	.	.
0.82	6	9	11	13	16	19	20	20	5	10	2	.	.	.	.	.	.	.	.
1.00	7	8	11	14	17	21	23	25	10	18	9	0	.	.	.	.	.	.	.
1.20	7	9	11	14	18	22	26	28	16	26	18	6	0	0	.	.	.	.	.
1.50	7	9	12	15	19	24	28	32	23	34	30	20	6	6	.	.	.	.	.
1.80	7	9	12	16	20	25	30	34	29	40	38	31	19	22	.	.	.	.	.
2.20	7	9	13	16	21	26	31	36	34	45	45	42	33	38	5	.	.	.	.
2.70	7	9	13	16	21	27	32	38	38	49	52	51	46	51	24	6	.	.	.
3.30	7	10	13	17	22	27	33	40	42	53	57	58	56	60	41	26	5	.	.
3.90	7	10	13	17	22	28	34	41	45	55	60	64	64	69	53	42	24	2	.
4.70	7	10	13	17	22	28	35	42	47	57	64	68	70	76	65	57	43	24	0
5.60	7	10	13	17	22	28	36	43	49	59	66	72	75	82	74	68	58	43	21
6.80	7	10	13	17	22	28	36	44	51	61	69	76	80	84	82	79	72	61	43
8.20	7	10	13	17	23	28	36	44	52	62	71	78	84	87	89	87	83	75	62
10.00	7	10	13	17	23	29	37	45	53	63	72	81	88	92	94	95	93	87	78
12.00	7	10	13	17	23	29	37	45	54	64	74	83	90	95	99	101	100	97	90
15.00	7	10	13	17	23	29	37	45	55	65	75	84	93	98	103	106	107	106	102
18.00	7	10	13	17	23	29	37	46	55	66	76	86	94	101	106	110	112	112	110
22.00	7	10	13	17	23	29	37	46	56	66	77	87	96	103	109	113	117	118	117
27.00	7	10	13	17	23	29	37	46	57	67	77	88	97	104	111	116	120	123	123
33.00	7	10	13	17	23	29	37	46	57	67	78	88	98	106	113	119	123	127	128

CASE # 2 COPYRIGHT 1982 STEVE SMITH DATA DATE 1/ 3/82 CHOKE/RESONANT CHARGE REACTOR DESIGN TABLE

$f(D) = U/(F \cdot G \cdot (D \cdot E)^2)$

P	Q	D	E	F	G	U	Um	DE	FG	f(D)	1/f(D)	1/(PQf(D))	DEUm
0.33	0.33	.885	0.107	0.058	0.223	2.166	0.729	0.0951	0.013	.1870E+5	.5347E-4	.4910E-3	.6934E-1
0.33	0.39	.888	0.109	0.056	0.281	2.170	0.845	0.0968	0.016	.1472E+5	.6794E-4	.5279E-3	.8181E-1
0.33	0.47	.892	0.111	0.054	0.359	2.176	1.000	0.0990	0.019	.1145E+5	.8735E-4	.5632E-3	.9905E-1
0.33	0.56	.894	0.112	0.053	0.448	2.179	1.178	0.1001	0.024	.9152E+4	.1093E-3	.5913E-3	.1179E+0
0.33	0.68	.895	0.112	0.053	0.568	2.180	1.417	0.1007	0.030	.7217E+4	.1386E-3	.6175E-3	.1426E+0
0.33	0.82	.897	0.114	0.051	0.707	2.183	1.694	0.1018	0.036	.5788E+4	.1728E-3	.6380E-3	.1725E+0
0.33	1.00	.898	0.114	0.051	0.886	2.184	2.053	0.1024	0.045	.4612E+4	.2168E-3	.6570E-3	.2102E+0
0.33	1.20	.899	0.114	0.051	1.086	2.186	2.452	0.1029	0.055	.3763E+4	.2657E-3	.6711E-3	.2524E+0
0.33	1.50	.899	0.114	0.051	1.386	2.186	3.052	0.1029	0.070	.2948E+4	.3392E-3	.6852E-3	.3141E+0
0.33	1.80	.900	0.115	0.050	1.685	2.187	3.651	0.1035	0.084	.2423E+4	.4127E-3	.6947E-3	.3778E+0
0.33	2.20	.900	0.115	0.050	2.085	2.187	4.451	0.1035	0.104	.1958E+4	.5106E-3	.7033E-3	.4606E+0
0.33	2.70	.901	0.115	0.050	2.585	2.187	5.451	0.1041	0.129	.1580E+4	.6331E-3	.7105E-3	.5641E+0
0.33	3.30	.901	0.116	0.049	3.184	2.189	6.649	0.1041	0.158	.1282E+4	.7800E-3	.7163E-3	.6920E+0
0.33	3.90	.901	0.116	0.049	3.785	2.189	7.849	0.1041	0.187	.1079E+4	.9270E-3	.7203E-3	.8169E+0
0.33	4.70	.901	0.116	0.049	4.585	2.189	9.449	0.1041	0.227	.8905E+3	.1123E-2	.7240E-3	.9834E+0
0.33	5.60	.901	0.116	0.049	5.485	2.189	11.249	0.1041	0.271	.7444E+3	.1343E-2	.7270E-3	.1171E+1
0.33	6.80	.901	0.116	0.049	6.685	2.189	13.649	0.1041	0.331	.6107E+3	.1637E-2	.7297E-3	.1420E+1
0.33	8.20	.901	0.116	0.049	8.085	2.189	16.449	0.1041	0.400	.5050E+3	.1980E-2	.7318E-3	.1712E+1
0.33	10.00	.901	0.116	0.049	9.885	2.189	20.049	0.1041	0.489	.4130E+3	.2421E-2	.7337E-3	.2086E+1
0.33	12.00	.902	0.116	0.049	11.884	2.190	24.048	0.1046	0.582	.3435E+3	.2911E-2	.7351E-3	.2516E+1
0.33	15.00	.902	0.116	0.049	14.884	2.190	30.048	0.1046	0.729	.2743E+3	.3646E-2	.7366E-3	.3144E+1
0.33	18.00	.902	0.116	0.049	17.884	2.190	36.048	0.1046	0.876	.2283E+3	.4381E-2	.7375E-3	.3772E+1
0.33	22.00	.902	0.116	0.049	21.884	2.190	44.048	0.1046	1.072	.1865E+3	.5361E-2	.7384E-3	.4609E+1
0.33	27.00	.902	0.116	0.049	26.884	2.190	54.048	0.1046	1.317	.1518E+3	.6585E-2	.7391E-3	.5655E+1
0.33	33.00	.902	0.116	0.049	32.884	2.190	66.048	0.1046	1.611	.1241E+3	.8055E-2	.7397E-3	.6911E+1
0.39	0.33	.867	0.126	0.070	0.204	2.191	0.745	0.1081	0.014	.1320E+5	.7573E-4	.5884E-3	.8052E-1
0.39	0.39	.867	0.128	0.067	0.262	2.200	0.858	0.1114	0.017	.1019E+5	.9811E-4	.6451E-3	.9557E-1
0.39	0.47	.872	0.131	0.064	0.339	2.207	1.012	0.1142	0.022	.7796E+4	.1283E-3	.6998E-3	.1156E+0
0.39	0.56	.875	0.132	0.063	0.428	2.211	1.188	0.1159	0.027	.6157E+4	.1624E-3	.7436E-3	.1377E+0
0.39	0.68	.878	0.134	0.061	0.546	2.216	1.424	0.1177	0.033	.4806E+4	.2081E-3	.7846E-3	.1676E+0
0.39	0.82	.880	0.135	0.060	0.685	2.218	1.702	0.1188	0.041	.3825E+4	.2615E-3	.8176E-3	.2022E+0
0.39	1.00	.881	0.136	0.060	0.865	2.220	2.061	0.1194	0.051	.3028E+4	.3302E-3	.8467E-3	.2460E+0
0.39	1.20	.882	0.136	0.059	1.064	2.221	2.460	0.1200	0.063	.2459E+4	.4066E-3	.8688E-3	.2950E+0
0.39	1.50	.883	0.136	0.058	1.364	2.223	3.058	0.1205	0.080	.1918E+4	.5213E-3	.8911E-3	.3686E+0
0.39	1.80	.884	0.137	0.058	1.663	2.224	3.657	0.1211	0.096	.1572E+4	.6360E-3	.9061E-3	.4429E+0
0.39	2.20	.885	0.137	0.058	2.063	2.226	4.456	0.1217	0.119	.1267E+4	.7890E-3	.9196E-3	.5425E+0
0.39	2.70	.885	0.137	0.058	2.563	2.226	5.456	0.1217	0.147	.1020E+4	.9803E-3	.9310E-3	.6639E+0

CHOKE/RESONANT CHARGE REACTOR DESIGN TABLE

$$f(D)=U/(F*G*(D*E)^2)$$

CASE # 2   COPYRIGHT 1982 STEVE SMITH   DATA DATE 1/3/82

P	Q	D	E	F	G	U	Um	DE	FG	f(D)	1/f(D)	1/(PQf(D))	DEUm
0.39	3.30	.885	0.137	0.058	3.163	2.226	6.656	0.1217	0.182	.8265E+3	.1210E-2	.9401E-3	.8100E+0
0.39	3.90	.886	0.138	0.057	3.762	2.227	7.855	0.1223	0.214	.6947E+3	.1439E-2	.9464E-3	.9604E+0
0.39	4.70	.886	0.138	0.057	4.562	2.227	9.455	0.1223	0.260	.5729E+3	.1746E-2	.9523E-3	.1156E+1
0.39	5.60	.886	0.138	0.057	5.462	2.227	11.255	0.1223	0.311	.4788E+3	.2090E-2	.9569E-3	.1376E+1
0.39	6.80	.886	0.138	0.057	6.662	2.227	13.655	0.1223	0.380	.3923E+3	.2549E-2	.9612E-3	.1670E+1
0.39	8.20	.886	0.138	0.057	8.062	2.227	16.455	0.1223	0.460	.3242E+3	.3085E-2	.9646E-3	.2012E+1
0.39	10.00	.886	0.138	0.057	9.862	2.227	20.055	0.1223	0.562	.2650E+3	.3773E-2	.9675E-3	.2452E+1
0.39	12.00	.887	0.139	0.057	11.862	2.227	24.055	0.1223	0.676	.2203E+3	.4539E-2	.9698E-3	.2941E+1
0.39	15.00	.887	0.139	0.056	14.862	2.229	30.054	0.1228	0.840	.1759E+3	.5687E-2	.9721E-3	.3692E+1
0.39	18.00	.887	0.139	0.056	17.862	2.229	36.054	0.1228	1.009	.1463E+3	.6834E-2	.9736E-3	.4429E+1
0.39	22.00	.887	0.139	0.056	21.862	2.229	44.054	0.1228	1.235	.1195E+3	.8365E-2	.9749E-3	.5412E+1
0.39	27.00	.887	0.139	0.056	26.862	2.229	54.054	0.1228	1.518	.9729E+2	.1028E-1	.9761E-3	.6640E+1
0.39	33.00	.887	0.139	0.056	32.862	2.229	66.054	0.1228	1.857	.7953E+2	.1257E-1	.9770E-3	.8115E+1
0.47	0.33	.837	0.149	0.086	0.182	2.223	0.769	0.1228	0.016	.9387E+4	.1065E-3	.6868E-3	.9447E-1
0.47	0.39	.837	0.154	0.082	0.237	2.237	0.877	0.1285	0.019	.7031E+4	.1422E-3	.7759E-3	.1127E+0
0.47	0.47	.845	0.158	0.077	0.313	2.248	1.077	0.1331	0.024	.5242E+4	.1908E-3	.8637E-3	.1367E+0
0.47	0.56	.851	0.161	0.074	0.400	2.257	1.200	0.1366	0.030	.4065E+4	.2460E-3	.9347E-3	.1639E+0
0.47	0.68	.855	0.163	0.072	0.518	2.263	1.435	0.1389	0.038	.3124E+4	.3201E-3	.1001E-2	.1994E+0
0.47	0.82	.858	0.164	0.071	0.656	2.267	1.712	0.1407	0.047	.2458E+4	.4068E-3	.1055E-2	.2408E+0
0.47	1.00	.860	0.165	0.070	0.835	2.270	2.069	0.1419	0.058	.1929E+4	.5185E-3	.1103E-2	.2936E+0
0.47	1.20	.862	0.166	0.069	1.034	2.273	2.467	0.1431	0.071	.1556E+4	.6428E-3	.1140E-2	.3530E+0
0.47	1.50	.863	0.166	0.069	1.334	2.274	3.066	0.1437	0.091	.1206E+4	.8293E-3	.1176E-2	.4405E+0
0.47	1.80	.864	0.167	0.068	1.633	2.276	3.664	0.1443	0.111	.9843E+3	.1016E-2	.1201E-2	.5287E+0
0.47	2.20	.865	0.168	0.067	2.033	2.277	4.463	0.1449	0.137	.7906E+3	.1265E-2	.1223E-2	.6466E+0
0.47	2.70	.866	0.168	0.067	2.532	2.278	5.462	0.1455	0.170	.6345E+3	.1576E-2	.1242E-2	.7946E+0
0.47	3.30	.866	0.168	0.067	3.132	2.278	6.662	0.1455	0.210	.5130E+3	.1949E-2	.1257E-2	.9692E+0
0.47	3.90	.866	0.168	0.067	3.732	2.278	7.862	0.1455	0.250	.4305E+3	.2323E-2	.1267E-2	.1144E+1
0.47	4.70	.867	0.168	0.067	4.532	2.280	9.461	0.1461	0.301	.3545E+3	.2821E-2	.1277E-2	.1382E+1
0.47	5.60	.867	0.168	0.067	5.431	2.280	11.261	0.1461	0.361	.2958E+3	.3381E-2	.1285E-2	.1645E+1
0.47	6.80	.867	0.168	0.067	6.632	2.280	13.661	0.1461	0.441	.2422E+3	.4128E-2	.1292E-2	.1996E+1
0.47	8.20	.867	0.168	0.067	8.032	2.280	16.461	0.1461	0.534	.2000E+3	.5000E-2	.1297E-2	.2405E+1
0.47	10.00	.868	0.169	0.066	9.831	2.281	20.059	0.1467	0.649	.1634E+3	.6120E-2	.1302E-2	.2943E+1
0.47	12.00	.868	0.169	0.066	11.831	2.281	24.059	0.1467	0.781	.1358E+3	.7365E-2	.1306E-2	.3529E+1
0.47	15.00	.868	0.169	0.066	14.831	2.281	30.059	0.1467	0.979	.1083E+3	.9233E-2	.1310E-2	.4449E+1
0.47	18.00	.868	0.169	0.066	17.831	2.281	36.059	0.1467	1.177	.9009E+2	.1110E-1	.1312E-2	.5290E+1
0.47	22.00	.868	0.169	0.066	21.831	2.281	44.059	0.1467	1.441	.7358E+2	.1359E-1	.1314E-2	.6463E+1

CHOKE/RESONANT CHARGE REACTOR DESIGN TABLE

$$f(D)=U/(F*G*(D*E)^2)$$

CASE # 2   COPYRIGHT 1982 STEVE SMITH   DATA DATE 1/ 3/82

P	Q	D	E	F	G	U	Um	DE	FG	f(D)	1/f(D)	1/(PQf(D))	DEUm
0.47	27.00	.868	0.169	0.066	26.831	2.281	54.059	0.1467	1.771	.5987E+2	.1670E-1	.1316E-2	.793E+1
0.47	33.00	.868	0.169	0.066	32.831	2.281	66.059	0.1467	2.167	.4893E+2	.2044E-1	.1318E-2	.969E+1
0.56	0.33	.783	0.172	0.109	0.159	2.250	0.803	0.1343	0.017	.7255E+4	.1378E-3	.7459E-3	.1079E+0
0.56	0.39	.801	0.181	0.100	0.210	2.276	0.902	0.1446	0.021	.5222E+4	.1915E-3	.8768E-3	.1303E+0
0.56	0.47	.815	0.188	0.093	0.283	2.296	1.044	0.1528	0.026	.3762E+4	.2658E-3	.1010E-2	.1596E+0
0.56	0.56	.824	0.192	0.088	0.368	2.308	1.214	0.1582	0.032	.2848E+4	.3511E-3	.1128E-2	.1920E+0
0.56	0.68	.830	0.195	0.085	0.485	2.317	1.446	0.1618	0.041	.2146E+4	.4661E-3	.1224E-2	.2341E+0
0.56	0.82	.835	0.198	0.083	0.623	2.324	1.720	0.1649	0.051	.1664E+4	.6009E-3	.1309E-2	.2837E+0
0.56	1.00	.840	0.199	0.081	0.801	2.328	2.077	0.1668	0.065	.1291E+4	.7749E-3	.1384E-2	.3463E+0
0.56	1.20	.840	0.200	0.080	1.000	2.331	2.474	0.1680	0.080	.1033E+4	.9685E-3	.1441E-2	.4157E+0
0.56	1.50	.842	0.201	0.079	1.299	2.334	3.072	0.1692	0.103	.7941E+3	.1259E-2	.1499E-2	.5199E+0
0.56	1.80	.844	0.202	0.078	1.598	2.337	3.669	0.1705	0.125	.6451E+3	.1550E-2	.1538E-2	.6256E+0
0.56	2.20	.845	0.203	0.077	1.998	2.338	4.468	0.1711	0.155	.5159E+3	.1938E-2	.1573E-2	.7645E+0
0.56	2.70	.846	0.203	0.077	2.497	2.340	5.467	0.1717	0.192	.4126E+3	.2424E-2	.1603E-2	.9389E+0
0.56	3.30	.846	0.203	0.077	3.097	2.340	6.667	0.1717	0.238	.3327E+3	.3006E-2	.1627E-2	.1145E+1
0.56	3.90	.847	0.204	0.077	3.697	2.341	7.866	0.1724	0.283	.2787E+3	.3588E-2	.1643E-2	.1356E+1
0.56	4.70	.847	0.204	0.077	4.496	2.341	9.466	0.1724	0.344	.2291E+3	.4365E-2	.1658E-2	.1632E+1
0.56	5.60	.848	0.204	0.076	5.396	2.343	11.264	0.1730	0.410	.1909E+3	.5239E-2	.1670E-2	.1949E+1
0.56	6.80	.848	0.204	0.076	6.596	2.343	13.664	0.1730	0.501	.1562E+3	.6404E-2	.1682E-2	.2364E+1
0.56	8.20	.848	0.204	0.076	7.996	2.343	16.464	0.1730	0.608	.1288E+3	.7763E-2	.1690E-2	.2848E+1
0.56	10.00	.848	0.204	0.076	9.796	2.343	20.064	0.1730	0.744	.1052E+3	.9510E-2	.1698E-2	.3471E+1
0.56	12.00	.848	0.204	0.076	11.796	2.343	24.064	0.1730	0.896	.8732E+2	.1145E-1	.1704E-2	.4163E+1
0.56	15.00	.849	0.204	0.076	14.796	2.344	30.063	0.1736	1.117	.6962E+2	.1436E-1	.1710E-2	.5220E+1
0.56	18.00	.849	0.204	0.076	17.796	2.344	36.063	0.1736	1.344	.5788E+2	.1728E-1	.1714E-2	.6261E+1
0.56	22.00	.849	0.204	0.076	21.796	2.344	44.063	0.1736	1.646	.4726E+2	.2116E-1	.1718E-2	.7650E+1
0.56	27.00	.849	0.204	0.076	26.795	2.344	54.063	0.1736	2.023	.3844E+2	.2601E-1	.1721E-2	.9386E+1
0.56	33.00	.849	0.204	0.076	32.796	2.344	66.063	0.1736	2.476	.3141E+2	.3184E-1	.1723E-2	.1147E+2
0.68	0.33	.714	0.197	0.143	0.133	2.271	0.861	0.1407	0.019	.6036E+4	.1657E-3	.7383E-3	.1212E+0
0.68	0.39	.746	0.213	0.127	0.177	2.317	0.943	0.1589	0.022	.4082E+4	.2450E-3	.9237E-3	.1498E+0
0.68	0.47	.771	0.226	0.114	0.244	2.353	1.072	0.1739	0.028	.2780E+4	.3597E-3	.1125E-2	.1864E+0
0.68	0.56	.787	0.234	0.107	0.327	2.376	1.233	0.1838	0.035	.2023E+4	.4943E-3	.1298E-2	.2265E+0
0.68	0.68	.798	0.239	0.101	0.441	2.391	1.459	0.1907	0.045	.1476E+4	.6775E-3	.1465E-2	.2783E+0
0.68	0.82	.805	0.243	0.098	0.578	2.401	1.731	0.1952	0.056	.1119E+4	.8936E-3	.1603E-2	.3379E+0
0.68	1.00	.811	0.246	0.095	0.755	2.410	2.084	0.1991	0.071	.8526E+3	.1173E-2	.1725E-2	.4149E+0
0.68	1.20	.814	0.247	0.093	0.953	2.414	2.480	0.2011	0.089	.6738E+3	.1484E-2	.1813E-2	.4986E+0
0.68	1.50	.817	0.248	0.092	1.252	2.418	3.076	0.2030	0.115	.5124E+3	.1952E-2	.1913E-2	.6246E+0

CASE # 2   COPYRIGHT 1982 STEVE SMITH   DATA DATE 1/ 3/82   CHOKE/RESONANT CHARGE REACTOR DESIGN TABLE

$$f(D)=U/(F*G*(D*E)^2)$$

P	Q	D	E	F	G	U	Um	DE	FG	f(D)	1/f(D)	1/(PQf(D))	DEUm
0.68	1.80	.819	0.250	0.091	1.550	2.421	3.674	0.2043	0.140	.4133E+3	.2420E-2	.1977E-2	.7507E+0
0.68	2.20	.821	0.251	0.090	1.950	2.424	4.471	0.2057	0.174	.3285E+3	.3044E-2	.2035E-2	.9196E+0
0.68	2.70	.822	0.251	0.089	2.449	2.426	5.470	0.2063	0.218	.2614E+3	.3825E-2	.2083E-2	.1129E+1
0.68	3.30	.823	0.252	0.088	3.049	2.427	6.669	0.2070	0.270	.2100E+3	.4762E-2	.2122E-2	.1380E+1
0.68	3.90	.824	0.252	0.088	3.648	2.428	7.868	0.2076	0.321	.1754E+3	.5700E-2	.2149E-2	.1634E+1
0.68	4.70	.824	0.252	0.088	4.448	2.428	9.468	0.2076	0.391	.1439E+3	.6950E-2	.2175E-2	.1966E+1
0.68	5.60	.825	0.253	0.088	5.348	2.430	11.267	0.2083	0.468	.1197E+3	.8356E-2	.2194E-2	.2347E+1
0.68	6.80	.825	0.253	0.088	6.548	2.430	13.667	0.2083	0.573	.9774E+2	.1023E-1	.2213E-2	.2847E+1
0.68	8.20	.825	0.253	0.088	7.948	2.430	16.467	0.2083	0.695	.8852E+2	.1242E-1	.2227E-2	.3430E+1
0.68	10.00	.826	0.253	0.087	9.747	2.431	20.065	0.2090	0.848	.6565E+2	.1523E-1	.2240E-2	.4193E+1
0.68	12.00	.826	0.253	0.087	11.747	2.431	24.065	0.2090	1.022	.5447E+2	.1836E-1	.2250E-2	.5029E+1
0.68	15.00	.826	0.253	0.087	14.747	2.431	30.065	0.2090	1.283	.4339E+2	.2305E-1	.2259E-2	.6283E+1
0.68	18.00	.826	0.253	0.087	17.747	2.431	36.065	0.2090	1.544	.3606E+2	.2773E-1	.2266E-2	.7537E+1
0.68	22.00	.826	0.253	0.087	21.747	2.431	44.065	0.2090	1.892	.2943E+2	.3398E-1	.2272E-2	.9209E+1
0.68	27.00	.826	0.253	0.087	26.747	2.431	54.065	0.2090	2.327	.2392E+2	.4180E-1	.2277E-2	.1130E+2
0.68	33.00	.826	0.253	0.087	32.747	2.431	66.065	0.2090	2.849	.1954E+2	.5117E-1	.2280E-2	.1381E+2
0.82	0.33	.619	0.220	0.191	0.110	2.275	0.947	0.1359	0.021	.5856E+4	.1708E-3	.6311E-3	.1286E+0
0.82	0.39	.669	0.245	0.166	0.146	2.347	1.006	0.1636	0.024	.3643E+4	.2745E-3	.8584E-3	.1646E+0
0.82	0.47	.714	0.267	0.143	0.203	2.411	1.111	0.1906	0.029	.2286E+4	.4375E-3	.1135E-2	.2119E+0
0.82	0.56	.742	0.281	0.129	0.279	2.451	1.257	0.2085	0.036	.1567E+4	.6383E-3	.1390E-2	.2622E+0
0.82	0.68	.762	0.291	0.119	0.389	2.480	1.473	0.2217	0.046	.1090E+4	.9178E-3	.1646E-2	.3266E+0
0.82	0.82	.774	0.297	0.113	0.523	2.497	1.739	0.2299	0.059	.7995E+3	.1251E-2	.1860E-2	.3996E+0
0.82	1.00	.782	0.301	0.109	0.699	2.508	2.089	0.2354	0.076	.5942E+3	.1683E-2	.2052E-2	.4917E+0
0.82	1.20	.788	0.304	0.106	0.896	2.517	2.482	0.2396	0.095	.4618E+3	.2165E-2	.2201E-2	.5945E+0
0.82	1.50	.792	0.306	0.104	1.194	2.523	3.077	0.2424	0.124	.3459E+3	.2891E-2	.2350E-2	.7456E+0
0.82	1.80	.795	0.308	0.103	1.493	2.527	3.673	0.2445	0.153	.2764E+3	.3618E-2	.2451E-2	.8979E+0
0.82	2.20	.797	0.309	0.102	1.892	2.530	4.471	0.2459	0.192	.2180E+3	.4588E-2	.2543E-2	.1099E+1
0.82	2.70	.799	0.310	0.100	2.391	2.533	5.468	0.2480	0.240	.1724E+3	.5801E-2	.2620E-2	.1352E+1
0.82	3.30	.800	0.311	0.100	2.990	2.534	6.667	0.2487	0.299	.1378E+3	.7257E-2	.2682E-2	.1653E+1
0.82	4.00	.801	0.311	0.100	3.590	2.536	7.866	0.2487	0.357	.1148E+3	.8713E-2	.2725E-2	.1956E+1
0.82	4.70	.801	0.311	0.100	4.390	2.536	9.466	0.2487	0.437	.9385E+2	.1065E-1	.2765E-2	.2354E+1
0.82	5.60	.802	0.311	0.099	5.289	2.537	11.265	0.2494	0.524	.7788E+2	.1284E-1	.2796E-2	.2818E+1
0.82	6.80	.803	0.311	0.099	6.489	2.538	13.663	0.2501	0.639	.6348E+2	.1575E-1	.2825E-2	.3418E+1
0.82	8.20	.803	0.311	0.099	7.889	2.538	16.463	0.2501	0.777	.5221E+2	.1915E-1	.2848E-2	.4118E+1
0.82	10.00	.803	0.311	0.099	9.689	2.538	20.063	0.2501	0.954	.4251E+2	.2352E-1	.2869E-2	.5019E+1
0.82	12.00	.804	0.312	0.098	11.688	2.540	24.062	0.2508	1.145	.3524E+2	.2838E-1	.2884E-2	.6036E+1

CASE # 2 COPYRIGHT 1982 STEVE SMITH DATA DATE 1/ 3/82 CHOKE/RESONANT CHARGE REACTOR DESIGN TABLE

$$f(D)=U/(F*G*(D*E)^2)$$

P	Q	D	E	F	G	U	Um	DE	FG	f(D)	1/f(D)	1/(PQf(D))	DEUm
0.82	15.00	.804	0.312	0.098	14.688	2.540	30.062	0.2508	1.439	.2804E+2	.3566E-1	.2899E-2	.7541E+1
0.82	18.00	.804	0.312	0.098	17.688	2.540	36.062	0.2508	1.733	.2329E+2	.4295E-1	.2910E-2	.9046E+1
0.82	22.00	.804	0.312	0.098	21.688	2.540	44.062	0.2508	2.125	.1899E+2	.5266E-1	.2919E-2	.1105E+2
0.82	27.00	.804	0.312	0.098	26.688	2.540	54.062	0.2508	2.615	.1543E+2	.6480E-1	.2927E-2	.1356E+2
0.82	33.00	.804	0.312	0.098	32.688	2.540	66.062	0.2508	3.203	.1260E+2	.7936E-1	.2933E-2	.1657E+2
1.00	0.33	.484	0.242	0.258	0.088	2.263	1.072	0.1171	0.023	.7264E+4	.1377E-3	.4172E-3	.1256E+0
1.00	0.39	.553	0.277	0.224	0.114	2.361	1.108	0.1529	0.025	.3981E+4	.2512E-3	.6441E-3	.1695E+0
1.00	0.47	.624	0.312	0.188	0.158	2.463	1.182	0.1947	0.030	.2187E+4	.4572E-3	.9727E-3	.2301E+0
1.00	0.56	.677	0.339	0.162	0.222	2.538	1.298	0.2292	0.036	.1351E+4	.7401E-3	.1322E-2	.2974E+0
1.00	0.68	.714	0.357	0.143	0.323	2.591	1.493	0.2549	0.046	.8635E+3	.1158E-2	.1703E-2	.3805E+0
1.00	0.82	.736	0.368	0.132	0.452	2.623	1.746	0.2708	0.060	.5992E+3	.1669E-2	.2035E-2	.4739E+0
1.00	1.00	.750	0.375	0.125	0.625	2.643	2.089	0.2813	0.078	.4276E+3	.2338E-2	.2338E-2	.5875E+0
1.00	1.20	.758	0.379	0.121	0.821	2.654	2.479	0.2873	0.099	.3237E+3	.3089E-2	.2574E-2	.7123E+0
1.00	1.50	.765	0.383	0.118	1.118	2.668	3.071	0.2926	0.131	.2370E+3	.4220E-2	.2813E-2	.8986E+0
1.00	1.80	.768	0.384	0.116	1.416	2.668	3.667	0.2949	0.164	.1868E+3	.5354E-2	.2974E-2	.1081E+1
1.00	2.20	.771	0.386	0.114	1.815	2.673	4.464	0.2972	0.208	.1456E+3	.6867E-2	.3121E-2	.1327E+1
1.00	2.70	.774	0.387	0.113	2.313	2.677	5.460	0.2995	0.261	.1142E+3	.8760E-2	.3244E-2	.1635E+1
1.00	3.30	.775	0.388	0.113	2.913	2.678	6.659	0.3003	0.328	.9064E+2	.1103E-1	.3343E-2	.2000E+1
1.00	3.90	.777	0.389	0.112	3.512	2.681	7.856	0.3019	0.392	.7515E+2	.1331E-1	.3412E-2	.2372E+1
1.00	4.70	.778	0.389	0.111	4.311	2.683	9.455	0.3026	0.479	.6121E+2	.1634E-1	.3476E-2	.2861E+1
1.00	5.60	.778	0.389	0.111	5.211	2.683	11.255	0.3026	0.578	.5064E+2	.1975E-1	.3526E-2	.3406E+1
1.00	6.80	.779	0.390	0.111	6.411	2.684	13.654	0.3034	0.708	.4116E+2	.2430E-1	.3573E-2	.4143E+1
1.00	8.20	.779	0.390	0.110	7.811	2.684	16.454	0.3034	0.863	.3378E+2	.2960E-1	.3610E-2	.4992E+1
1.00	10.00	.780	0.390	0.110	9.610	2.686	20.053	0.3042	1.057	.2745E+2	.3642E-1	.3642E-2	.6100E+1
1.00	12.00	.780	0.390	0.110	11.610	2.686	24.053	0.3042	1.277	.2272E+2	.4401E-1	.3667E-2	.7317E+1
1.00	15.00	.781	0.391	0.109	14.610	2.687	30.051	0.3050	1.600	.1806E+2	.5538E-1	.3692E-2	.9165E+1
1.00	18.00	.781	0.391	0.109	17.610	2.687	36.051	0.3050	1.928	.1498E+2	.6675E-1	.3708E-2	.1099E+2
1.00	22.00	.781	0.391	0.109	21.610	2.687	44.051	0.3050	2.366	.1221E+2	.8191E-1	.3723E-2	.1343E+2
1.00	27.00	.781	0.391	0.109	26.610	2.687	54.051	0.3050	2.914	.9915E+1	.1009E+0	.3736E-2	.1648E+2
1.00	33.00	.781	0.391	0.109	32.610	2.687	66.051	0.3050	3.571	.8090E+1	.1236E+0	.3746E-2	.2014E+2
1.20	0.33	.330	0.265	0.335	0.065	2.242	1.216	0.0875	0.022	.1347E+5	.7426E-4	.1875E-3	.1060E+0
1.20	0.39	.410	0.305	0.295	0.085	2.357	1.239	0.1251	0.025	.6010E+4	.1663E-3	.3555E-3	.1549E+0
1.20	0.47	.503	0.352	0.248	0.118	2.490	1.286	0.1768	0.029	.2705E+4	.3697E-3	.6555E-3	.2274E+0
1.20	0.56	.586	0.393	0.207	0.167	2.608	1.365	0.2303	0.035	.1423E+4	.7029E-3	.1046E-2	.3144E+0
1.20	0.68	.654	0.427	0.173	0.253	2.705	1.523	0.2793	0.044	.7926E+3	.1262E-2	.1546E-2	.4252E+0
1.20	0.82	.694	0.447	0.153	0.373	2.763	1.754	0.3102	0.057	.5030E+3	.1988E-2	.2020E-2	.5442E+0

CASE # 2   COPYRIGHT 1982 STEVE SMITH   DATA DATE 1/ 3/82   CHOKE/RESONANT CHARGE REACTOR DESIGN TABLE

$f(D) = U/(F \cdot G \cdot (D \cdot E)^2)$

P	Q	D	E	F	G	U	Um	DE	FG	f(D)	1/f(D)	1/(PQf(D))	DEUm
1.20	1.00	.717	0.459	0.142	0.542	2.796	2.086	0.3287	0.077	.3376E+3	.2962E-2	.2468E-2	.6858E+0
1.20	1.20	.730	0.465	0.135	0.735	2.814	2.470	0.3395	0.099	.2461E+3	.4063E-2	.2821E-2	.8386E+0
1.20	1.50	.740	0.470	0.130	1.030	2.828	3.058	0.3478	0.134	.1746E+3	.5727E-2	.3181E-2	.1064E+1
1.20	1.80	.745	0.473	0.128	1.328	2.836	3.652	0.3520	0.169	.1355E+3	.7396E-2	.3424E-2	.1286E+1
1.20	2.20	.749	0.475	0.126	1.726	2.841	4.447	0.3554	0.217	.1039E+3	.9627E-2	.3647E-2	.1581E+1
1.20	2.70	.752	0.476	0.124	2.224	2.846	5.444	0.3580	0.276	.8053E+2	.1242E-1	.3833E-2	.1949E+1
1.20	3.30	.754	0.477	0.123	2.823	2.848	6.641	0.3597	0.347	.6342E+2	.1577E-1	.3982E-2	.2389E+1
1.20	3.90	.756	0.478	0.122	3.422	2.851	7.839	0.3614	0.417	.5230E+2	.1912E-1	.4086E-2	.2833E+1
1.20	4.70	.757	0.479	0.122	4.222	2.853	9.438	0.3622	0.513	.4239E+2	.2359E-1	.4183E-2	.3419E+1
1.20	5.60	.758	0.479	0.121	5.121	2.854	11.236	0.3631	0.620	.3494E+2	.2862E-1	.4259E-2	.4080E+1
1.20	6.80	.759	0.480	0.121	6.321	2.856	13.635	0.3639	0.762	.2831E+2	.3533E-1	.4329E-2	.4962E+1
1.20	8.20	.759	0.480	0.121	7.721	2.856	16.435	0.3639	0.930	.2317E+2	.4315E-1	.4385E-2	.5981E+1
1.20	10.00	.760	0.480	0.120	9.520	2.857	20.034	0.3648	1.142	.1879E+2	.5321E-1	.4434E-2	.7308E+1
1.20	12.00	.760	0.480	0.120	11.520	2.857	24.033	0.3648	1.382	.1553E+2	.6439E-1	.4472E-2	.8768E+1
1.20	15.00	.761	0.481	0.120	14.520	2.858	30.033	0.3657	1.735	.1232E+2	.8116E-1	.4509E-2	.1098E+2
1.20	18.00	.761	0.481	0.120	17.520	2.858	36.033	0.3657	2.094	.1021E+2	.9793E-1	.4534E-2	.1318E+2
1.20	22.00	.761	0.481	0.120	21.520	2.858	42.033	0.3657	2.572	.8313E+1	.1203E+0	.4556E-2	.1610E+2
1.20	27.00	.761	0.481	0.120	26.519	2.858	54.033	0.3657	3.169	.6746E+1	.1482E+0	.4575E-2	.1976E+2
1.20	33.00	.761	0.481	0.120	32.520	2.858	66.033	0.3657	3.886	.5501E+1	.1818E+0	.4590E-2	.2415E+2
1.50	0.33	.109	0.305	0.446	0.025	2.227	1.420	0.0332	0.011	.1779E+6	.5621E-5	.1135E-4	.4714E-1
1.50	0.39	.193	0.347	0.404	0.044	2.347	1.438	0.0669	0.018	.2989E+5	.3345E-4	.5718E-4	.9618E-1
1.50	0.47	.300	0.400	0.350	0.070	2.500	1.468	0.1200	0.024	.7085E+4	.1411E-3	.2002E-3	.1762E+0
1.50	0.56	.412	0.456	0.294	0.104	2.660	1.512	0.1879	0.031	.2461E+4	.4058E-3	.4831E-3	.2841E+0
1.50	0.68	.532	0.516	0.234	0.164	2.831	1.607	0.2745	0.038	.9779E+3	.1021E-2	.1001E-2	.4410E+0
1.50	0.82	.618	0.559	0.191	0.261	2.954	1.782	0.3455	0.050	.4965E+3	.2014E-2	.1637E-2	.6156E+0
1.50	1.00	.669	0.585	0.166	0.416	3.027	2.080	0.3910	0.069	.2879E+3	.3474E-2	.2316E-2	.8134E+0
1.50	1.20	.694	0.597	0.153	0.603	3.063	2.450	0.4143	0.092	.1934E+3	.5171E-2	.2873E-2	.1015E+1
1.50	1.50	.710	0.605	0.145	0.895	3.086	3.030	0.4296	0.130	.1289E+3	.7760E-2	.3449E-2	.1302E+1
1.50	1.80	.718	0.612	0.141	1.191	3.097	3.621	0.4373	0.168	.9645E+2	.1037E-1	.3840E-2	.1583E+1
1.50	2.20	.724	0.614	0.138	1.588	3.106	4.413	0.4431	0.219	.7218E+2	.1385E-1	.4198E-2	.1955E+1
1.50	2.70	.728	0.614	0.136	2.086	3.111	5.408	0.4470	0.284	.5489E+2	.1822E-1	.4498E-2	.2418E+1
1.50	3.30	.731	0.616	0.135	2.684	3.116	6.605	0.4499	0.361	.4262E+2	.2346E-1	.4746E-2	.2972E+1
1.50	3.90	.733	0.617	0.134	3.284	3.118	7.802	0.4519	0.438	.3484E+2	.2871E-1	.4907E-2	.3526E+1
1.50	4.70	.734	0.617	0.133	4.083	3.120	9.401	0.4529	0.543	.2801E+2	.3570E-1	.5064E-2	.4258E+1
1.50	5.60	.735	0.618	0.132	4.983	3.121	11.200	0.4539	0.660	.2295E+2	.4357E-1	.5187E-2	.5083E+1
1.50	6.80	.736	0.618	0.132	6.182	3.123	13.599	0.4548	0.816	.1850E+2	.5406E-1	.5300E-2	.6185E+1

CHOKE/RESONANT CHARGE REACTOR DESIGN TABLE

CASE # 2   COPYRIGHT 1982 STEVE SMITH   DATA DATE 1/ 3/82

$f(D) = U/(F*G*(D*E)^2)$

P	Q	D	E	F	G	U	Um	DE	FG	f(D)	1/f(D)	1/(PQf(D))	DEUm
1.50	8.20	.737	0.619	0.132	7.582	3.124	16.398	0.4558	0.997	.1508E+2	.6631E-1	.5391E-2	.7475E+1
1.50	10.00	.738	0.619	0.131	9.381	3.126	19.996	0.4568	1.229	.1229E+2	.8205E-1	.5470E-2	.9135E+1
1.50	12.00	.738	0.619	0.131	11.381	3.126	23.996	0.4568	1.491	.1005E+2	.9954E-1	.5530E-2	.1096E+2
1.50	15.00	.739	0.620	0.131	14.380	3.127	29.995	0.4578	1.877	.7950E+1	.1258E+0	.5590E-2	.1373E+2
1.50	18.00	.739	0.620	0.131	17.381	3.127	35.995	0.4578	2.268	.6578E+1	.1520E+0	.5631E-2	.1648E+2
1.50	22.00	.739	0.620	0.131	21.381	3.127	43.995	0.4578	2.790	.5347E+1	.1870E+0	.5667E-2	.2014E+2
1.50	27.00	.740	0.620	0.130	26.380	3.128	53.994	0.4588	3.429	.4334E+1	.2307E+0	.5598E-2	.2477E+2
1.50	33.00	.740	0.620	0.130	32.380	3.128	65.994	0.4588	4.209	.3531E+1	.2832E+0	.5722E-2	.3028E+2
1.80	0.47	.094	0.447	0.453	0.023	2.505	1.654	0.0420	0.010	.1362E+6	.7343E-5	.8679E-5	.6950E-1
1.80	0.56	.215	0.508	0.393	0.053	2.678	1.687	0.1091	0.021	.1092E+5	.9161E-4	.9088E-4	.1841E+1
1.80	0.68	.365	0.583	0.318	0.098	2.892	1.745	0.2126	0.031	.2067E+4	.4838E-3	.3953E-3	.3710E+0
1.80	0.82	.506	0.653	0.247	0.167	3.040	1.854	0.3304	0.041	.6870E+3	.1456E-2	.9861E-3	.6125E+0
1.80	1.00	.610	0.705	0.195	0.295	3.243	2.087	0.4301	0.058	.3048E+3	.3281E-2	.1823E-2	.8977E+0
1.80	1.20	.658	0.729	0.171	0.471	3.311	2.429	0.4797	0.081	.1787E+3	.5597E-2	.2591E-2	.1165E+1
1.80	1.50	.685	0.743	0.158	0.758	3.350	2.996	0.5086	0.119	.1085E+3	.9213E-2	.3412E-2	.1524E+1
1.80	1.80	.697	0.749	0.151	1.052	3.367	3.582	0.5217	0.159	.7765E+2	.1288E-1	.3975E-2	.1869E+1
1.80	2.20	.705	0.753	0.148	1.448	3.378	4.372	0.5305	0.214	.5622E+2	.1779E-1	.4492E-2	.2319E+1
1.80	2.70	.711	0.756	0.145	1.945	3.387	5.365	0.5372	0.281	.4178E+2	.2394E-1	.4925E-2	.2882E+1
1.80	3.30	.714	0.757	0.143	2.543	3.391	6.561	0.5405	0.364	.3192E+2	.3133E-1	.5274E-2	.3546E+1
1.80	3.90	.717	0.759	0.142	3.142	3.396	7.757	0.5434	0.445	.2583E+2	.3872E-1	.5516E-2	.4219E+1
1.80	4.70	.719	0.760	0.141	3.941	3.398	9.355	0.5461	0.554	.2058E+2	.4858E-1	.5742E-2	.5109E+1
1.80	5.60	.720	0.760	0.140	4.840	3.400	11.154	0.5472	0.678	.1676E+2	.5968E-1	.5920E-2	.6103E+1
1.80	6.80	.721	0.761	0.140	6.040	3.401	13.553	0.5483	0.843	.1343E+2	.7447E-1	.6084E-2	.7431E+1
1.80	8.20	.722	0.761	0.139	7.439	3.403	16.351	0.5494	1.034	.1090E+2	.9174E-1	.6215E-2	.8984E+1
1.80	10.00	.723	0.762	0.139	9.239	3.404	19.950	0.5506	1.280	.8777E+1	.1139E+0	.6330E-2	.1098E+2
1.80	12.00	.723	0.762	0.139	11.239	3.404	23.950	0.5506	1.557	.7215E+1	.1386E+0	.6417E-2	.1319E+2
1.80	15.00	.724	0.762	0.138	14.238	3.406	29.949	0.5517	1.965	.5695E+1	.1756E+0	.6504E-2	.1652E+2
1.80	18.00	.724	0.762	0.138	17.238	3.406	35.949	0.5517	2.379	.4704E+1	.2126E+0	.6562E-2	.1983E+2
1.80	22.00	.724	0.762	0.138	21.238	3.406	43.949	0.5517	2.931	.3818E+1	.2619E+0	.6615E-2	.2425E+2
1.80	27.00	.725	0.763	0.137	26.238	3.407	53.948	0.5528	3.608	.3090E+1	.3236E+0	.6658E-2	.2982E+2
1.80	33.00	.725	0.763	0.137	32.238	3.407	65.948	0.5528	4.433	.2515E+1	.3976E+0	.6694E-2	.3646E+2
2.20	0.68	.107	0.654	0.447	0.027	2.924	1.973	0.0699	0.012	.5054E+5	.1979E-4	.1323E-4	.1379E+0
2.20	0.82	.286	0.743	0.357	0.077	3.180	2.035	0.2125	0.027	.2562E+4	.3904E-3	.2164E-3	.4325E+0
2.20	1.00	.475	0.838	0.263	0.162	3.450	2.166	0.3978	0.043	.5110E+3	.1957E-2	.8895E-3	.8615E+0
2.20	1.20	.592	0.896	0.204	0.304	3.617	2.423	0.5304	0.062	.2073E+3	.4824E-2	.1827E-2	.1285E+1
2.20	1.50	.653	0.927	0.174	0.574	3.704	2.949	0.6050	0.100	.1017E+3	.9833E-2	.2980E-2	.1784E+1

CASE # 2   COPYRIGHT 1982 STEVE SMITH   DATA DATE 1/ 3/82

CHOKE/RESONANT CHARGE REACTOR DESIGN TABLE

$f(D)=U/(F*G*(D*E)^2)$

P	Q	D	E	F	G	U	Um	DE	FG	f(D)	1/f(D)	1/(PQf(D))	DEUm
2.20	1.80	.674	0.937	0.163	0.863	3.734	3.524	0.6315	0.141	.665E+2	.1502E-1	.3794E-2	.2225E+1
2.20	2.20	.687	0.944	0.157	1.257	3.753	4.308	0.6482	0.197	.4542E+2	.2202E-1	.4549E-2	.2792E+1
2.20	2.70	.694	0.947	0.153	1.753	3.763	5.300	0.6572	0.268	.324BE+2	.3079E-1	.5183E-2	.3483E+1
2.20	3.30	.699	0.950	0.151	2.351	3.770	6.493	0.6637	0.354	.2419E+2	.4134E-1	.5694E-2	.4310E+1
2.20	3.90	.702	0.951	0.149	2.949	3.774	7.690	0.6676	0.439	.1927E+2	.5189E-1	.6048E-2	.5134E+1
2.20	4.70	.704	0.952	0.148	3.748	3.777	9.287	0.6702	0.555	.1516E+2	.6597E-1	.6380E-2	.6224E+1
2.20	5.60	.706	0.953	0.147	4.647	3.780	11.085	0.6728	0.683	.1222E+2	.8181E-1	.6641E-2	.7458E+1
2.20	6.80	.707	0.954	0.146	5.847	3.781	13.484	0.6741	0.858	.9715E+1	.1029E+0	.6881E-2	.9090E+1
2.20	8.20	.708	0.954	0.146	7.246	3.783	16.283	0.6754	1.058	.7838E+1	.1276E+0	.7073E-2	.1100E+2
2.20	10.00	.709	0.955	0.146	9.046	3.784	19.881	0.6767	1.316	.6278E+1	.1593E+0	.7240E-2	.1345E+2
2.20	12.00	.710	0.955	0.145	11.045	3.786	23.880	0.6781	1.602	.5141E+1	.1945E+0	.7368E-2	.1619E+2
2.20	15.00	.710	0.955	0.145	14.045	3.786	29.880	0.6781	2.037	.4043E+1	.2473E+0	.7495E-2	.2026E+2
2.20	18.00	.711	0.956	0.145	17.045	3.787	35.879	0.6794	2.463	.3331E+1	.3002E+0	.7580E-2	.2437E+2
2.20	22.00	.711	0.956	0.145	21.045	3.787	43.879	0.6794	3.041	.2690E+1	.3706E+0	.7657E-2	.2981E+2
2.20	27.00	.711	0.956	0.145	26.045	3.787	53.879	0.6794	3.763	.2180E+1	.4587E+0	.7722E-2	.3660E+2
2.20	33.00	.711	0.956	0.145	32.045	3.787	65.879	0.6794	4.630	.1772E+1	.5643E+0	.7773E-2	.4476E+2
2.70	1.00	.196	0.948	0.402	0.052	3.551	2.397	0.1858	0.021	.4920E+4	.2032E-3	.7528E-4	.4454E+0
2.70	1.20	.421	1.061	0.290	0.140	3.872	2.524	0.4465	0.040	.4810E+3	.2079E-2	.6416E-3	.1127E+1
2.70	1.50	.595	1.148	0.202	0.352	4.121	2.912	0.6828	0.071	.1239E+3	.8074E-2	.1994E-2	.1989E+1
2.70	1.80	.645	1.173	0.178	0.627	4.193	3.452	0.7563	0.111	.6582E+2	.1519E-1	.3126E-2	.2610E+1
2.70	2.20	.668	1.184	0.166	1.016	4.226	4.224	0.7909	0.169	.4005E+2	.2497E-1	.4203E-2	.3341E+1
2.70	2.70	.679	1.190	0.161	1.511	4.241	5.210	0.8077	0.242	.2682E+2	.3729E-1	.5115E-2	.4208E+1
2.70	3.30	.686	1.193	0.157	2.107	4.251	6.402	0.8184	0.331	.1919E+2	.5212E-1	.5849E-2	.5239E+1
2.70	3.90	.689	1.195	0.156	2.706	4.256	7.598	0.8230	0.421	.1493E+2	.6696E-1	.6359E-2	.6253E+1
2.70	4.70	.692	1.196	0.154	3.504	4.260	9.195	0.8276	0.540	.1152E+2	.8677E-1	.6838E-2	.7610E+1
2.70	5.60	.694	1.197	0.153	4.403	4.263	10.992	0.8307	0.674	.9169E+1	.1091E+0	.7213E-2	.9131E+1
2.70	6.80	.696	1.198	0.152	5.602	4.266	13.390	0.8338	0.852	.7205E+1	.1388E+0	.7559E-2	.1116E+2
2.70	8.20	.697	1.199	0.151	7.001	4.267	16.189	0.8354	1.061	.5765E+1	.1735E+0	.7835E-2	.1352E+2
2.70	10.00	.698	1.199	0.151	8.801	4.268	19.787	0.8369	1.329	.4586E+1	.2181E+0	.8077E-2	.1656E+2
2.70	12.00	.698	1.199	0.151	10.801	4.268	23.787	0.8369	1.631	.3737E+1	.2676E+0	.8260E-2	.1991E+2
2.70	15.00	.699	1.200	0.151	13.801	4.270	29.786	0.8385	2.077	.2924E+1	.3420E+0	.8444E-2	.2497E+2
2.70	18.00	.700	1.200	0.150	16.800	4.270	35.785	0.8400	2.520	.2402E+1	.4163E+0	.8566E-2	.3006E+2
2.70	22.00	.700	1.200	0.150	20.800	4.271	43.785	0.8400	3.120	.1940E+1	.5154E+0	.8677E-2	.3678E+2
2.70	27.00	.700	1.200	0.150	25.800	4.271	53.785	0.8400	3.870	.1564E+1	.6393E+0	.8778E-2	.4518E+2
2.70	33.00	.700	1.200	0.150	31.800	4.271	65.785	0.8400	4.708	.1269E+1	.7880E+0	.8844E-2	.5526E+2
3.30	1.20	.066	1.183	0.467	0.017	3.965	2.826	0.0781	0.008	.8193E+5	.1221E-4	.3082E-5	.2207E+0

CASE # 2   COPYRIGHT 1982 STEVE SMITH   DATA DATE 1/ 3/82   CHOKE/RESONANT CHARGE REACTOR DESIGN TABLE

$$f(D) = U/(F*G*(D*E)^2)$$

P	Q	D	E	F	G	DE	FG	Um	U	f(D)	1/f(D)	1/(PQf(D))	DEUm
3.30	1.50	.416	1.358	0.292	0.142	0.5649	0.041	3.001	4.465	.3374E+3	.2963E-2	.5987E-3	.1695E+1
3.30	1.80	.585	1.443	0.208	0.358	0.8439	0.074	3.396	4.767	.8910E+2	.1112E-1	.1889E-2	.2866E+1
3.30	2.20	.643	1.472	0.179	0.729	0.9462	0.130	4.125	4.790	.4114E+2	.2430E-1	.3348E-2	.3903E+1
3.30	2.70	.664	1.482	0.168	1.218	0.9840	0.205	5.100	4.820	.2432E+2	.4111E-1	.4614E-2	.5019E+1
3.30	3.30	.674	1.487	0.163	1.813	1.0022	0.296	6.288	4.834	.1628E+2	.6141E-1	.5639E-2	.6302E+1
3.30	3.90	.679	1.490	0.161	2.411	1.0114	0.387	7.482	4.841	.1223E+2	.8174E-1	.6351E-2	.7567E+1
3.30	4.70	.683	1.492	0.159	3.209	1.0187	0.509	9.077	4.847	.9184E+1	.1089E+0	.7020E-2	.9247E+1
3.30	5.60	.685	1.493	0.158	4.108	1.0224	0.647	10.874	4.850	.7172E+1	.1394E+0	.7545E-2	.1112E+2
3.30	6.80	.687	1.494	0.157	5.306	1.0260	0.830	13.272	4.853	.5551E+1	.1802E+0	.8029E-2	.1362E+2
3.30	8.20	.688	1.494	0.156	6.706	1.0279	1.046	16.071	4.854	.4392E+1	.2277E+0	.8415E-2	.1652E+2
3.30	10.00	.689	1.495	0.155	8.506	1.0297	1.323	19.670	4.856	.3462E+1	.2888E+0	.8752E-2	.2025E+2
3.30	12.00	.690	1.495	0.155	10.505	1.0316	1.628	23.668	4.857	.2803E+1	.3567E+0	.9008E-2	.2442E+2
3.30	15.00	.691	1.496	0.155	13.505	1.0334	2.086	29.667	4.858	.2180E+1	.4586E+0	.9265E-2	.3066E+2
3.30	18.00	.691	1.496	0.155	16.505	1.0334	2.550	35.667	4.858	.1784E+1	.5605E+0	.9436E-2	.3688E+2
3.30	22.00	.692	1.496	0.154	20.504	1.0352	3.158	43.666	4.860	.1436E+1	.6963E+0	.9591E-2	.4520E+2
3.30	27.00	.692	1.496	0.154	25.504	1.0352	3.928	53.666	4.860	.1155E+1	.8661E+0	.9721E-2	.5555E+2
3.30	33.00	.692	1.496	0.154	31.504	1.0352	4.852	65.666	4.860	.9347E+0	.1070E+1	.9825E-2	.6798E+2
3.90	1.50	.066	1.483	0.467	0.017	0.0979	0.008	3.297	4.565	.6002E+5	.1666E-4	.2848E-5	.3228E+0
3.90	1.80	.412	1.656	0.294	0.144	0.6823	0.042	3.477	5.060	.2567E+3	.3895E-2	.5548E-3	.2372E+1
3.90	2.20	.602	1.751	0.199	0.449	1.0541	0.089	4.046	5.397	.5377E+2	.1862E-1	.2170E-2	.4265E+1
3.90	2.70	.648	1.774	0.176	0.926	1.1496	0.163	4.991	5.420	.2506E+2	.3991E-1	.3798E-2	.5737E+1
3.90	3.30	.664	1.782	0.168	1.518	1.1832	0.255	6.171	5.437	.1518E+2	.6588E-1	.5119E-2	.7302E+1
3.90	3.90	.671	1.786	0.165	2.115	1.1981	0.348	7.363	5.441	.1088E+2	.9195E-1	.6045E-2	.8821E+1
3.90	4.70	.676	1.788	0.162	2.912	1.2087	0.472	8.957	5.444	.7889E+1	.1268E+0	.6915E-2	.1083E+2
3.90	5.60	.679	1.790	0.161	3.811	1.2151	0.612	10.753	5.446	.6026E+1	.1659E+0	.7598E-2	.1307E+2
3.90	6.80	.681	1.791	0.160	5.010	1.2193	0.799	13.151	5.448	.4583E+1	.2182E+0	.8228E-2	.1603E+2
3.90	8.20	.682	1.791	0.159	6.409	1.2215	1.019	15.949	5.448	.3582E+1	.2792E+0	.8730E-2	.1948E+2
3.90	10.00	.684	1.792	0.158	8.208	1.2257	1.297	19.547	5.450	.2796E+1	.3576E+0	.9170E-2	.2396E+2
3.90	12.00	.684	1.792	0.158	10.208	1.2257	1.613	23.546	5.451	.2248E+1	.4448E+0	.9503E-2	.2866E+2
3.90	15.00	.685	1.793	0.157	13.208	1.2279	2.080	29.546	5.451	.1738E+1	.5755E+0	.9837E-2	.3628E+2
3.90	18.00	.686	1.793	0.157	16.207	1.2300	2.544	35.544	5.451	.1416E+1	.7062E+0	.1006E-1	.4372E+2
3.90	22.00	.686	1.793	0.157	20.207	1.2300	3.172	43.544	5.451	.1136E+1	.8805E+0	.1026E-1	.5356E+2
3.90	27.00	.686	1.793	0.157	25.207	1.2300	3.957	53.544	5.451	.9105E+0	.1098E+1	.1043E-1	.6586E+2
3.90	33.00	.687	1.794	0.157	31.207	1.2321	4.884	65.543	5.453	.7354E+0	.1360E+1	.1057E-1	.8076E+2
4.70	2.20	.409	2.054	0.296	0.146	0.8403	0.043	4.109	5.855	.1929E+3	.5185E-2	.5014E-3	.3453E+1
4.70	2.70	.612	2.156	0.194	0.544	1.3195	0.106	4.863	6.145	.3345E+2	.2990E-1	.2356E-2	.6416E+1

CASE # 2   DATA DATE 1/ 3/82

CHOKE/RESONANT CHARGE REACTOR DESIGN TABLE

$$f(D)=U/(F*G*(D*E)^2)$$

P	Q	D	E	F	G	U	Um	DE	FG	f(D)	1/f(D)	1/(PQf(D))	DEUm
4.70	3.30	.650	2.175	0.175	1.125	6.200	6.016	1.4138	0.197	.1576E+2	.6347E-1	.4092E-2	.8506E+1
4.70	3.90	.662	2.181	0.169	1.719	6.202	7.202	1.4438	0.291	.1027E+2	.9741E-1	.5314E-2	.1040E+2
4.70	4.70	.669	2.185	0.166	2.516	6.227	8.793	1.4614	0.416	.7003E+1	.1428E+0	.6464E-2	.1285E+2
4.70	5.60	.673	2.186	0.164	3.414	6.233	10.589	1.4715	0.558	.5157E+1	.1939E+0	.7367E-2	.1558E+2
4.70	6.80	.675	2.188	0.163	4.613	6.236	12.986	1.4766	0.750	.3816E+1	.2621E+0	.8200E-2	.1918E+2
4.70	8.20	.677	2.189	0.162	6.011	6.238	15.784	1.4816	0.971	.2927E+1	.3416E+0	.8864E-2	.2339E+2
4.70	10.00	.679	2.189	0.161	7.811	6.241	19.381	1.4867	1.254	.2253E+1	.4439E+0	.9445E-2	.2882E+2
4.70	12.00	.679	2.189	0.161	9.811	6.241	23.381	1.4867	1.575	.1793E+1	.5576E+0	.9887E-2	.3476E+2
4.70	15.00	.680	2.190	0.160	12.810	6.243	29.380	1.4892	2.050	.1373E+1	.7281E+0	.1033E-1	.4375E+2
4.70	18.00	.681	2.191	0.160	15.810	6.244	35.379	1.4917	2.522	.1113E+1	.8987E+0	.1062E-1	.5278E+2
4.70	22.00	.681	2.191	0.160	19.810	6.244	43.379	1.4917	3.160	.8881E+0	.1126E+1	.1089E-1	.6471E+2
4.70	27.00	.682	2.191	0.159	24.809	6.246	53.378	1.4943	3.945	.7091E+0	.1410E+1	.1111E-1	.7976E+2
4.70	33.00	.682	2.191	0.159	30.809	6.246	65.378	1.4943	4.899	.5710E+0	.1751E+1	.1129E-1	.9769E+2
5.60	2.70	.448	2.524	0.277	0.176	6.811	4.869	1.1308	0.049	.1097E+3	.9119E-2	.6031E-3	.5506E+1
5.60	3.30	.624	2.612	0.190	0.688	7.063	5.855	1.6299	0.131	.2055E+2	.4865E-1	.2633E-2	.9543E+1
5.60	3.90	.651	2.626	0.174	1.276	7.101	7.022	1.7092	0.222	.1093E+2	.9149E-1	.4189E-2	.1200E+2
5.60	4.70	.662	2.631	0.167	2.096	7.117	8.603	1.7417	0.350	.6709E+1	.1490E+0	.5663E-2	.1498E+2
5.60	5.60	.667	2.634	0.165	2.997	7.124	10.403	1.7565	0.494	.4673E+1	.2139E+0	.6821E-2	.1827E+2
5.60	6.80	.671	2.636	0.163	4.205	7.130	12.798	1.7684	0.685	.3328E+1	.3005E+0	.7891E-2	.2263E+2
5.60	8.20	.673	2.637	0.162	5.617	7.133	15.595	1.7744	0.910	.2491E+1	.4015E+0	.8744E-2	.2767E+2
5.60	10.00	.675	2.638	0.162	7.383	7.136	19.192	1.7803	1.196	.1882E+1	.5314E+0	.9490E-2	.3417E+2
5.60	12.00	.676	2.638	0.161	9.422	7.137	23.192	1.7833	1.517	.1480E+1	.6758E+0	.1006E-1	.4136E+2
5.60	15.00	.677	2.639	0.161	12.399	7.138	29.191	1.7863	1.996	.1121E+1	.8924E+0	.1062E-1	.5215E+2
5.60	18.00	.677	2.639	0.161	15.410	7.138	35.191	1.7863	2.481	.9018E+0	.1109E+1	.1100E-1	.6287E+2
5.60	22.00	.678	2.639	0.161	19.360	7.140	43.189	1.7892	3.117	.7155E+0	.1398E+1	.1134E-1	.7728E+2
5.60	27.00	.678	2.639	0.161	24.360	7.140	53.189	1.7892	3.922	.5686E+0	.1759E+1	.1163E-1	.9517E+2
5.60	33.00	.678	2.639	0.161	30.360	7.140	65.189	1.7892	4.888	.4563E+0	.2192E+1	.1186E-1	.1166E+3
6.80	3.30	.446	3.123	0.277	0.177	8.008	5.814	1.3929	0.049	.8419E+2	.1188E-1	.5293E-3	.8098E+1
6.80	3.90	.621	3.211	0.190	0.690	8.258	6.801	1.9937	0.131	.1590E+2	.6289E-1	.2371E-2	.1356E+2
6.80	4.70	.651	3.226	0.174	1.474	8.301	8.365	2.0998	0.257	.7317E+1	.1367E+0	.4276E-2	.1756E+2
6.80	5.60	.661	3.231	0.169	2.370	8.316	10.152	2.1354	0.402	.4541E+1	.2202E+0	.5783E-2	.2168E+2
6.80	6.80	.666	3.233	0.167	3.567	8.323	12.546	2.1532	0.596	.3014E+1	.3318E+0	.7176E-2	.2701E+2
6.80	8.20	.669	3.235	0.166	4.966	8.327	15.343	2.1639	0.822	.2164E+1	.4621E+0	.8287E-2	.3320E+2
6.80	10.00	.671	3.236	0.165	6.765	8.330	18.940	2.1710	1.113	.1588E+1	.6296E+0	.9259E-2	.4112E+2
6.80	12.00	.673	3.237	0.164	8.764	8.333	22.938	2.1782	1.433	.1226E+1	.8158E+0	.9998E-2	.4996E+2
6.80	15.00	.674	3.237	0.163	11.763	8.334	28.937	2.1817	1.917	.9132E+0	.1095E+1	.1074E-1	.6313E+2

CASE # 2   COPYRIGHT 1982 STEVE SMITH   DATA DATE 1/ 3/82   CHOKE/RESONANT CHARGE REACTOR DESIGN TABLE

$$f(D)=U/(F*G*(D*E)^2)$$

P	Q	D	E	F	G	U	Um	DE	FG	f(D)	1/f(D)	1/(PQf(D))	DEUm
6.80	18.00	.674	3.237	0.163	14.763	8.334	34.937	2.1817	2.406	.7276E+0	.1374E+1	.1123E-1	.7622E+2
6.80	22.00	.675	3.238	0.163	18.763	8.336	42.935	2.1853	3.049	.5725E+0	.1747E+1	.1168E-1	.9383E+2
6.80	27.00	.675	3.238	0.163	23.763	8.336	52.935	2.1853	3.861	.4520E+0	.2212E+1	.1205E-1	.1157E+3
8.20	33.00	.675	3.238	0.163	29.763	8.336	64.935	2.1853	4.836	.3609E+0	.2771E+1	.1235E-1	.1419E+3
8.20	3.90	.358	3.779	0.321	0.121	6.820	6.820	1.3529	0.039	.1306E+3	.7659E-2	.2395E-3	.9227E+1
8.20	4.70	.625	3.913	0.188	0.787	9.664	8.096	2.4453	0.148	.1095E+2	.9136E-1	.2371E-1	.1980E+2
8.20	5.60	.652	3.926	0.174	1.674	9.703	9.863	2.5598	0.291	.5084E+1	.1967E+0	.4284E-2	.2525E+2
8.20	6.80	.662	3.931	0.169	2.869	9.717	12.251	2.6023	0.485	.2959E+1	.3379E+0	.6060E-2	.3188E+2
8.20	8.20	.666	3.933	0.167	4.267	9.723	15.046	2.6194	0.713	.1989E+1	.5029E+0	.7479E-2	.3941E+2
8.20	10.00	.668	3.934	0.166	6.066	9.726	18.644	2.6279	1.007	.1399E+1	.7159E+0	.8720E-2	.4999E+2
8.20	12.00	.670	3.935	0.165	8.065	9.728	22.641	2.6365	1.331	.1052E+1	.9508E+0	.9663E-2	.5969E+2
8.20	15.00	.671	3.936	0.165	11.065	9.730	28.640	2.6407	1.820	.7666E+0	.1304E+1	.1061E-1	.7563E+2
8.20	18.00	.672	3.936	0.164	14.064	9.731	34.639	2.6450	2.306	.6031E+0	.1658E+1	.1123E-1	.9162E+2
8.20	22.00	.672	3.936	0.164	18.064	9.732	42.637	2.6450	2.962	.4695E+0	.2130E+1	.1181E-1	.1128E+3
8.20	27.00	.673	3.937	0.164	23.063	9.733	52.637	2.6493	3.771	.3677E+0	.2719E+1	.1228E-1	.1395E+3
8.20	33.00	.673	3.937	0.164	29.064	9.733	64.637	2.6493	4.752	.2918E+0	.3427E+1	.1266E-1	.1712E+3
10.00	4.70	.251	4.626	0.375	0.074	8.164	8.164	1.1610	0.028	.2906E+3	.3441E-2	.7321E-4	.9478E+1
10.00	5.60	.623	4.812	0.188	0.788	11.461	9.512	2.9976	0.149	.8582E+1	.1165E+0	.2081E-2	.2851E+2
10.00	6.80	.654	4.827	0.173	1.973	11.505	11.874	3.1569	0.341	.3382E+1	.2957E+0	.4348E-2	.3749E+2
10.00	8.20	.662	4.831	0.169	3.369	11.517	14.665	3.1981	0.569	.1978E+1	.5056E+0	.6166E-2	.4690E+2
10.00	10.00	.666	4.833	0.167	5.167	11.523	18.260	3.2188	0.863	.1289E+1	.7759E+0	.7759E-1	.5877E+2
10.00	12.00	.668	4.834	0.166	7.166	11.526	22.257	3.2291	1.190	.9292E+0	.1076E+1	.8968E-2	.7187E+2
10.00	15.00	.669	4.835	0.166	10.166	11.527	28.256	3.2343	1.682	.6555E+0	.1527E+1	.1018E-1	.9139E+2
10.00	18.00	.670	4.835	0.165	13.165	11.528	34.255	3.2395	2.172	.5057E+0	.1977E+1	.1099E-1	.1110E+3
10.00	22.00	.671	4.836	0.165	17.165	11.530	42.254	3.2446	2.824	.3879E+0	.2578E+1	.1172E-1	.1371E+3
10.00	27.00	.671	4.836	0.165	22.165	11.530	52.254	3.2446	3.646	.3004E+0	.3329E+1	.1233E-1	.1695E+3
10.00	33.00	.671	4.836	0.165	28.165	11.530	64.254	3.2446	4.633	.2364E+0	.4230E+1	.1282E-1	.2085E+3

CASE # 3 COPYRIGHT 1982 STEVE SMITH DATA DATE 1/ 3/82 CHOKE/RESONANT CHARGE REACTOR DESIGN TABLE

SCALE BY $1.000E-04$ FUNCTION $1/f(D)$

$$f(D)=U/(F*G*(D*E)^2)$$

Q/P	0.33	0.39	0.47	0.56	0.68	0.82	1.00	1.20	1.50	1.80	2.20	2.70	3.30	3.90	4.70	5.60	6.80	8.20	10.00
0.33	0	0	0	0	0	0													
0.39	0	0	0	1	1	1	0												
0.47	0	0	0	2	2	2	2												
0.56	0	1	1	3	3	4	4				1								
0.68	1	1	2	4	5	6	7		6		3								
0.82	1	2	3	5	7	10	10	13	11	8	9								
1.00	2	3	4	7	11	15	18	21	21	17	19								
1.20	2	4	6	10	14	20	27	34	36	31	47	4							
1.50	3	5	9	13	20	29	42	54	65	64	93	18	0						
1.80	4	7	11	16	26	38	56	76	98	106	176	51	8						
2.20	5	8	14	21	33	51	76	98	146	176	262	130	47	3					
2.70	7	10	17	27	43	66	101	144	208	262	372	279	170	50					
3.30	8	13	21	34	54	84	131	190	284	372	461	499	420	240	32				
3.90	10	16	26	40	66	103	161	237	360	484	627	736	733	573	227	6			
4.70	12	19	31	50	81	128	201	298	463	635	850	1061	1182	1129	790	245			
5.60	15	23	38	60	98	155	247	378	578	805	1104	1431	1702	1801	1629	1043	142		
6.80	18	28	47	74	121	192	307	469	732	1031	1442	1927	2403	2717	2832	2511	1392	85	
8.20	22	34	57	90	148	236	378	570	912	1296	1838	2507	3225	3795	4264	4340	3653	1807	5
10.00	26	42	70	111	183	291	469	710	1143	1638	2348	3254	4284	5185	6116	6727	6749	5580	2350
12.00	32	51	84	134	221	353	570	866	1401	2017	2914	4084	5462	6733	8180	9391	10230	10009	7771
15.00	40	63	105	168	278	446	721	1099	1786	2585	3764	5331	7230	9057	11280	13394	15469	16711	16350
18.00	48	76	127	203	336	539	872	1332	2172	3154	4615	6578	8999	11383	14382	17402	21016	23434	24988
22.00	59	94	156	249	412	662	1074	1643	2687	3913	5749	8241	11358	14484	18519	22747	27776	32403	36524
27.00	73	115	192	306	508	817	1326	2032	3330	4861	7166	10320	14308	18361	23692	29430	36468	43619	50954
33.00	89	141	235	375	623	1002	1629	2499	4102	5999	8867	12815	17847	23014	29900	37451	46973	57082	68275

CASE # 3   COPYRIGHT 1982 STEVE SMITH   DATA DATE 1/ 3/82   CHOKE/RESONANT CHARGE REACTOR DESIGN TABLE

SCALE BY 1.00E-04   FUNCTION 1/[P*Q*f(D)]

$$f(D)=U/(F*G*(D*E)^2)$$

Q/P	0.33	0.39	0.47	0.56	0.68	0.82	1.00	1.20	1.50	1.80	2.20	2.70	3.30	3.90	4.70	5.60	6.80	8.20	10.00
0.33	3	4	4	4	4	5	3	2	1	0	0	.	.	.	.	.	.	.	.
0.39	4	4	5	5	5	6	4	3	2	0	0	.	.	.	.	.	.	.	.
0.47	5	5	6	7	7	9	6	4	3	1	0	.	.	.	.	.	.	.	.
0.56	5	5	7	8	9	11	8	6	4	2	0	.	.	.	.	.	.	.	.
0.68	6	6	9	10	11	14	11	9	6	3	2	.	.	.	.	.	.	.	.
0.82	6	7	10	11	14	18	14	13	9	5	4	.	.	.	.	.	.	.	.
1.00	6	8	11	13	16	21	19	18	14	9	7	0	.	.	.	.	.	.	.
1.20	7	8	11	14	18	24	23	23	20	14	14	0	.	.	.	.	.	.	.
1.50	7	9	12	15	20	26	28	30	29	24	23	1	0	.	.	.	.	.	.
1.80	7	9	13	16	21	28	31	35	36	33	36	10	1	.	.	.	.	.	.
2.20	8	9	13	17	22	30	35	40	44	44	50	22	7	0	.	.	.	.	.
2.70	8	10	13	18	23	31	37	44	51	54	63	38	19	5	.	.	.	.	.
3.30	8	10	14	18	24	32	40	48	57	63	73	56	39	19	2	.	.	.	.
3.90	8	10	14	19	25	33	41	51	62	69	82	70	57	38	12	.	.	.	.
4.70	8	10	14	19	25	34	43	53	66	75	90	84	76	62	36	9	.	.	.
5.60	8	10	15	19	26	35	44	55	69	80	96	95	92	82	62	33	.	.	.
6.80	8	11	15	20	27	36	45	57	72	84	102	105	107	102	89	66	4	2	.
8.20	8	11	15	20	27	36	46	58	74	88	107	113	119	119	111	95	30	27	0
10.00	8	11	15	20	27	36	47	59	76	91	110	121	130	133	130	120	66	68	24
12.00	8	11	15	20	27	37	48	60	78	93	114	126	138	144	145	145	99	102	65
15.00	8	11	15	20	28	37	48	61	79	96	117	132	146	155	160	159	125	136	109
18.00	8	11	15	20	28	37	49	62	80	97	119	135	151	162	170	173	152	159	139
22.00	8	11	15	20	28	37	49	62	81	99	120	139	156	169	179	185	169	180	166
27.00	8	11	15	20	28	37	49	63	82	100	121	142	161	174	187	187	185	197	189
33.00	8	11	15	20	28	37	49	63	83	101	122	144	164	179	193	203	209	211	207

CASE # 3   COPYRIGHT 1982 STEVE SMITH   DATA DATE 1/ 3/82

CHOKE/RESONANT CHARGE REACTOR DESIGN TABLE

$f(D) = U/(F*G*(D*E)^2)$

P	Q	D	E	F	G	U	Um	DE	FG	f(D)	1/f(D)	1/(PQf(D))	DEUm
0.33	0.33	.919	0.084	0.081	0.162	2.133	0.750	0.0772	0.013	.2728E+5	.3666E-4	.3366E-3	.5789E-1
0.33	0.39	.926	0.091	0.074	0.208	2.150	0.850	0.0843	0.015	.1967E+5	.5083E-4	.3949E-3	.7162E-1
0.33	0.47	.931	0.096	0.069	0.278	2.162	0.996	0.0894	0.019	.1411E+5	.7086E-4	.4569E-3	.8898E-1
0.33	0.56	.935	0.100	0.065	0.360	2.172	1.164	0.0935	0.023	.1062E+5	.9418E-4	.5096E-3	.1088E+0
0.33	0.68	.938	0.103	0.062	0.474	2.179	1.396	0.0966	0.029	.7945E+4	.1259E-3	.5609E-3	.1348E+0
0.33	0.82	.940	0.105	0.060	0.610	2.184	1.670	0.0987	0.037	.6126E+4	.1632E-3	.6032E-3	.1648E+0
0.33	1.00	.942	0.107	0.058	0.786	2.189	2.024	0.1008	0.046	.4727E+4	.2116E-3	.6411E-3	.2040E+0
0.33	1.20	.943	0.108	0.057	0.984	2.192	2.421	0.1018	0.056	.3767E+4	.2655E-3	.6703E-3	.2466E+0
0.33	1.50	.944	0.109	0.056	1.282	2.194	3.018	0.1029	0.072	.2886E+4	.3465E-3	.6999E-3	.3106E+0
0.33	1.80	.944	0.109	0.056	1.582	2.194	3.618	0.1029	0.089	.2339E+4	.4275E-3	.7197E-3	.3723E+0
0.33	2.10	.945	0.110	0.055	1.980	2.196	4.416	0.1039	0.109	.1867E+4	.5358E-3	.7380E-3	.4590E+0
0.33	2.70	.945	0.110	0.055	2.480	2.196	5.416	0.1039	0.136	.1490E+4	.6710E-3	.7531E-3	.5629E+0
0.33	3.30	.946	0.111	0.054	3.078	2.199	6.613	0.1050	0.166	.1200E+4	.8335E-3	.7654E-3	.6944E+0
0.33	3.90	.946	0.111	0.054	3.678	2.199	7.813	0.1050	0.199	.1004E+4	.9960E-3	.7739E-3	.8204E+0
0.33	4.70	.946	0.111	0.054	4.478	2.199	9.413	0.1050	0.242	.8247E+3	.1213E-2	.7818E-3	.9884E+0
0.33	5.60	.946	0.111	0.054	5.378	2.199	11.213	0.1050	0.290	.6867E+3	.1456E-2	.7880E-3	.1177E+1
0.33	6.80	.946	0.111	0.054	6.578	2.199	13.613	0.1050	0.355	.5614E+3	.1781E-2	.7938E-3	.1429E+1
0.33	8.20	.946	0.111	0.054	7.978	2.199	16.413	0.1050	0.431	.4629E+3	.2160E-2	.7984E-3	.1723E+1
0.33	10.00	.947	0.112	0.053	9.776	2.201	20.010	0.1061	0.518	.3777E+3	.2648E-2	.8024E-3	.2122E+1
0.33	12.00	.947	0.112	0.053	11.776	2.201	24.010	0.1061	0.624	.3135E+3	.3190E-2	.8055E-3	.2547E+1
0.33	15.00	.947	0.112	0.053	14.776	2.201	30.010	0.1061	0.783	.2499E+3	.4002E-2	.8085E-3	.3183E+1
0.33	18.00	.947	0.112	0.053	17.776	2.201	36.010	0.1061	0.942	.2077E+3	.4815E-2	.8106E-3	.3819E+1
0.33	22.00	.947	0.112	0.053	21.776	2.201	44.010	0.1061	1.154	.1695E+3	.5898E-2	.8124E-3	.4668E+1
0.33	27.00	.947	0.112	0.053	26.776	2.201	54.010	0.1061	1.419	.1379E+3	.7252E-2	.8140E-3	.5729E+1
0.33	33.00	.947	0.112	0.053	32.776	2.201	66.010	0.1061	1.737	.1126E+3	.8878E-2	.8152E-3	.7001E+1
0.39	0.33	.896	0.091	0.104	0.148	2.137	0.790	0.0815	0.015	.2089E+5	.4788E-4	.3720E-3	.6440E-1
0.39	0.39	.905	0.100	0.095	0.190	2.159	0.884	0.0905	0.018	.1461E+5	.6847E-4	.4501E-3	.8002E-1
0.39	0.47	.914	0.109	0.086	0.252	2.181	1.018	0.0996	0.022	.1014E+5	.9862E-4	.5380E-3	.1015E+0
0.39	0.56	.920	0.115	0.080	0.330	2.196	1.181	0.1058	0.026	.7430E+4	.1346E-3	.6162E-3	.1250E+0
0.39	0.68	.925	0.120	0.075	0.440	2.208	1.407	0.1110	0.033	.5430E+4	.1842E-3	.6944E-3	.1562E+0
0.39	0.82	.928	0.123	0.072	0.574	2.215	1.678	0.1141	0.041	.4114E+4	.2431E-3	.7601E-3	.1916E+0
0.39	1.00	.930	0.125	0.070	0.750	2.220	2.033	0.1163	0.053	.3129E+4	.3196E-3	.8195E-3	.2363E+0
0.39	1.20	.932	0.127	0.068	0.946	2.225	2.427	0.1184	0.064	.2469E+4	.4051E-3	.8656E-3	.2873E+0
0.39	1.50	.933	0.128	0.067	1.244	2.227	3.024	0.1194	0.083	.1874E+4	.5337E-3	.9123E-3	.3612E+0
0.39	1.80	.933	0.129	0.066	1.542	2.230	3.621	0.1204	0.102	.1509E+4	.6626E-3	.9439E-3	.4363E+0
0.39	2.20	.935	0.130	0.065	1.940	2.232	4.418	0.1216	0.126	.1198E+4	.8347E-3	.9728E-3	.5371E+0

CASE # 3 COPYRIGHT 1982 STEVE SMITH DATA DATE 1/ 3/82 CHOKE/RESONANT CHARGE REACTOR DESIGN TABLE

$$f(D)=U/(F*G*(D*E)^2)$$

P	Q	D	E	F	G	U	Um	DE	FG	f(D)	1/f(D)	1/(PQf(D))	DEUm
0.39	2.70	.936	0.131	0.064	2.438	2.235	5.416	0.1226	0.156	.9525E+3	.1050E-2	.9970E-3	.6640E+0
0.39	3.30	.936	0.131	0.064	3.038	2.235	6.616	0.1226	0.194	.7644E+3	.1308E-2	.1016E-2	.8112E+0
0.39	3.90	.936	0.131	0.064	3.638	2.235	7.816	0.1226	0.233	.6383E+3	.1567E-2	.1030E-2	.9583E+0
0.39	4.70	.937	0.132	0.063	4.436	2.237	9.413	0.1237	0.279	.5232E+3	.1911E-2	.1043E-2	.1164E+1
0.39	5.60	.937	0.132	0.063	5.336	2.237	11.213	0.1237	0.336	.4350E+3	.2299E-2	.1053E-2	.1387E+1
0.39	6.80	.937	0.132	0.063	6.536	2.237	13.613	0.1237	0.412	.3551E+3	.2816E-2	.1062E-2	.1684E+1
0.39	8.20	.937	0.132	0.063	7.936	2.237	16.413	0.1237	0.500	.2925E+3	.3419E-2	.1069E-2	.2030E+1
0.39	10.00	.937	0.132	0.063	9.736	2.237	20.013	0.1237	0.613	.2384E+3	.4195E-2	.1076E-2	.2475E+1
0.39	12.00	.937	0.132	0.063	11.736	2.237	24.013	0.1237	0.739	.1978E+3	.5056E-2	.1080E-2	.2970E+1
0.39	15.00	.937	0.132	0.063	14.736	2.237	30.013	0.1237	0.928	.1575E+3	.6349E-2	.1085E-2	.3712E+1
0.39	18.00	.937	0.132	0.063	17.736	2.237	36.013	0.1237	1.117	.1309E+3	.7641E-2	.1088E-2	.4454E+1
0.39	22.00	.938	0.133	0.062	21.734	2.239	44.010	0.1248	1.348	.1068E+3	.9365E-2	.1091E-2	.5490E+1
0.39	27.00	.938	0.133	0.062	26.734	2.239	54.010	0.1248	1.658	.8681E+2	.1152E-1	.1094E-2	.6738E+1
0.39	33.00	.938	0.133	0.062	32.734	2.239	66.010	0.1248	2.030	.7090E+2	.1411E-1	.1096E-2	.8235E+1
0.47	0.33	.862	0.097	0.138	0.136	2.135	0.853	0.0836	0.019	.1627E+5	.6146E-4	.3963E-3	.7130E-1
0.47	0.39	.874	0.109	0.126	0.172	2.164	0.938	0.0953	0.022	.1140E+5	.9089E-4	.4959E-3	.8940E-1
0.47	0.47	.886	0.121	0.114	0.228	2.193	1.064	0.1072	0.026	.7341E+4	.1362E-3	.6166E-3	.1141E+0
0.47	0.56	.896	0.131	0.104	0.298	2.217	1.216	0.1174	0.031	.5193E+4	.1926E-3	.7316E-3	.1427E+0
0.47	0.68	.905	0.140	0.095	0.400	2.239	1.430	0.1267	0.038	.3671E+4	.2724E-3	.8524E-3	.1812E+0
0.47	0.82	.909	0.145	0.090	0.530	2.251	1.696	0.1320	0.048	.2711E+4	.3689E-3	.9571E-3	.2237E+0
0.47	1.00	.914	0.149	0.086	0.702	2.261	2.044	0.1362	0.060	.2019E+4	.4952E-3	.1054E-2	.2784E+0
0.47	1.20	.917	0.152	0.083	0.896	2.268	2.436	0.1394	0.074	.1570E+4	.6369E-3	.1129E-2	.3395E+0
0.47	1.50	.919	0.154	0.081	1.192	2.273	3.024	0.1415	0.097	.1175E+4	.8507E-3	.1207E-2	.4288E+0
0.47	1.80	.921	0.156	0.079	1.488	2.278	3.624	0.1437	0.118	.9388E+3	.1065E-2	.1259E-2	.5207E+0
0.47	2.20	.922	0.157	0.078	1.886	2.281	4.421	0.1448	0.147	.7398E+3	.1352E-2	.1307E-2	.6400E+0
0.47	2.70	.923	0.158	0.077	2.384	2.283	5.418	0.1458	0.184	.5848E+3	.1710E-2	.1348E-2	.7902E+0
0.47	3.30	.923	0.158	0.077	2.984	2.283	6.618	0.1458	0.230	.4672E+3	.2140E-2	.1380E-2	.9652E+0
0.47	3.90	.924	0.159	0.076	3.582	2.285	7.816	0.1469	0.272	.3889E+3	.2571E-2	.1403E-2	.1148E+1
0.47	4.70	.924	0.159	0.076	4.382	2.285	9.416	0.1469	0.333	.3179E+3	.3145E-2	.1424E-2	.1383E+1
0.47	5.60	.925	0.160	0.075	5.280	2.288	11.213	0.1480	0.396	.2638E+3	.3791E-2	.1440E-2	.1659E+1
0.47	6.80	.925	0.160	0.075	6.480	2.288	13.613	0.1480	0.486	.2149E+3	.4653E-2	.1456E-2	.2015E+1
0.47	8.20	.925	0.160	0.075	7.680	2.288	16.413	0.1480	0.591	.1767E+3	.5658E-2	.1468E-2	.2429E+1
0.47	10.00	.925	0.160	0.075	9.680	2.288	20.013	0.1480	0.726	.1439E+3	.6951E-2	.1479E-2	.2962E+1
0.47	12.00	.925	0.160	0.075	11.680	2.288	24.013	0.1480	0.876	.1192E+3	.8387E-2	.1487E-2	.3554E+1
0.47	15.00	.925	0.160	0.075	14.680	2.288	30.013	0.1480	1.101	.9487E+2	.1054E-1	.1495E-2	.4442E+1
0.47	18.00	.926	0.161	0.074	17.678	2.290	36.010	0.1491	1.308	.7877E+2	.1270E-1	.1501E-2	.5369E+1

CASE # 3   DATA DATE 1/ 3/82   CHOKE/RESONANT CHARGE REACTOR DESIGN TABLE

$$f(D) = U/(F \cdot G \cdot (D \cdot E)^2)$$

P	Q	D	E	F	G	U	Um	DE	FG	f(D)	1/f(D)	1/(PQf(D))	DEUm
0.47	22.00	.926	.161	0.074	21.678	2.290	44.010	0.1491	1.604	.6423E+2	.1557E-1	.1506E-2	.6561E+1
0.47	27.00	.926	.161	0.074	26.678	2.290	54.010	0.1491	1.974	.5219E+2	.1916E-1	.1510E-2	.8052E+1
0.47	33.00	.926	.161	0.074	32.678	2.290	66.010	0.1491	2.418	.4261E+2	.2347E-1	.1513E-2	.9841E+1
0.56	0.33	.821	.101	0.179	0.128	2.125	0.931	0.0829	0.023	.1349E+5	.7413E-4	.4011E-3	.7722E-1
0.56	0.39	.835	.115	0.165	0.160	2.159	1.011	0.0960	0.026	.8870E+4	.1127E-3	.5162E-3	.9711E-1
0.56	0.47	.852	.132	0.148	0.206	2.200	1.123	0.1125	0.030	.5706E+4	.1752E-3	.6658E-3	.1263E+0
0.56	0.56	.866	.146	0.134	0.268	2.234	1.263	0.1264	0.036	.3892E+4	.2569E-3	.8193E-3	.1596E+0
0.56	0.68	.879	.159	0.121	0.362	2.266	1.466	0.1398	0.044	.2649E+4	.3776E-3	.9915E-3	.2048E+0
0.56	0.82	.888	.168	0.112	0.484	2.288	1.720	0.1492	0.054	.1896E+4	.5273E-3	.1148E-2	.2566E+0
0.56	1.00	.895	.175	0.105	0.650	2.305	2.060	0.1566	0.068	.1377E+4	.7264E-3	.1297E-2	.3226E+0
0.56	1.20	.899	.179	0.101	0.842	2.315	2.448	0.1609	0.085	.1051E+4	.9514E-3	.1416E-2	.3940E+0
0.56	1.50	.903	.183	0.097	1.134	2.324	3.037	0.1652	0.110	.7738E+3	.1292E-2	.1538E-2	.5018E+0
0.56	1.80	.905	.185	0.095	1.430	2.329	3.631	0.1674	0.136	.6117E+3	.1635E-2	.1622E-2	.6080E+0
0.56	2.20	.908	.187	0.092	1.826	2.334	4.425	0.1696	0.170	.4778E+3	.2093E-2	.1699E-2	.7506E+0
0.56	2.70	.908	.188	0.092	2.324	2.337	5.423	0.1707	0.214	.3750E+3	.2667E-2	.1764E-2	.9257E+0
0.56	3.30	.909	.189	0.091	2.922	2.339	6.620	0.1718	0.266	.2980E+3	.3355E-2	.1816E-2	.1137E+1
0.56	3.90	.910	.191	0.090	3.520	2.341	7.817	0.1729	0.317	.2472E+3	.4045E-2	.1852E-2	.1352E+1
0.56	4.70	.911	.191	0.089	4.328	2.344	9.414	0.1740	0.384	.2016E+3	.4964E-2	.1886E-2	.1638E+1
0.56	5.60	.911	.191	0.089	5.218	2.344	11.214	0.1740	0.464	.1667E+3	.5999E-2	.1913E-2	.1951E+1
0.56	6.80	.912	.191	0.088	6.418	2.344	13.614	0.1740	0.571	.1355E+3	.7379E-2	.1938E-2	.2360E+1
0.56	8.20	.912	.192	0.088	7.816	2.346	16.411	0.1751	0.688	.1113E+3	.8989E-2	.1957E-2	.2874E+1
0.56	10.00	.912	.192	0.089	9.616	2.346	20.011	0.1751	0.846	.9043E+2	.1106E-1	.1975E-2	.3504E+1
0.56	12.00	.913	.192	0.088	11.616	2.346	24.011	0.1751	1.022	.7486E+2	.1336E-1	.1988E-2	.4204E+1
0.56	15.00	.913	.192	0.088	14.616	2.346	30.011	0.1751	1.286	.5949E+2	.1681E-1	.2001E-2	.5255E+1
0.56	18.00	.913	.192	0.088	17.616	2.346	36.011	0.1762	1.550	.4936E+2	.2026E-1	.2010E-2	.6300E+1
0.56	22.00	.913	.193	0.087	21.614	2.349	44.008	0.1762	1.880	.4023E+2	.2486E-1	.2018E-2	.7755E+1
0.56	27.00	.913	.193	0.087	26.614	2.349	54.008	0.1762	2.315	.3267E+2	.3061E-1	.2024E-2	.9517E+1
0.56	33.00	.913	.193	0.087	32.614	2.349	66.008	0.1762	2.837	.2666E+2	.3751E-1	.2030E-2	.1163E+2
0.68	0.33	.764	.104	0.236	0.122	2.107	1.043	0.0795	0.029	.1159E+5	.8628E-4	.3845E-3	.8285E-1
0.68	0.39	.781	.121	0.219	0.148	2.148	1.114	0.0945	0.032	.7421E+4	.1348E-3	.5081E-3	.1053E+0
0.68	0.47	.801	.141	0.199	0.188	2.197	1.217	0.1129	0.037	.4603E+4	.2173E-3	.6798E-3	.1374E+0
0.68	0.56	.820	.160	0.180	0.240	2.243	1.343	0.1312	0.043	.3016E+4	.3316E-3	.8707E-3	.1762E+0
0.68	0.68	.839	.179	0.161	0.322	2.289	1.528	0.1502	0.052	.1958E+4	.5108E-3	.1105E-2	.2295E+0
0.68	0.82	.855	.195	0.145	0.430	2.328	1.763	0.1667	0.062	.1343E+4	.7446E-3	.1335E-2	.2939E+0
0.68	1.00	.867	.207	0.133	0.586	2.357	2.088	0.1795	0.078	.9389E+3	.1065E-2	.1566E-2	.3748E+0
0.68	1.20	.875	.215	0.125	0.770	2.376	2.465	0.1881	0.096	.6976E+3	.1433E-2	.1757E-2	.4630E+0

CHOKE/RESONANT CHARGE REACTOR DESIGN TABLE

f(D) =U/(F*G*(D*E)^2)

CASE # 3 / COPYRIGHT 1982 STEVE SMITH   DATA DATE 1/ 3/82

P	Q	D	E	F	G	U	Um	DE	FG	f(D)	1/f(D)	1/(PQf(D))	DEUm
0.68	1.50	.881	.221	0.119	1.058	2.391	3.048	0.1947	0.126	.5010E+3	.1996E-2	.1957E-2	.5935E+0
0.68	1.80	.885	.225	0.115	1.350	2.401	3.637	0.1991	0.155	.3900E+3	.2564E-2	.2095E-2	.7242E+0
0.68	2.20	.887	.227	0.113	1.746	2.406	4.431	0.2013	0.197	.3007E+3	.3325E-2	.2223E-2	.8922E+0
0.68	2.70	.890	.230	0.110	2.240	2.413	5.423	0.2047	0.246	.2337E+3	.4279E-2	.2331E-2	.1110E+1
0.68	3.30	.891	.231	0.109	2.838	2.415	6.620	0.2058	0.309	.1843E+3	.5426E-2	.2418E-2	.1362E+1
0.68	3.90	.892	.232	0.108	3.436	2.418	7.817	0.2069	0.371	.1521E+3	.6573E-2	.2479E-2	.1618E+1
0.68	4.70	.893	.233	0.107	4.234	2.420	9.414	0.2081	0.453	.1234E+3	.8104E-2	.2536E-2	.1959E+1
0.68	5.60	.894	.234	0.106	5.132	2.422	11.211	0.2090	0.544	.1018E+3	.9827E-2	.2581E-2	.2345E+1
0.68	6.80	.894	.234	0.106	6.332	2.423	13.611	0.2092	0.671	.8247E+2	.1213E-1	.2622E-2	.2847E+1
0.68	8.20	.895	.235	0.105	7.730	2.425	16.408	0.2103	0.812	.6754E+2	.1481E-1	.2655E-2	.3451E+1
0.68	10.00	.895	.235	0.105	9.530	2.425	20.008	0.2103	1.001	.5478E+2	.1825E-1	.2684E-2	.4208E+1
0.68	12.00	.895	.235	0.105	11.530	2.425	24.008	0.2103	1.211	.4528E+2	.2209E-1	.2707E-2	.5050E+1
0.68	15.00	.896	.236	0.104	14.528	2.427	30.005	0.2115	1.511	.3593E+2	.2783E-1	.2729E-2	.6345E+1
0.68	18.00	.896	.236	0.104	17.528	2.427	36.005	0.2115	1.823	.2978E+2	.3358E-1	.2743E-2	.7618E+1
0.68	21.00	.896	.236	0.104	21.528	2.427	44.005	0.2115	2.239	.2425E+2	.4124E-1	.2757E-2	.9305E+1
0.68	27.00	.896	.236	0.104	26.528	2.427	54.005	0.2115	2.759	.1968E+2	.5082E-1	.2768E-2	.1142E+2
0.68	33.00	.896	.236	0.104	32.528	2.427	66.005	0.2115	3.383	.1605E+2	.6232E-1	.2777E-2	.1396E+2
0.82	0.33	.697	.107	0.303	0.116	2.084	1.174	0.0746	0.035	.1066E+5	.9381E-4	.3467E-3	.8757E-1
0.82	0.39	.715	.125	0.285	0.140	2.128	1.243	0.0894	0.040	.6676E+4	.1498E-3	.4684E-3	.1111E+0
0.82	0.47	.737	.147	0.263	0.176	2.181	1.340	0.1083	0.046	.4015E+4	.2491E-3	.6463E-3	.1452E+0
0.82	0.56	.760	.170	0.240	0.220	2.237	1.454	0.1292	0.053	.2538E+4	.3940E-3	.8580E-3	.1879E+0
0.82	0.68	.787	.197	0.213	0.286	2.303	1.617	0.1550	0.061	.1572E+4	.6359E-3	.1140E-2	.2507E+0
0.82	0.82	.810	.220	0.190	0.380	2.358	1.831	0.1782	0.072	.1029E+4	.9721E-3	.1446E-2	.3263E+0
0.82	1.00	.830	.240	0.170	0.520	2.407	2.134	0.1992	0.088	.6862E+3	.1457E-2	.1777E-2	.4251E+0
0.82	1.20	.843	.253	0.157	0.694	2.439	2.497	0.2133	0.109	.4920E+3	.2032E-2	.2065E-2	.5325E+0
0.82	1.50	.854	.264	0.146	0.972	2.465	3.065	0.2255	0.142	.3418E+3	.2926E-2	.2379E-2	.6911E+0
0.82	1.80	.860	.270	0.140	1.260	2.480	3.648	0.2322	0.176	.2607E+3	.3835E-2	.2598E-2	.8471E+0
0.82	2.20	.865	.275	0.135	1.650	2.492	4.434	0.2379	0.223	.1977E+3	.5058E-2	.2804E-2	.1055E+1
0.82	2.70	.868	.278	0.132	2.144	2.499	5.425	0.2413	0.283	.1517E+3	.6593E-2	.2978E-2	.1309E+1
0.82	3.30	.871	.281	0.128	2.738	2.507	6.617	0.2448	0.353	.1185E+3	.8441E-2	.3119E-2	.1619E+1
0.82	3.90	.872	.282	0.128	3.336	2.509	7.814	0.2459	0.427	.9711E+2	.1029E-1	.3218E-2	.1921E+1
0.82	4.70	.873	.283	0.127	4.134	2.511	9.411	0.2471	0.525	.7837E+2	.1276E-1	.3311E-2	.2325E+1
0.82	5.60	.874	.284	0.126	5.032	2.514	11.208	0.2482	0.634	.6435E+2	.1554E-1	.3384E-2	.2782E+1
0.82	6.80	.875	.285	0.125	6.230	2.516	13.605	0.2494	0.779	.5196E+2	.1925E-1	.3452E-2	.3393E+1
0.82	8.00	.876	.286	0.124	7.628	2.519	16.402	0.2505	0.946	.4242E+2	.2357E-1	.3506E-2	.4109E+1
0.82	10.00	.876	.286	0.124	9.428	2.519	20.002	0.2505	1.169	.3432E+2	.2913E-1	.3553E-2	.5011E+1

CASE # 3   DATA DATE 1/ 3/82   CHOKE/RESONANT CHARGE REACTOR DESIGN TABLE

$f(D) = U/(F*G*(D*E)^2)$

P	Q	D	E	F	G	U	Um	DE	FG	f(D)	1/f(D)	1/(PQf(D))	DEUm
0.82	12.00	.877	0.287	0.123	11.426	2.521	24.000	0.2517	1.405	.2832E+2	.3531E-1	.3589E-2	.6041E+1
0.82	15.00	.877	0.287	0.123	14.426	2.521	30.000	0.2517	1.774	.2243E+2	.4459E-1	.3625E-2	.7551E+1
0.82	18.00	.877	0.287	0.123	17.426	2.521	36.000	0.2517	2.143	.1857E+2	.5386E-1	.3649E-2	.9061E+1
0.82	22.00	.878	0.288	0.122	21.424	2.524	43.997	0.2529	2.614	.1510E+2	.6622E-1	.3671E-2	.1113E+2
0.82	27.00	.878	0.288	0.122	26.424	2.524	53.997	0.2529	3.224	.1224E+2	.8168E-1	.3689E-2	.1365E+2
0.82	33.00	.878	0.288	0.122	32.424	2.524	65.997	0.2529	3.956	.9978E+1	.1002E+0	.3704E-2	.1669E+2
1.00	0.33	.609	0.109	0.391	0.112	2.050	1.348	0.0664	0.044	.1062E+5	.9412E-4	.2852E-3	.8951E-1
1.00	0.39	.628	0.128	0.372	0.134	2.096	1.414	0.0804	0.050	.6508E+4	.1536E-3	.3940E-3	.1137E+0
1.00	0.47	.653	0.153	0.347	0.164	2.157	1.503	0.0999	0.057	.3797E+4	.2633E-3	.5603E-3	.1501E+0
1.00	0.56	.679	0.179	0.321	0.202	2.220	1.608	0.1215	0.065	.2318E+4	.4314E-3	.7704E-3	.1955E+0
1.00	0.68	.711	0.211	0.289	0.258	2.298	1.757	0.1500	0.075	.1369E+4	.7303E-3	.1074E-2	.2636E+0
1.00	0.82	.743	0.243	0.257	0.334	2.376	1.945	0.1805	0.086	.8490E+3	.1178E-2	.1436E-2	.3512E+0
1.00	1.00	.775	0.275	0.225	0.450	2.453	2.214	0.2131	0.101	.5335E+3	.1875E-2	.1875E-2	.4718E+0
1.00	1.20	.798	0.298	0.202	0.604	2.509	2.548	0.2378	0.122	.3637E+3	.2750E-2	.2291E-2	.6060E+0
1.00	1.50	.818	0.318	0.182	0.864	2.558	3.091	0.2601	0.157	.2404E+3	.4160E-2	.2773E-2	.8041E+0
1.00	1.80	.828	0.328	0.172	1.144	2.582	3.662	0.2716	0.197	.1779E+3	.5621E-2	.3123E-2	.9947E+0
1.00	2.20	.836	0.336	0.164	1.528	2.602	4.440	0.2800	0.251	.1316E+3	.7600E-2	.3455E-2	.1247E+1
1.00	2.70	.842	0.342	0.158	2.016	2.616	5.422	0.2880	0.319	.9995E+2	.1010E-1	.3739E-2	.1561E+1
1.00	3.30	.845	0.345	0.155	2.610	2.623	6.614	0.2915	0.405	.7630E+2	.1311E-1	.3971E-2	.1928E+1
1.00	3.90	.848	0.348	0.152	3.204	2.631	7.805	0.2951	0.487	.6203E+2	.1612E-1	.4134E-2	.2303E+1
1.00	4.70	.850	0.350	0.150	4.000	2.636	9.400	0.2975	0.600	.4963E+2	.2015E-1	.4287E-2	.2796E+1
1.00	5.60	.851	0.351	0.149	4.898	2.638	11.197	0.2987	0.730	.4051E+2	.2468E-1	.4408E-2	.3344E+1
1.00	6.80	.853	0.352	0.147	6.096	2.640	13.594	0.2999	0.902	.3254E+2	.3073E-1	.4519E-2	.4077E+1
1.00	8.20	.854	0.353	0.146	7.494	2.643	16.391	0.3011	1.102	.2646E+2	.3779E-1	.4609E-2	.4935E+1
1.00	10.00	.854	0.354	0.146	9.292	2.645	19.988	0.3023	1.357	.2134E+2	.4687E-1	.4687E-2	.6043E+1
1.00	12.00	.855	0.355	0.145	11.290	2.648	23.985	0.3035	1.637	.1756E+2	.5696E-1	.4747E-2	.7280E+1
1.00	15.00	.855	0.355	0.145	14.290	2.648	29.985	0.3035	2.072	.1387E+2	.7210E-1	.4806E-2	.9101E+1
1.00	18.00	.856	0.356	0.144	17.288	2.650	35.982	0.3047	2.489	.1146E+2	.8723E-1	.4846E-2	.1097E+2
1.00	22.00	.856	0.356	0.144	21.288	2.650	43.982	0.3047	3.065	.9316E+1	.1074E+0	.4883E-2	.1340E+2
1.00	27.00	.856	0.356	0.144	26.288	2.650	53.982	0.3047	3.785	.7539E+1	.1326E+0	.4913E-2	.1645E+2
1.00	33.00	.856	0.356	0.144	32.288	2.650	65.982	0.3047	4.649	.6138E+1	.1629E+0	.4937E-2	.2011E+2
1.20	0.33	.511	0.111	0.489	0.108	2.061	1.543	0.0567	0.050	.1184E+5	.8444E-4	.2132E-3	.8750E-1
1.20	0.39	.531	0.131	0.469	0.128	2.094	1.606	0.0696	0.060	.7094E+4	.1410E-3	.3012E-3	.1117E+0
1.20	0.47	.557	0.157	0.443	0.156	2.124	1.691	0.0874	0.069	.4019E+4	.2488E-3	.4412E-3	.1479E+0
1.20	0.56	.586	0.186	0.414	0.188	2.194	1.788	0.1090	0.078	.2373E+4	.4214E-3	.6271E-3	.1949E+0
1.20	0.68	.622	0.222	0.378	0.236	2.222	1.925	0.1381	0.089	.1341E+4	.7455E-3	.9135E-3	.2659E+0

CHOKE/RESONANT CHARGE REACTOR DESIGN TABLE

CASE # 3   COPYRIGHT 1982 STEVE SMITH   DATA DATE 1/ 3/82

$f(D)=U/(F*G*(D*E)^2)$

P	Q	D	E	F	G	U	Um	DE	FG	f(D)	1/f(D)	1/(PQf(D))	DEUm
1.20	0.82	.661	0.261	0.339	0.298	2.377	2.094	0.1725	0.101	.7904E+3	.1265E-2	.1286E-2	.3613E+0
1.20	1.00	.703	0.303	0.297	0.394	2.479	2.334	0.2130	0.117	.4668E+3	.2142E-2	.1785E-2	.4971E+0
1.20	1.20	.739	0.339	0.261	0.522	2.566	2.631	0.2505	0.136	.3001E+3	.3332E-2	.2314E-2	.6591E+0
1.20	1.50	.772	0.372	0.228	0.756	2.646	3.137	0.2872	0.172	.1861E+3	.5372E-2	.2985E-2	.9008E+0
1.20	1.80	.791	0.391	0.209	1.018	2.692	3.682	0.3093	0.213	.1323E+3	.7559E-2	.3500E-2	.1139E+1
1.20	2.20	.804	0.404	0.196	1.392	2.724	4.445	0.3248	0.273	.9463E+2	.1057E-1	.4003E-2	.1444E+1
1.20	2.70	.813	0.413	0.187	1.874	2.746	5.419	0.3358	0.350	.6950E+2	.1439E-1	.4441E-2	.1820E+1
1.20	3.00	.819	0.419	0.181	2.462	2.760	6.680	0.3432	0.446	.5260E+2	.1901E-1	.4801E-2	.2266E+1
1.20	3.90	.822	0.422	0.178	3.056	2.768	7.794	0.3469	0.544	.4228E+2	.2365E-1	.5054E-2	.2704E+1
1.20	4.70	.825	0.425	0.175	3.850	2.775	9.385	0.3506	0.674	.3350E+2	.2985E-1	.5292E-2	.3291E+1
1.20	5.60	.828	0.428	0.172	4.744	2.782	11.574	0.3544	0.816	.2715E+2	.3683E-1	.5481E-2	.3961E+1
1.20	6.80	.829	0.429	0.171	5.942	2.785	13.574	0.3556	1.016	.2167E+2	.4615E-1	.5656E-2	.4827E+1
1.20	8.20	.831	0.431	0.169	7.338	2.789	16.368	0.3582	1.240	.1753E+2	.5703E-1	.5796E-2	.5862E+1
1.20	10.00	.832	0.432	0.168	9.136	2.792	19.965	0.3594	1.535	.1408E+2	.7102E-1	.5918E-2	.7176E+1
1.20	12.00	.833	0.433	0.167	11.114	2.794	23.962	0.3607	1.859	.1155E+2	.8657E-1	.6012E-2	.8643E+1
1.20	15.00	.833	0.433	0.167	14.134	2.794	29.962	0.3607	2.360	.9100E+1	.1099E+0	.6105E-2	.1081E+2
1.20	18.00	.834	0.434	0.166	17.132	2.797	35.959	0.3620	2.844	.7506E+1	.1332E+0	.6168E-2	.1302E+2
1.20	22.00	.834	0.434	0.166	21.132	2.797	43.959	0.3620	3.508	.6085E+1	.1643E+0	.6224E-2	.1591E+2
1.20	27.00	.835	0.435	0.165	26.130	2.799	53.957	0.3632	4.311	.4921E+1	.2032E+0	.6272E-2	.1960E+2
1.20	33.00	.835	0.435	0.165	32.130	2.799	65.957	0.3632	5.301	.4002E+1	.2499E+0	.6310E-2	.2396E+2
1.50	0.33	.365	0.115	0.635	0.100	1.957	1.831	0.0420	0.063	.1750E+5	.5716E-4	.1155E-3	.7687E-1
1.50	0.39	.386	0.136	0.614	0.118	2.008	1.891	0.0525	0.072	.1006E+5	.9941E-4	.1699E-3	.9928E-1
1.50	0.45	.414	0.164	0.586	0.142	2.076	1.970	0.0679	0.083	.5411E+4	.1847E-3	.2620E-3	.1338E+0
1.50	0.56	.445	0.195	0.555	0.170	2.152	2.063	0.0868	0.094	.3029E+4	.3302E-3	.3931E-3	.1790E+0
1.50	0.68	.486	0.236	0.514	0.208	2.251	2.185	0.1147	0.107	.1601E+4	.6247E-3	.6125E-3	.2507E+0
1.50	0.82	.531	0.281	0.469	0.258	2.361	2.337	0.1492	0.121	.8763E+3	.1141E-2	.9278E-3	.3487E+0
1.50	1.00	.584	0.334	0.416	0.332	2.489	2.545	0.1951	0.138	.4738E+3	.2111E-2	.1407E-2	.4965E+0
1.50	1.20	.635	0.385	0.365	0.430	2.613	2.800	0.2445	0.157	.2786E+3	.3590E-2	.1994E-2	.6844E+0
1.50	1.50	.693	0.443	0.307	0.614	2.754	3.234	0.3070	0.188	.1550E+3	.6450E-2	.2867E-2	.9928E+0
1.50	1.80	.729	0.479	0.271	0.842	2.842	3.731	0.3492	0.228	.1021E+3	.9791E-2	.3626E-2	.1303E+1
1.50	2.20	.755	0.505	0.245	1.190	2.905	4.457	0.3813	0.292	.6854E+2	.1459E-1	.4421E-2	.1699E+1
1.50	2.70	.772	0.522	0.228	1.556	2.946	5.408	0.4030	0.378	.4805E+2	.2081E-1	.5139E-2	.2179E+1
1.50	3.30	.783	0.533	0.217	2.234	2.973	6.576	0.4173	0.485	.3521E+2	.2840E-1	.5738E-2	.2745E+1
1.50	3.90	.789	0.539	0.211	2.822	2.987	7.759	0.4253	0.595	.2774E+2	.3605E-1	.6162E-2	.3300E+1
1.50	4.70	.793	0.543	0.207	3.614	2.997	9.348	0.4306	0.748	.2161E+2	.4628E-1	.6565E-2	.4025E+1
1.50	5.60	.797	0.547	0.203	4.506	3.007	11.136	0.4360	0.915	.1730E+2	.5782E-1	.6883E-2	.4855E+1

CASE # 3 COPYRIGHT 1982 STEVE SMITH DATA DATE 1/ 3/82 CHOKE/RESONANT CHARGE REACTOR DESIGN TABLE

$$f(D)=U/(F*G*(D*E)^2)$$

P	Q	D	E	F	G	U	Um	DE	FG	f(D)	1/f(D)	1/(PQf(D))	DEUm
1.50	6.80	.799	0.549	0.201	5.702	3.012	13.531	0.4387	1.146	.1366E+2	.7322E-1	.7179E-2	.5935E+1
1.50	8.20	.801	0.551	0.199	7.098	3.017	16.325	0.4414	1.413	.1096E+2	.9121E-1	.7415E-2	.7205E+1
1.50	10.00	.803	0.553	0.197	8.894	3.021	19.919	0.4441	1.752	.8745E+1	.1143E+0	.7623E-2	.8845E+1
1.50	12.00	.804	0.554	0.196	10.892	3.024	23.916	0.4454	2.135	.7140E+1	.1401E+0	.7781E-2	.1065E+2
1.50	15.00	.805	0.555	0.195	13.890	3.026	29.914	0.4468	2.709	.5598E+1	.1786E+0	.7940E-2	.1336E+2
1.50	18.00	.806	0.556	0.194	16.888	3.029	35.911	0.4481	3.276	.4603E+1	.2172E+0	.8046E-2	.1699E+2
1.50	22.00	.807	0.557	0.193	20.886	3.031	43.908	0.4495	4.031	.3722E+1	.2687E+0	.8142E-2	.1974E+2
1.50	27.00	.807	0.557	0.193	25.886	3.031	53.908	0.4495	4.996	.3003E+1	.3330E+0	.8223E-2	.2423E+2
1.50	33.00	.808	0.558	0.192	31.884	3.034	65.905	0.4509	6.122	.2438E+1	.4102E+0	.8287E-2	.2971E+2
1.80	0.33	.221	0.121	0.779	0.088	1.908	2.114	0.0267	0.069	.3892E+5	.2570E-4	.4326E-4	.5653E-1
1.80	0.44	.244	0.144	0.756	0.102	1.956	2.116	0.0351	0.077	.2063E+5	.4848E-4	.6906E-4	.7619E-1
1.80	0.47	.274	0.174	0.726	0.122	2.036	2.243	0.0477	0.089	.1012E+5	.9886E-4	.1169E-3	.1069E+0
1.80	0.56	.307	0.207	0.693	0.146	2.117	2.328	0.0635	0.101	.5180E+4	.1931E-3	.1915E-3	.1480E+0
1.80	0.68	.339	0.259	0.650	0.180	2.180	2.445	0.0875	0.117	.2479E+4	.4033E-3	.3295E-3	.2140E+0
1.80	0.82	.399	0.299	0.601	0.222	2.340	2.585	0.1193	0.133	.1232E+4	.8111E-3	.5498E-3	.3084E+0
1.80	1.00	.459	0.359	0.541	0.282	2.486	2.774	0.1648	0.153	.6001E+3	.1666E-2	.9258E-3	.4571E+0
1.80	1.20	.521	0.421	0.479	0.358	2.636	2.997	0.2193	0.171	.3196E+3	.3129E-2	.1449E-2	.6573E+0
1.80	1.50	.599	0.499	0.401	0.502	2.826	3.374	0.2989	0.201	.1571E+3	.6364E-2	.2357E-2	.1008E+1
1.80	1.80	.656	0.556	0.344	0.688	2.964	3.811	0.3647	0.237	.9415E+2	.1062E-1	.3278E-2	.1390E+1
1.80	2.20	.703	0.603	0.297	0.994	3.079	4.476	0.4239	0.295	.5803E+2	.1723E-1	.4352E-2	.1898E+1
1.80	2.70	.732	0.632	0.268	1.436	3.149	5.393	0.4626	0.385	.3823E+2	.2616E-1	.5382E-2	.2495E+1
1.80	3.30	.749	0.649	0.251	2.004	3.190	6.545	0.4861	0.503	.2687E+2	.3722E-1	.6266E-2	.3181E+1
1.80	3.90	.759	0.659	0.241	2.582	3.215	7.716	0.5002	0.622	.2065E+2	.4843E-1	.6899E-2	.3860E+1
1.80	4.70	.766	0.666	0.234	3.368	3.232	9.296	0.5102	0.788	.1576E+2	.6347E-1	.7503E-2	.4743E+1
1.80	5.60	.771	0.671	0.229	4.258	3.244	11.082	0.5173	0.975	.1243E+2	.8045E-1	.7982E-2	.5733E+1
1.80	6.80	.774	0.674	0.226	5.452	3.251	13.473	0.5217	1.232	.9695E+1	.1031E+0	.8427E-2	.7029E+1
1.80	8.20	.777	0.677	0.223	6.846	3.258	16.265	0.5260	1.527	.7713E+1	.1296E+0	.8784E-2	.8556E+1
1.80	10.00	.779	0.679	0.221	8.642	3.263	19.859	0.5289	1.910	.6107E+1	.1638E+0	.9097E-2	.1050E+2
1.80	12.00	.781	0.681	0.219	10.638	3.268	23.851	0.5319	2.330	.4959E+1	.2017E+0	.9336E-2	.1269E+2
1.80	15.00	.782	0.682	0.218	13.636	3.270	29.851	0.5333	2.973	.3868E+1	.2585E+0	.9575E-2	.1592E+2
1.80	18.00	.783	0.683	0.217	16.634	3.273	35.848	0.5348	3.610	.3170E+1	.3154E+0	.9735E-2	.1917E+2
1.80	22.00	.784	0.684	0.216	20.632	3.275	43.842	0.5363	4.457	.2556E+1	.3913E+0	.9881E-2	.2351E+2
1.80	27.00	.785	0.685	0.215	25.630	3.277	53.842	0.5377	5.510	.2057E+1	.4861E+0	.1000E-1	.2895E+2
1.80	33.00	.785	0.685	0.215	31.630	3.278	65.842	0.5377	6.800	.1667E+1	.5999E+0	.1010E-1	.3540E+2
2.20	0.33	.047	0.147	0.953	0.036	1.885	2.440	0.0069	0.034	.1151E+7	.8688E-6	.1197E-5	.1686E-1
2.20	0.39	.069	0.169	0.931	0.052	1.938	2.497	0.0117	0.048	.2945E+6	.3396E-5	.3958E-5	.2912E-1

CHOKE/RESONANT CHARGE REACTOR DESIGN TABLE

CASE # 3   COPYRIGHT 1982 STEVE SMITH   DATA DATE 1/ 3/82

$f(D)=U/(F*G*(D*E)^2)$

P	Q	D	E	F	G	U	Um	DE	FG	f(D)	1/f(D)	1/(PQf(D))	DEUm
2.20	0.47	.099	0.199	0.901	0.072	2.011	2.571	0.0197	0.065	.7988E+5	.1252E-4	.1211E-4	.5065E-1
2.20	0.56	.133	0.233	0.867	0.094	2.094	2.654	0.0310	0.081	.2675E+5	.3738E-4	.3034E-4	.8224E-1
2.20	0.68	.178	0.278	0.822	0.124	2.203	2.765	0.0495	0.102	.8827E+4	.1133E-3	.7572E-4	.1368E+0
2.20	0.82	.230	0.330	0.770	0.160	2.330	2.897	0.0759	0.123	.3282E+4	.3047E-3	.1689E-3	.2199E+0
2.20	1.00	.294	0.394	0.706	0.212	2.485	3.074	0.1158	0.150	.1237E+4	.8082E-3	.3674E-3	.3561E+0
2.20	1.20	.363	0.463	0.637	0.274	2.653	3.277	0.1681	0.175	.5380E+3	.1859E-2	.7040E-3	.5507E+0
2.20	1.50	.459	0.559	0.541	0.382	2.886	3.602	0.2566	0.207	.2121E+3	.4715E-2	.1429E-2	.9242E+0
2.20	1.80	.541	0.641	0.459	0.518	3.085	3.968	0.3468	0.238	.1079E+3	.9268E-2	.2340E-2	.1376E+1
2.20	2.20	.620	0.720	0.380	0.760	3.277	4.542	0.4464	0.289	.5694E+2	.1756E-1	.3629E-2	.2028E+1
2.20	2.70	.676	0.776	0.324	1.148	3.413	5.382	0.5246	0.372	.3334E+2	.2999E-1	.5049E-2	.2823E+1
2.20	3.30	.708	0.808	0.292	1.684	3.491	6.490	0.5721	0.492	.2169E+2	.4610E-1	.6350E-2	.3713E+1
2.20	3.90	.723	0.823	0.277	2.254	3.527	7.648	0.5950	0.624	.1596E+2	.6267E-1	.7305E-2	.4551E+1
2.20	4.70	.735	0.835	0.265	3.030	3.556	9.213	0.6137	0.803	.1176E+2	.8504E-1	.8225E-2	.5654E+1
2.20	5.60	.742	0.842	0.258	3.916	3.573	10.993	0.6248	1.010	.9061E+1	.1104E+0	.8958E-2	.6868E+1
2.20	6.80	.748	0.848	0.252	5.104	3.588	13.376	0.6343	1.286	.6933E+1	.1442E+0	.9641E-2	.8485E+1
2.20	8.20	.752	0.852	0.248	6.496	3.598	16.165	0.6407	1.611	.5440E+1	.1838E+0	.1019E-1	.1036E+2
2.20	10.00	.755	0.855	0.245	8.290	3.605	19.756	0.6455	2.031	.4259E+1	.2348E+0	.1067E-1	.1275E+2
2.20	12.00	.757	0.857	0.243	10.286	3.610	23.750	0.6487	2.499	.3431E+1	.2914E+0	.1104E-1	.1541E+2
2.20	15.00	.759	0.859	0.241	13.282	3.615	29.745	0.6520	3.201	.2656E+1	.3764E+0	.1141E-1	.1939E+2
2.20	18.00	.760	0.860	0.240	16.280	3.617	35.742	0.6536	3.907	.2167E+1	.4615E+0	.1155E-1	.2336E+2
2.20	22.00	.761	0.861	0.239	20.278	3.619	43.739	0.6552	4.846	.1740E+1	.5749E+0	.1188E-1	.2866E+2
2.20	27.00	.762	0.862	0.238	25.276	3.622	53.736	0.6568	6.016	.1395E+1	.7166E+0	.1206E-1	.3530E+2
2.20	33.00	.762	0.862	0.238	31.276	3.622	65.736	0.6568	7.444	.1128E+1	.8867E+0	.1221E-1	.4318E+2
2.70	0.82	.041	0.391	0.959	0.038	2.370	3.222	0.0160	0.036	.2531E+6	.3951E-5	.1785E-5	.5166E-1
2.70	1.00	.104	0.454	0.896	0.092	2.523	3.402	0.0472	0.082	.1373E+5	.7283E-4	.2697E-4	.1606E+0
2.70	1.20	.174	0.524	0.826	0.152	2.693	3.602	0.0912	0.126	.2581E+4	.3875E-3	.1196E-3	.3284E+0
2.70	1.50	.278	0.628	0.722	0.244	2.946	3.905	0.1746	0.176	.5487E+3	.1823E-2	.4500E-3	.6817E+0
2.70	1.80	.375	0.725	0.625	0.350	3.182	4.228	0.2719	0.219	.1968E+3	.5082E-2	.1046E-2	.1149E+1
2.70	2.20	.488	0.838	0.512	0.524	3.456	4.705	0.4089	0.268	.7703E+2	.1298E-1	.2185E-2	.1924E+1
2.70	2.70	.589	0.939	0.411	0.822	3.702	5.416	0.5531	0.338	.3582E+2	.2792E-1	.3830E-2	.2995E+1
2.70	3.30	.653	1.003	0.347	1.294	3.857	6.433	0.6550	0.449	.2002E+2	.4994E-1	.5605E-2	.4213E+1
2.70	3.70	.683	1.033	0.317	1.834	3.970	7.547	0.7055	0.581	.1358E+2	.7364E-1	.6993E-2	.5325E+1
2.70	4.70	.703	1.053	0.297	2.594	3.979	9.090	0.7403	0.770	.9424E+1	.1061E+0	.8362E-2	.6729E+1
2.70	5.60	.714	1.064	0.286	3.472	4.005	10.859	0.7597	0.993	.6999E+1	.1431E+0	.9463E-2	.8249E+1
2.70	6.80	.722	1.072	0.278	4.656	4.025	13.236	0.7740	1.294	.5191E+1	.1927E+0	.1049E-1	.1024E+2
2.70	8.20	.728	1.078	0.272	6.044	4.039	16.019	0.7848	1.644	.3989E+1	.2507E+0	.1113E-1	.1257E+2

CASE # 3   COPYRIGHT 1982 STEVE SMITH   DATA DATE 1/ 3/82   CHOKE/RESONANT CHARGE REACTOR DESIGN TABLE

$$f(D) = U / (F * G * (D * E)^2)$$

P	Q	D	E	F	G	U	Um	DE	FG	f(D)	1/f(D)	1/(PQf(D))	DEUm
2.70	10.00	.732	1.082	0.268	7.836	4.049	19.607	0.7920	2.100	.3074E+1	.3254E+0	.1205E-1	.1553E+2
2.70	12.00	.734	1.084	0.266	9.832	4.054	23.601	0.7957	2.615	.2448E+1	.4084E+0	.1261E-1	.1878E+2
2.70	15.00	.737	1.087	0.263	12.826	4.061	29.593	0.8011	3.373	.1876E+1	.5331E+0	.1316E-1	.2371E+2
2.70	18.00	.738	1.088	0.262	15.824	4.064	35.590	0.8029	4.146	.1520E+1	.6578E+0	.1353E-1	.2858E+2
2.70	22.00	.740	1.090	0.260	19.820	4.068	43.584	0.8066	5.153	.1213E+1	.8241E+0	.1387E-1	.3516E+2
2.70	27.00	.741	1.091	0.259	24.818	4.071	53.581	0.8084	6.428	.9690E+0	.1032E+1	.1416E-1	.4332E+2
2.70	33.00	.741	1.091	0.259	30.818	4.071	65.581	0.8084	7.982	.7804E+0	.1281E+1	.1438E-1	.5302E+2
3.30	1.50	.067	0.717	0.933	0.066	3.034	4.251	0.0480	0.062	.2135E+5	.4684E-4	.9464E-5	.2042E+0
3.30	1.80	.169	0.819	0.831	0.162	3.281	4.559	0.1384	0.135	.1272E+4	.7860E-3	.1323E-3	.6310E+0
3.30	2.20	.300	0.950	0.700	0.300	3.600	4.985	0.2850	0.210	.2110E+3	.4739E-2	.6527E-3	.1421E+1
3.30	2.70	.444	1.094	0.556	0.512	3.944	5.573	0.4857	0.285	.5880E+2	.1701E-1	.1969E-2	.2707E+1
3.30	3.30	.566	1.216	0.434	0.868	4.246	6.424	0.6883	0.377	.2379E+2	.4203E-1	.3859E-2	.4421E+1
3.30	3.90	.629	1.279	0.371	1.342	4.399	7.444	0.8045	0.498	.1365E+2	.7326E-1	.5692E-2	.5999E+1
3.30	4.70	.667	1.317	0.333	2.066	4.492	8.935	0.8784	0.688	.8460E+1	.1182E+0	.7621E-2	.7849E+1
3.30	5.80	.687	1.337	0.313	2.926	4.540	10.678	0.9185	0.916	.5875E+1	.1702E+0	.9210E-2	.9808E+1
3.30	6.80	.699	1.349	0.301	4.102	4.569	13.044	0.9430	1.235	.4162E+1	.2403E+0	.1071E-1	.1230E+2
3.30	8.20	.707	1.357	0.293	5.486	4.588	15.821	0.9594	1.607	.3101E+1	.3225E+0	.1192E-1	.1518E+2
3.30	10.00	.712	1.362	0.288	7.276	4.600	19.407	0.9697	2.095	.2335E+1	.4284E+0	.1298E-1	.1882E+2
3.30	12.00	.716	1.366	0.284	9.268	4.610	23.395	0.9781	2.632	.1831E+1	.5462E+0	.1379E-1	.2288E+2
3.30	15.00	.719	1.369	0.281	12.262	4.617	29.387	0.9843	3.446	.1383E+1	.7230E+0	.1461E-1	.2893E+2
3.30	18.00	.721	1.371	0.279	15.258	4.622	35.381	0.9885	4.257	.1111E+1	.8999E+0	.1515E-1	.3497E+2
3.30	22.00	.722	1.372	0.278	19.256	4.625	43.378	0.9906	5.353	.8804E+0	.1136E+1	.1564E-1	.4297E+2
3.30	27.00	.723	1.373	0.277	24.254	4.627	53.375	0.9927	6.773	.6989E+0	.1431E+1	.1606E-1	.5298E+2
3.30	33.00	.724	1.374	0.276	30.252	4.630	65.373	0.9948	8.350	.5603E+0	.1785E+1	.1639E-1	.6503E+2
3.90	2.20	.100	1.050	0.900	0.100	3.714	5.299	0.1050	0.090	.3743E+4	.2672E-3	.3114E-4	.5564E+0
3.90	2.70	.263	1.213	0.737	0.274	4.251	5.833	0.3190	0.202	.2000E+3	.5001E-2	.4749E-3	.1861E+1
3.90	3.30	.434	1.384	0.566	0.532	4.525	6.544	0.6007	0.301	.4165E+2	.2401E-1	.1865E-2	.3931E+1
3.90	3.90	.551	1.501	0.449	0.898	4.809	7.410	0.8271	0.403	.1744E+2	.5735E-1	.3770E-2	.6128E+1
3.90	4.70	.627	1.577	0.373	1.546	4.994	8.792	0.9888	0.577	.8858E+1	.1129E+0	.6159E-2	.8694E+1
3.90	5.80	.661	1.611	0.331	2.434	5.077	10.495	1.0649	0.806	.5553E+1	.1801E+0	.8245E-2	.1118E+2
3.90	6.80	.680	1.630	0.320	3.540	5.123	12.841	1.1084	1.133	.3681E+1	.2717E+0	.1024E-1	.1423E+2
3.90	8.20	.691	1.641	0.309	4.918	5.149	15.609	1.1339	1.520	.2635E+1	.3795E+0	.1187E-1	.1770E+2
3.90	10.00	.698	1.648	0.302	6.704	5.166	19.189	1.1503	2.025	.1929E+1	.5185E+0	.1330E-1	.2207E+2
3.90	12.00	.702	1.652	0.298	8.696	5.176	23.176	1.1597	2.591	.1485E+1	.6733E+0	.1439E-1	.2688E+2
3.90	15.00	.706	1.656	0.294	11.684	5.186	29.166	1.1691	3.436	.1104E+1	.9057E+0	.1548E-1	.3410E+2
3.90	18.00	.708	1.658	0.292	14.684	5.191	35.161	1.1739	4.288	.8785E+0	.1138E+1	.1621E-1	.4127E+2

CHOKE/RESONANT CHARGE REACTOR DESIGN TABLE

$f(D)=U/(F*G*(D*E)^2)$

CASE # 3   COPYRIGHT 1982 STEVE SMITH   DATA DATE 1/ 3/82

P	Q	D	E	F	G	U	Um	DE	FG	f(D)	1/f(D)	1/(PQf(D))	DEUm
3.90	22.00	.710	1.660	0.290	18.680	5.196	43.155	1.1786	5.417	.6904E+0	.1448E+1	.1688E-1	.5086E+2
3.90	27.00	.711	1.661	0.289	23.678	5.198	53.152	1.1810	6.843	.5446E+0	.1836E+1	.1744E-1	.6277E+2
3.90	33.00	.712	1.662	0.288	29.676	5.200	65.149	1.1833	8.547	.4345E+0	.2301E+1	.1788E-1	.7709E+2
4.70	3.30	.197	1.547	0.803	0.206	4.749	6.878	0.3048	0.165	.3091E+3	.3235E-2	.2086E-3	.2096E+1
4.70	3.90	.374	1.724	0.626	0.452	5.179	7.572	0.6448	0.283	.4403E+2	.2271E-1	.1239E-2	.4882E+1
4.70	4.70	.539	1.889	0.461	0.922	5.580	8.700	1.0182	0.425	.1266E+2	.7896E-1	.3575E-2	.8859E+1
4.70	5.60	.619	1.969	0.381	1.662	5.774	10.272	1.2188	0.633	.6139E+1	.1629E+0	.6189E-2	.1252E+2
4.70	6.80	.656	2.006	0.344	2.788	5.864	12.566	1.3159	0.959	.3531E+1	.2832E+0	.8861E-2	.1654E+2
4.70	8.20	.674	2.024	0.326	4.152	5.908	15.315	1.3642	1.354	.2345E+1	.4264E+0	.1106E-1	.2089E+2
4.70	10.00	.684	2.034	0.316	5.932	5.932	18.886	1.3913	1.875	.1635E+1	.6116E+0	.1301E-1	.2628E+2
4.70	12.00	.689	2.039	0.311	7.922	5.945	22.872	1.4049	2.464	.1222E+1	.8180E+0	.1450E-1	.3213E+2
4.70	15.00	.694	2.044	0.306	10.912	5.957	28.857	1.4185	3.339	.8865E+0	.1128E+1	.1600E-1	.4094E+2
4.70	18.00	.697	2.047	0.303	13.906	5.964	34.849	1.4268	4.214	.6953E+0	.1438E+1	.1700E-1	.4972E+2
4.70	22.00	.699	2.049	0.301	17.902	5.969	42.843	1.4323	5.389	.5400E+0	.1852E+1	.1791E-1	.6136E+2
4.70	27.00	.700	2.050	0.300	22.900	5.971	52.840	1.4350	6.870	.4221E+0	.2369E+1	.1867E-1	.7583E+2
4.70	33.00	.702	2.052	0.298	28.896	5.976	64.835	1.4405	8.611	.3345E+0	.2990E+1	.1928E-1	.9339E+2
5.60	3.90	.099	1.899	0.901	0.102	5.411	7.972	0.1880	0.092	.1666E+4	.6003E-3	.2748E-4	.1499E+1
5.60	4.70	.343	2.143	0.657	0.414	6.004	8.874	0.7350	0.272	.4085E+2	.2448E-1	.9300E-3	.6523E+1
5.60	5.60	.530	2.330	0.470	0.954	6.458	10.140	1.2349	0.442	.9586E+1	.1043E+0	.3327E-2	.1252E+2
5.60	6.80	.623	2.423	0.377	1.954	6.684	12.274	1.5095	0.737	.3982E+1	.2511E+0	.6595E-2	.1853E+2
5.60	8.20	.655	2.455	0.345	3.290	6.762	14.983	1.6080	1.135	.2304E+1	.4340E+0	.9452E-2	.2409E+2
5.60	10.00	.671	2.471	0.329	5.058	6.801	18.537	1.6580	1.664	.1487E+1	.6727E+0	.1201E-1	.3073E+2
5.60	12.00	.679	2.479	0.321	7.042	6.847	22.514	1.6832	2.260	.1065E+1	.9391E+0	.1397E-1	.3790E+2
5.60	15.00	.685	2.485	0.315	10.030	6.835	28.497	1.7022	3.159	.7466E+0	.1339E+1	.1595E-1	.4851E+2
5.60	18.00	.688	2.488	0.312	13.024	6.842	34.488	1.7117	4.063	.5747E+0	.1740E+1	.1726E-1	.5904E+2
5.60	22.00	.690	2.490	0.310	17.020	6.847	42.483	1.7181	5.276	.4396E+0	.2275E+1	.1846E-1	.7299E+2
5.60	27.00	.692	2.492	0.308	22.016	6.852	52.477	1.7245	6.781	.3398E+0	.2943E+1	.1946E-1	.9049E+2
5.60	33.00	.693	2.493	0.307	28.014	6.854	64.474	1.7276	8.600	.2670E+0	.3745E+1	.2027E-1	.1114E+3
6.80	5.60	.255	2.655	0.745	0.290	6.990	10.411	0.6770	0.216	.7059E+2	.1417E-1	.3720E-3	.7044E+1
6.80	6.80	.523	2.923	0.477	0.954	7.641	12.045	1.5287	0.455	.7185E+1	.1392E+0	.3010E-2	.1841E+2
6.80	8.20	.622	3.022	0.378	2.156	7.882	14.562	1.8797	0.815	.2737E+1	.3653E+0	.6552E-2	.2737E+2
6.80	10.00	.654	3.054	0.346	3.892	7.959	18.070	1.9973	1.347	.1482E+1	.6749E+0	.9925E-2	.3609E+2
6.80	12.00	.667	3.067	0.333	5.866	7.991	22.033	2.0457	1.953	.9776E+0	.1023E+1	.1254E-1	.4507E+2
6.80	15.00	.676	3.076	0.324	8.848	8.013	28.008	2.0794	2.867	.6465E+0	.1547E+1	.1517E-1	.5824E+2
6.80	18.00	.680	3.080	0.320	11.840	8.030	33.996	2.0944	3.789	.4827E+0	.2072E+1	.1692E-1	.7120E+2
6.80	22.00	.683	3.083	0.317	15.834	8.030	41.988	2.1057	5.019	.3608E+0	.2772E+1	.1853E-1	.8841E+2

CASE # 3   COPYRIGHT 1982 STEVE SMITH   DATA DATE 1/ 3/82   CHOKE/RESONANT CHARGE REACTOR DESIGN TABLE

$f(D)=U/(F*G*(D*E)^2)$

P	Q	D	E	F	G	U	Um	DE	FG	f(D)	1/f(D)	1/(PQf(D))	DEUm
6.80	27.00	.685	3.085	0.315	20.830	8.035	51.982	2.1132	6.561	.2742E+0	.3647E+1	.1986E-1	.1098E+3
6.80	33.00	.686	3.086	0.314	26.828	8.037	62.979	2.1170	8.424	.2129E+0	.4697E+1	.2293E-1	.1354E+3
8.20	6.80	.194	3.294	0.806	0.212	8.242	12.384	0.6390	0.171	.1181E+3	.8466E-2	.1518E-3	.7914E+1
8.20	8.20	.518	3.618	0.482	0.964	9.029	14.258	1.8741	0.465	.5533E+1	.1807E+0	.2688E-2	.2672E+2
8.20	10.00	.627	3.727	0.373	2.546	9.294	17.547	2.3368	0.950	.1792E+1	.5580E+0	.6805E-2	.4100E+2
8.20	12.00	.654	3.754	0.346	4.492	9.359	21.470	2.4551	1.554	.9991E+0	.1001E+1	.1017E-1	.5271E+2
8.20	15.00	.667	3.767	0.333	7.466	9.391	27.432	2.5126	2.486	.5983E+0	.1671E+1	.1359E-1	.6893E+2
8.20	18.00	.672	3.772	0.328	10.456	9.403	33.418	2.5348	3.430	.4267E+0	.2343E+1	.1588E-1	.8471E+2
8.20	22.00	.676	3.776	0.324	14.448	9.413	41.407	2.5526	4.681	.3086E+0	.3240E+1	.1796E-1	.1057E+3
8.20	27.00	.679	3.779	0.321	19.442	9.420	51.398	2.5659	6.241	.2293E+0	.4362E+1	.1970E-1	.1319E+3
8.20	33.00	.681	3.781	0.319	25.438	9.425	63.392	2.5749	8.115	.1752E+0	.5708E+1	.2109E-1	.1632E+3
10.00	8.20	.066	4.066	0.934	0.068	9.731	14.778	0.2684	0.064	.2128E+4	.4700E-3	.5732E-5	.3966E+1
10.00	10.00	.514	4.514	0.486	0.972	10.819	17.097	2.3202	0.472	.4255E+1	.2350E+0	.2350E-2	.3967E+2
10.00	12.00	.627	4.627	0.373	2.746	11.094	20.774	2.9011	1.024	.1287E+1	.7771E+0	.6476E-2	.6027E+2
10.00	15.00	.656	4.656	0.344	5.688	11.164	26.691	3.0543	1.957	.6116E+0	.1635E+1	.1090E-1	.8152E+2
10.00	18.00	.665	4.665	0.335	8.670	11.186	32.666	3.1022	2.904	.4002E+0	.2499E+1	.1388E-1	.1013E+3
10.00	22.00	.671	4.671	0.329	12.658	11.201	40.648	3.1342	4.164	.2738E+0	.3652E+1	.1660E-1	.1274E+3
10.00	27.00	.674	4.674	0.326	17.652	11.208	50.640	3.1503	5.755	.1963E+0	.5095E+1	.1887E-1	.1595E+3
10.00	33.00	.676	4.676	0.324	23.648	11.213	62.634	3.1610	7.662	.1465E+0	.6828E+1	.2069E-1	.1980E+3

CASE # 4   COPYRIGHT 1982 STEVE SMITH   DATA DATE 1/ 3/82   CHOKE/RESONANT CHARGE REACTOR DESIGN TABLE

$f(D) = U/(F*G*(D*E)^2)$

SCALE BY 1.000E-04   FUNCTION 1/f(D)

Q/P	0.33	0.39	0.47	0.56	0.68	0.82	1.00	1.20	1.50	1.80	2.20	2.70	3.30	3.90	4.70	5.60	6.80	8.20	10.00
0.33	0	0	0	0	0	1	0	0	0	0	0	0	.	.	.	.	.	.	.
0.39	0	0	0	0	0	1	1	1	1	1	1	0	.	.	.	.	.	.	.
0.47	0	0	0	1	1	2	2	2	2	2	1	0	.	.	.	.	.	.	.
0.56	0	0	1	1	1	2	3	4	4	4	3	0	.	.	.	.	.	.	.
0.68	0	0	1	2	2	3	5	6	9	7	6	2	1	.	.	.	.	.	.
0.82	0	0	2	2	3	5	7	9	14	13	12	8	3	.	.	.	.	.	.
1.00	0	1	2	3	5	7	10	14	19	21	21	17	8	2	.	.	.	.	.
1.20	1	1	3	4	6	10	14	19	28	33	36	31	19	6	.	.	.	.	.
1.50	1	1	4	5	8	12	19	28	42	54	64	64	47	24	3	.	.	.	.
1.80	2	1	5	6	10	15	25	36	56	76	96	106	93	60	16	.	.	.	.
2.20	2	2	7	7	13	20	32	48	76	106	142	172	176	142	69	9	.	.	.
2.70	3	3	8	9	16	25	41	62	101	144	201	262	300	290	206	79	.	.	.
3.30	3	4	10	12	19	32	52	80	131	190	274	372	461	499	455	295	61	.	.
3.90	4	5	13	14	24	38	63	97	161	237	346	484	627	722	750	632	299	.	.
4.70	5	6	14	17	28	47	77	120	201	298	444	635	850	1026	1165	1162	882	17	.
5.60	6	7	17	20	32	56	94	146	247	368	554	805	1104	1372	1643	1792	1694	318	209
6.80	7	9	20	25	41	69	115	181	307	462	700	1031	1442	1835	2285	2648	2838	2589	1545
8.20	7	11	21	30	51	84	141	222	378	570	872	1296	1838	2376	3038	3654	4196	4361	3774
10.00	8	14	23	37	62	104	174	274	469	710	1092	1638	2348	3074	4008	4951	5951	6667	6793
12.00	11	16	28	45	76	125	210	332	570	866	1337	2017	2914	3850	5087	6394	7905	9241	10182
15.00	13	21	35	56	96	158	265	419	721	1099	1704	2585	3764	5014	6706	8561	10841	13110	15283
18.00	15	25	42	68	115	190	319	507	872	1332	2071	3154	4615	6178	8326	10729	13779	16981	20391
22.00	19	30	51	83	141	233	392	623	1074	1643	2561	3913	5749	7731	10487	13620	17697	22145	27205
27.00	23	37	63	102	174	287	483	768	1326	2032	3173	4861	7166	9672	13188	17235	22595	28602	35726
33.00	28	45	77	125	213	352	592	942	1629	2499	3908	5999	8867	12002	16429	21573	28474	36351	45952

CASE # 4   COPYRIGHT 1982 STEVE SMITH   DATA DATE 1/ 3/82   CHOKE/RESONANT CHARGE REACTOR DESIGN TABLE

SCALE BY 1.00E-04   FUNCTION 1/[P*Q*f(D)] $f(D)=U/[F*G*(D*E)^2]$

Q/P	0.33	0.39	0.47	0.56	0.68	0.82	1.00	1.20	1.50	1.80	2.20	2.70	3.30	3.90	4.70	5.60	6.80	8.20	10.00
0.33	2	2	2	2															
0.39	2	2	3	3															
0.47	2	2	3	3															
0.56	2	2	3	4							1		1						
0.68	2	3	3	5	3					1	2	1	2						
0.82	2	3	4	5	3	3			2	2	3	2	5						
1.00	2	3	4	6	4	3	3		3	3	4	4	10						
1.20	2	3	4	6	5	4	3	2	4	4	6	6	16	1					
1.50	2	3	4	6	5	5	5	4	6	6	10	10	24	4					
1.80	2	3	4	6	6	6	6	6	7	9	14	16	29	9	2				
2.20	2	3	5	6	7	7	7	8	10	12	19	22	36	17	7				
2.70	2	3	5	6	8	8	9	10	12	15	24	29	42	27	16	5			
3.30	3	3	5	6	8	9	10	12	15	20	29	34	46	39	29	16	3		
3.90	3	3	5	7	8	10	11	14	18	23	34	42	55	47	41	29	11		
4.70	3	3	5	7	9	10	13	15	21	27	38	46	56	56	53	44	28		
5.60	3	3	5	7	9	11	14	17	25	32	43	50	60	63	62	57	44	8	4
6.80	3	3	5	7	9	12	15	19	26	34	45	53	64	69	72	70	61	25	23
8.20	3	3	5	7	9	12	16	20	28	35	47	56	68	74	79	80	75	46	46
10.00	3	3	5	7	9	12	16	21	29	37	48	59	71	79	85	88	88	65	68
12.00	3	3	5	7	9	12	17	22	30	38	50	61	74	82	90	95	97	81	85
15.00	3	3	5	7	9	13	17	23	31	39	51	62	76	86	95	102	106	94	104
18.00	3	4	5	7	9	13	18	23	32	40	52	65	78	88	98	106	113	107	113
22.00	3	4	5	7	9	13	18	23	32	41	52	66	79	90	101	111	118	115	124
27.00	3	4	5	7	9	13	18	24	33	41	53	67	80	92	104	114	123	129	132
33.00	3	4	5	7	9	13	18	24	33	42	54	67	81	93	106	117	127	134	139

CASE # 4 COPYRIGHT 1982 STEVE SMITH DATA DATE 1/ 3/82

CHOKE/RESONANT CHARGE REACTOR DESIGN TABLE

$f(D)=U/(F*G*(D*E)^2)$

P	Q	D	E	F	G	U	Um	DE	FG	f(D)	1/f(D)	1/(PQf(D))	DEUm
0.33	0.33	.955	0.065	0.045	0.200	2.111	1.114	0.0621	0.009	.6086E+5	.1643E-4	.1509E-3	.6916E-1
0.33	0.39	.957	0.067	0.045	0.256	2.116	1.284	0.0641	0.011	.4675E+5	.2139E-4	.1662E-3	.8236E-1
0.33	0.47	.959	0.069	0.041	0.332	2.120	1.515	0.0662	0.014	.3558E+5	.2811E-4	.1812E-3	.1002E+0
0.33	0.56	.960	0.070	0.040	0.420	2.123	1.780	0.0672	0.017	.2798E+5	.3574E-4	.1934E-3	.1196E+0
0.33	0.68	.961	0.070	0.039	0.538	2.125	2.135	0.0682	0.021	.2176E+5	.4596E-4	.2048E-3	.1457E+0
0.33	0.82	.961	0.071	0.039	0.678	2.125	2.555	0.0682	0.026	.1726E+5	.5792E-4	.2141E-3	.1743E+0
0.33	1.00	.962	0.072	0.038	0.856	2.128	3.090	0.0693	0.033	.1363E+5	.7333E-4	.2223E-3	.2140E+0
0.33	1.20	.962	0.072	0.037	1.056	2.128	3.590	0.0693	0.040	.1105E+5	.9048E-4	.2285E-3	.2556E+0
0.33	1.50	.963	0.073	0.037	1.354	2.130	4.585	0.0703	0.050	.8664E+4	.1162E-3	.2348E-3	.3223E+0
0.33	1.80	.963	0.073	0.037	1.654	2.130	5.485	0.0703	0.061	.7043E+4	.1420E-3	.2390E-3	.3856E+0
0.33	2.20	.963	0.073	0.037	2.054	2.130	6.685	0.0703	0.076	.5672E+4	.1763E-3	.2429E-3	.4700E+0
0.33	2.70	.963	0.073	0.037	2.554	2.130	8.185	0.0703	0.094	.4561E+4	.2192E-3	.2461E-3	.5754E+0
0.33	3.30	.964	0.074	0.036	3.152	2.133	9.980	0.0713	0.113	.3693E+4	.2708E-3	.2486E-3	.7120E+0
0.33	3.90	.964	0.074	0.036	3.752	2.133	11.780	0.0713	0.135	.3103E+4	.3223E-3	.2504E-3	.8404E+0
0.33	4.70	.964	0.074	0.036	4.552	2.133	14.180	0.0713	0.164	.2557E+4	.3910E-3	.2521E-3	.1012E+1
0.33	5.60	.964	0.074	0.036	5.452	2.133	16.880	0.0713	0.196	.2135E+4	.4684E-3	.2534E-3	.1204E+1
0.33	6.80	.964	0.074	0.036	6.652	2.133	20.480	0.0713	0.239	.1750E+4	.5714E-3	.2547E-3	.1461E+1
0.33	8.20	.964	0.074	0.036	8.052	2.133	24.680	0.0713	0.290	.1446E+4	.6917E-3	.2556E-3	.1761E+1
0.33	10.00	.964	0.074	0.036	9.852	2.133	30.080	0.0713	0.355	.1182E+4	.8463E-3	.2565E-3	.2146E+1
0.33	12.00	.964	0.074	0.036	11.852	2.133	36.080	0.0713	0.427	.9822E+3	.1018E-2	.2571E-3	.2574E+1
0.33	15.00	.964	0.074	0.036	14.852	2.133	45.080	0.0713	0.535	.7838E+3	.1276E-2	.2578E-3	.3216E+1
0.33	18.00	.964	0.074	0.036	17.852	2.133	54.080	0.0713	0.643	.6521E+3	.1534E-2	.2582E-3	.3858E+1
0.33	22.00	.964	0.074	0.036	21.852	2.133	66.080	0.0713	0.787	.5327E+3	.1877E-2	.2586E-3	.4714E+1
0.33	27.00	.964	0.074	0.036	26.852	2.133	81.080	0.0713	0.967	.4335E+3	.2307E-2	.2589E-3	.5784E+1
0.33	33.00	.964	0.074	0.036	32.852	2.133	99.080	0.0713	1.183	.3543E+3	.2822E-2	.2592E-3	.7068E+1
0.39	0.33	.943	0.073	0.057	0.184	2.122	1.155	0.0688	0.010	.4269E+5	.2343E-4	.1820E-3	.7953E-1
0.39	0.39	.947	0.077	0.053	0.236	2.131	1.316	0.0729	0.013	.3205E+5	.3121E-4	.2052E-3	.9595E-1
0.39	0.47	.949	0.079	0.051	0.312	2.136	1.546	0.0750	0.016	.2388E+5	.4187E-4	.2284E-3	.1159E+0
0.39	0.56	.951	0.081	0.049	0.398	2.141	1.806	0.0770	0.020	.1850E+5	.5405E-4	.2475E-3	.1392E+0
0.39	0.68	.953	0.083	0.047	0.514	2.146	2.157	0.0791	0.024	.1420E+5	.7044E-4	.2656E-3	.1706E+0
0.39	0.82	.954	0.084	0.046	0.652	2.148	2.572	0.0801	0.030	.1115E+5	.8966E-4	.2803E-3	.2061E+0
0.39	1.00	.955	0.085	0.045	0.830	2.151	3.147	0.0812	0.037	.8739E+4	.1144E-3	.2934E-3	.2522E+0
0.39	1.20	.955	0.085	0.045	1.030	2.151	3.707	0.0812	0.046	.7042E+4	.1420E-3	.3034E-3	.3009E+0
0.39	1.50	.956	0.086	0.044	1.328	2.153	4.602	0.0822	0.058	.5451E+4	.1833E-3	.3136E-3	.3784E+0
0.39	1.80	.956	0.086	0.044	1.628	2.153	5.502	0.0822	0.072	.4447E+4	.2249E-3	.3223E-3	.4524E+0
0.39	2.20	.957	0.087	0.043	2.026	2.156	6.697	0.0833	0.087	.3569E+4	.2802E-3	.3265E-3	.5576E+0

CASE # 4   COPYRIGHT 1982 STEVE SMITH   DATA DATE 1/ 3/82   CHOKE/RESONANT CHARGE REACTOR DESIGN TABLE

$f(D)=U/(F*G*(D*E)^2)$

P	Q	D	E	F	G	U	Um	DE	FG	f(D)	1/f(D)	1/(PQf(D))	DEUm
0.39	2.70	.957	.087	.043	2.526	2.156	8.197	.0833	0.109	.2863E+4	.3493E-3	.3317E-3	.6825E+0
0.39	3.10	.957	.087	.043	3.126	2.156	9.397	.0833	0.134	.2313E+4	.4323E-3	.3359E-3	.8324E+0
0.39	3.90	.957	.087	.043	3.726	2.156	11.797	.0833	0.160	.1941E+4	.5152E-3	.3388E-3	.9822E+0
0.39	4.70	.957	.087	.043	4.526	2.156	14.197	.0833	0.195	.1598E+4	.6259E-3	.3414E-3	.1182E+1
0.39	5.40	.957	.087	.042	5.426	2.156	16.897	.0833	0.233	.1333E+4	.7503E-3	.3436E-3	.1407E+1
0.39	6.80	.958	.088	.042	6.624	2.158	20.492	.0843	0.278	.1091E+4	.9163E-3	.3455E-3	.1728E+1
0.39	8.20	.958	.088	.042	8.024	2.158	24.692	.0843	0.337	.9010E+3	.1110E-2	.3471E-3	.2082E+1
0.39	10.00	.958	.088	.042	9.824	2.158	30.092	.0843	0.413	.7359E+3	.1359E-2	.3484E-3	.2537E+1
0.39	12.00	.958	.088	.042	11.824	2.158	36.092	.0843	0.497	.6114E+3	.1636E-2	.3495E-3	.3043E+1
0.39	15.00	.958	.088	.042	14.824	2.158	45.092	.0843	0.623	.4877E+3	.2051E-2	.3505E-3	.3801E+1
0.39	18.00	.958	.088	.042	17.824	2.158	54.092	.0843	0.749	.4056E+3	.2466E-2	.3512E-3	.4560E+1
0.39	21.00	.958	.088	.042	21.824	2.158	66.092	.0843	0.917	.3313E+3	.3019E-2	.3518E-3	.5572E+1
0.39	27.00	.958	.088	.042	26.824	2.158	81.092	.0843	1.127	.2695E+3	.3710E-2	.3524E-3	.6836E+1
0.39	33.00	.958	.088	.042	32.824	2.158	99.092	.0843	1.379	.2202E+3	.4540E-2	.3528E-3	.8354E+1
0.47	0.33	.925	.082	.075	.167	2.131	1.220	.0755	0.012	.2988E+5	.3347E-4	.2158E-3	.9215E-1
0.47	0.39	.931	.088	.069	.215	2.146	1.371	.0816	0.015	.2175E+5	.4598E-4	.2599E-3	.1119E+0
0.47	0.47	.936	.093	.064	.285	2.158	1.586	.0867	0.018	.1574E+5	.6352E-4	.2875E-3	.1376E+0
0.47	0.56	.939	.096	.061	.369	2.165	1.842	.0898	0.022	.1193E+5	.8382E-4	.3185E-3	.1655E+0
0.47	0.68	.942	.099	.058	.483	2.172	2.187	.0929	0.028	.8983E+4	.1113E-3	.3483E-3	.2033E+0
0.47	0.82	.944	.101	.056	.619	2.177	2.598	.0950	0.035	.6959E+4	.1437E-3	.3728E-3	.2468E+0
0.47	1.00	.945	.102	.055	.797	2.180	3.133	.0961	0.044	.5389E+4	.1855E-3	.3948E-3	.3010E+0
0.47	1.20	.946	.103	.054	.995	2.182	3.728	.0971	0.054	.4307E+4	.2322E-3	.4117E-3	.3621E+0
0.47	1.50	.947	.104	.053	1.293	2.185	4.623	.0982	0.069	.3308E+4	.3023E-3	.4287E-3	.4539E+0
0.47	1.80	.948	.104	.053	1.593	2.185	5.523	.0982	0.084	.2665E+4	.3724E-3	.4442E-3	.5422E+0
0.47	2.20	.948	.105	.052	1.991	2.187	6.718	.0992	0.104	.2146E+4	.4660E-3	.4507E-3	.6666E+0
0.47	2.70	.948	.105	.052	2.491	2.187	8.218	.0992	0.130	.1715E+4	.5830E-3	.4595E-3	.8154E+0
0.47	3.30	.949	.105	.052	3.091	2.187	10.018	.0992	0.161	.1382E+4	.7235E-3	.4665E-3	.9940E+0
0.47	3.90	.949	.106	.051	3.689	2.189	11.813	.1003	0.188	.1157E+4	.8640E-3	.4714E-3	.1185E+1
0.47	4.70	.949	.106	.051	4.489	2.189	14.213	.1003	0.229	.9511E+3	.1051E-2	.4760E-3	.1425E+1
0.47	5.60	.949	.106	.051	5.389	2.189	16.913	.1003	0.275	.7923E+3	.1262E-2	.4796E-3	.1696E+1
0.47	6.80	.949	.106	.051	6.589	2.189	20.513	.1003	0.336	.6480E+3	.1543E-2	.4829E-3	.2057E+1
0.47	8.00	.949	.106	.051	7.989	2.189	24.713	.1003	0.407	.5344E+3	.1871E-2	.4855E-3	.2478E+1
0.47	10.00	.949	.106	.051	9.789	2.189	30.113	.1003	0.499	.4361E+3	.2293E-2	.4878E-3	.3020E+1
0.47	12.00	.949	.106	.051	11.789	2.189	36.113	.1003	0.601	.3622E+3	.2761E-2	.4896E-3	.3621E+1
0.47	15.00	.949	.106	.051	14.789	2.189	45.113	.1003	0.754	.2887E+3	.3464E-2	.4913E-3	.4524E+1
0.47	18.00	.949	.106	.051	17.789	2.189	54.113	.1003	0.907	.2400E+3	.4167E-2	.4925E-3	.5426E+1

CASE # 4   COPYRIGHT 1982 STEVE SMITH   DATA DATE 1/ 3/82   CHOKE/RESONANT CHARGE REACTOR DESIGN TABLE

$f(D)=U/(F*G*(D*E)^2)$

P	Q	D	E	F	G	U	Um	DE	FG	f(D)	1/f(D)	1/(Pqf(D))	DEUm
0.47	22.00	.949	0.106	0.051	21.789	2.189	66.113	0.1003	1.111	.1959E+3	.5104E-2	.4936E-3	.6630E+1
0.47	27.00	.949	0.106	0.051	26.789	2.189	81.113	0.1003	1.366	.1594E+3	.6275E-2	.4945E-3	.8134E+1
0.47	33.00	.949	0.106	0.051	32.789	2.189	99.113	0.1003	1.672	.1302E+3	.7680E-2	.4952E-3	.9933E+1
0.56	0.33	.903	0.090	0.097	0.151	2.138	1.301	0.0810	0.015	.2231E+5	.4482E-4	.2425E-3	.1053E+0
0.56	0.39	.911	0.098	0.089	0.195	2.157	1.442	0.0890	0.017	.1573E+5	.6358E-4	.2911E-3	.1283E+0
0.56	0.47	.919	0.106	0.081	0.259	2.177	1.643	0.0971	0.021	.1102E+5	.9077E-4	.3449E-3	.1596E+0
0.56	0.56	.924	0.111	0.076	0.339	2.189	1.889	0.1023	0.026	.8133E+4	.1230E-3	.3921E-3	.1932E+0
0.56	0.68	.928	0.115	0.072	0.451	2.198	2.230	0.1064	0.032	.5983E+4	.1671E-3	.4389E-3	.2372E+0
0.56	0.82	.931	0.118	0.069	0.585	2.206	2.635	0.1095	0.040	.4556E+4	.2195E-3	.4780E-3	.2887E+0
0.56	1.00	.934	0.121	0.066	0.759	2.213	3.160	0.1127	0.050	.3480E+4	.2874E-3	.5132E-3	.3562E+0
0.56	1.20	.935	0.122	0.065	0.957	2.215	3.756	0.1138	0.062	.2753E+4	.3632E-3	.5405E-3	.4272E+0
0.56	1.50	.936	0.123	0.064	1.255	2.218	4.651	0.1148	0.080	.2095E+4	.4773E-3	.5682E-3	.5340E+0
0.56	1.80	.937	0.124	0.063	1.553	2.220	5.546	0.1159	0.098	.1690E+4	.5916E-3	.5869E-3	.6426E+0
0.56	2.20	.938	0.125	0.062	1.951	2.223	6.741	0.1169	0.121	.1344E+4	.7440E-3	.6039E-3	.7883E+0
0.56	2.70	.938	0.125	0.062	2.451	2.223	8.241	0.1169	0.152	.1070E+4	.9348E-3	.6182E-3	.9637E+0
0.56	3.30	.939	0.126	0.061	3.049	2.225	10.036	0.1180	0.186	.8593E+3	.1164E-2	.6297E-3	.1184E+1
0.56	3.90	.939	0.126	0.061	3.649	2.225	11.836	0.1180	0.223	.7180E+3	.1393E-2	.6377E-3	.1397E+1
0.56	4.70	.939	0.126	0.061	4.449	2.225	14.236	0.1180	0.271	.5889E+3	.1698E-2	.6452E-3	.1688E+1
0.56	5.60	.940	0.126	0.061	5.349	2.225	16.936	0.1180	0.326	.4998E+3	.2042E-2	.6510E-3	.1998E+1
0.56	6.60	.940	0.127	0.060	6.547	2.228	20.531	0.1191	0.393	.4000E+3	.2500E-2	.6565E-3	.2445E+1
0.56	8.20	.940	0.127	0.060	7.947	2.228	24.731	0.1191	0.477	.3295E+3	.3034E-2	.6608E-3	.2945E+1
0.56	10.00	.940	0.127	0.060	9.747	2.228	30.131	0.1191	0.585	.2687E+3	.3722E-2	.6646E-3	.3588E+1
0.56	12.00	.940	0.127	0.060	11.747	2.228	36.131	0.1191	0.705	.2229E+3	.4486E-2	.6675E-3	.4302E+1
0.56	15.00	.940	0.127	0.060	14.747	2.228	45.131	0.1191	0.885	.1776E+3	.5631E-2	.6704E-3	.5374E+1
0.56	18.00	.940	0.127	0.060	17.747	2.228	54.131	0.1191	1.065	.1476E+3	.6777E-2	.6723E-3	.6445E+1
0.56	22.00	.940	0.127	0.060	21.747	2.228	66.131	0.1191	1.305	.1204E+3	.8304E-2	.6740E-3	.7874E+1
0.56	27.00	.940	0.127	0.060	26.747	2.228	81.131	0.1191	1.605	.9791E+2	.1021E-1	.6755E-3	.9660E+1
0.56	33.00	.940	0.127	0.060	32.747	2.228	99.131	0.1191	1.965	.7997E+2	.1250E-1	.6766E-3	.1180E+2
0.68	0.33	.869	0.096	0.131	0.139	2.135	1.432	0.0831	0.018	.1701E+5	.5880E-4	.2620E-3	.1190E+0
0.68	0.47	.881	0.108	0.119	0.175	2.164	1.554	0.0949	0.021	.1157E+5	.8641E-4	.3258E-3	.1474E+0
0.68	0.56	.892	0.119	0.108	0.233	2.191	1.740	0.1059	0.025	.7782E+4	.1285E-3	.4021E-3	.1842E+0
0.68	0.68	.902	0.129	0.098	0.303	2.215	1.962	0.1161	0.030	.5445E+4	.1803E-3	.4736E-3	.2277E+0
0.68	0.82	.909	0.136	0.091	0.409	2.232	2.288	0.1233	0.037	.3947E+4	.2534E-3	.5479E-3	.2821E+0
0.68	0.82	.914	0.141	0.086	0.539	2.254	2.683	0.1286	0.046	.2931E+4	.3412E-3	.6119E-3	.3450E+0
0.68	1.00	.918	0.145	0.082	0.711	2.255	3.204	0.1338	0.058	.2193E+4	.4560E-3	.6705E-3	.4255E+0
0.68	1.20	.920	0.147	0.080	0.907	2.259	3.794	0.1349	0.073	.1711E+4	.5846E-3	.7164E-3	.5120E+0

CHOKE/RESONANT CHARGE REACTOR DESIGN TABLE

$$f(D) = U/(F \ast G \ast (D \ast E)^2)$$

CASE # 4   COPYRIGHT 1982 STEVE SMITH   DATA DATE 1/ 3/82

P	Q	D	E	F	G	U	Um	DE	FG	f(D)	1/f(D)	1/(PQf(D))	DEUm
0.68	1.50	.922	0.149	0.078	1.203	2.264	4.684	0.1371	0.094	.1284E+4	.7785E-3	.7633E-3	.6421E+0
0.68	1.80	.926	0.150	0.077	1.501	2.266	5.580	0.1381	0.116	.1028E+4	.9730E-3	.7949E-3	.7708E+0
0.68	2.20	.925	0.152	0.075	1.897	2.271	6.770	0.1403	0.142	.8112E+3	.1233E-2	.8240E-3	.9497E+0
0.68	2.70	.925	0.152	0.075	2.397	2.271	8.270	0.1403	0.180	.6420E+3	.1558E-2	.8484E-3	.1160E+1
0.68	3.30	.926	0.153	0.074	2.995	2.274	10.465	0.1414	0.222	.5134E+3	.1948E-2	.8681E-3	.1423E+1
0.68	3.90	.926	0.153	0.074	3.595	2.274	11.865	0.1414	0.266	.4277E+3	.2338E-2	.8817E-3	.1677E+1
0.68	4.70	.927	0.154	0.073	4.393	2.276	14.260	0.1424	0.321	.3493E+3	.2859E-2	.8945E-3	.2031E+1
0.68	5.60	.927	0.154	0.073	5.293	2.276	16.960	0.1424	0.386	.2903E+3	.3445E-2	.9046E-3	.2416E+1
0.68	6.80	.927	0.154	0.073	6.493	2.276	20.560	0.1424	0.474	.2367E+3	.4226E-2	.9138E-3	.2929E+1
0.68	8.20	.928	0.155	0.072	7.891	2.278	24.755	0.1435	0.568	.1947E+3	.5137E-2	.9212E-3	.3553E+1
0.68	10.00	.928	0.155	0.072	9.691	2.278	30.155	0.1435	0.698	.1585E+3	.6309E-2	.9277E-3	.4328E+1
0.68	12.00	.928	0.155	0.072	11.691	2.278	36.155	0.1435	0.842	.1314E+3	.7611E-2	.9327E-3	.5189E+1
0.68	15.00	.928	0.155	0.072	14.691	2.278	45.155	0.1435	1.058	.1046E+3	.9564E-2	.9376E-3	.6481E+1
0.68	18.00	.928	0.155	0.072	17.691	2.278	54.155	0.1435	1.274	.8683E+2	.1152E-1	.9409E-3	.7773E+1
0.68	22.00	.928	0.155	0.072	21.691	2.278	66.155	0.1435	1.562	.7082E+2	.1412E-1	.9439E-3	.9495E+1
0.68	27.00	.928	0.155	0.072	26.691	2.278	81.155	0.1435	1.922	.5755E+2	.1738E-1	.9464E-3	.1165E+2
0.68	33.00	.928	0.155	0.072	32.691	2.278	99.155	0.1435	2.354	.4699E+2	.2128E-1	.9484E-3	.1423E+2
0.82	0.33	.827	0.100	0.173	0.129	2.126	1.596	0.0830	0.022	.1380E+5	.7245E-4	.2677E-3	.1324E+0
0.82	0.39	.841	0.114	0.159	0.161	2.160	1.708	0.0962	0.026	.9109E+4	.1098E-3	.3433E-3	.1642E+0
0.82	0.47	.857	0.130	0.143	0.209	2.178	1.870	0.1111	0.030	.5889E+4	.1698E-3	.4406E-3	.2089E+0
0.82	0.56	.871	0.144	0.129	0.271	2.233	2.072	0.1257	0.035	.4037E+4	.2477E-3	.5394E-3	.2605E+0
0.82	0.68	.883	0.156	0.117	0.367	2.262	2.374	0.1380	0.043	.2763E+4	.3620E-3	.6492E-3	.3277E+0
0.82	0.82	.892	0.165	0.102	0.489	2.284	2.750	0.1475	0.053	.1987E+4	.5032E-3	.7483E-3	.4056E+0
0.82	1.00	.898	0.171	0.100	0.657	2.299	3.261	0.1539	0.067	.1448E+4	.6904E-3	.8420E-3	.5017E+0
0.82	1.20	.902	0.175	0.098	0.849	2.309	3.841	0.1582	0.083	.1109E+4	.9018E-3	.9164E-3	.6075E+0
0.82	1.50	.906	0.179	0.094	1.141	2.318	4.722	0.1625	0.107	.8186E+3	.1222E-2	.9932E-3	.7672E+0
0.82	1.80	.908	0.181	0.092	1.437	2.323	5.612	0.1647	0.132	.6481E+3	.1543E-2	.1045E-2	.9241E+0
0.82	2.20	.909	0.182	0.091	1.835	2.326	6.807	0.1657	0.167	.5069E+3	.1973E-2	.1094E-2	.1128E+1
0.82	2.70	.910	0.183	0.090	2.333	2.328	8.303	0.1668	0.210	.3983E+3	.2511E-2	.1134E-2	.1385E+1
0.82	3.30	.911	0.184	0.089	2.931	2.330	10.098	0.1679	0.261	.3168E+3	.3157E-2	.1167E-2	.1696E+1
0.82	3.90	.912	0.185	0.088	3.529	2.333	11.893	0.1690	0.311	.2629E+3	.3803E-2	.1189E-2	.2010E+1
0.82	4.70	.913	0.186	0.087	4.327	2.333	14.288	0.1701	0.376	.2143E+3	.4666E-2	.1211E-2	.2431E+1
0.82	5.60	.913	0.186	0.087	5.227	2.335	16.988	0.1701	0.455	.1774E+3	.5636E-2	.1227E-2	.2890E+1
0.82	6.80	.913	0.186	0.087	6.427	2.335	20.588	0.1701	0.559	.1443E+3	.6930E-2	.1243E-2	.3502E+1
0.82	8.20	.914	0.187	0.086	7.825	2.338	24.783	0.1712	0.673	.1185E+3	.8440E-2	.1255E-2	.4243E+1
0.82	10.00	.914	0.187	0.086	9.625	2.338	30.183	0.1712	0.828	.9633E+2	.1038E-1	.1266E-2	.5168E+1

CASE # 4   COPYRIGHT 1982 STEVE SMITH   DATA DATE 1/ 3/82   CHOKE/RESONANT CHARGE REACTOR DESIGN TABLE

$$f(D)=U/(F*G*(D*E)^2)$$

P	Q	D	E	F	G	U	Um	DE	FG	f(D)	1/f(D)	1/(PQf(D))	DEUm
0.82	12.00	.914	0.187	0.086	11.625	2.338	36.183	0.1712	1.000	.7976E+2	.1254E-1	.1274E-2	.6195E+1
0.82	15.00	.914	0.187	0.086	14.625	2.338	45.183	0.1712	1.258	.6340E+2	.1577E-1	.1282E-2	.7736E+1
0.82	18.00	.914	0.187	0.086	17.625	2.338	54.183	0.1712	1.516	.5261E+2	.1901E-1	.1288E-2	.9277E+1
0.82	22.00	.914	0.187	0.086	21.625	2.338	66.183	0.1712	1.860	.4288E+2	.2332E-1	.1293E-2	.1133E+2
0.82	27.00	.915	0.188	0.085	26.623	2.340	81.178	0.1723	2.263	.3482E+2	.2872E-1	.1297E-2	.1399E+2
0.82	33.00	.915	0.188	0.085	32.623	2.340	99.178	0.1723	2.773	.2842E+2	.3519E-1	.1300E-2	.1709E+2
1.00	0.33	.770	0.103	0.230	0.123	2.108	1.821	0.0796	0.028	.1174E+5	.8519E-4	.2582E-3	.1449E+0
1.00	0.39	.787	0.120	0.213	0.149	2.149	1.919	0.0947	0.032	.7534E+4	.1327E-3	.3403E-3	.1817E+0
1.00	0.47	.807	0.140	0.193	0.189	2.198	2.062	0.1132	0.037	.4690E+4	.2132E-3	.4537E-3	.2335E+0
1.00	0.56	.825	0.158	0.175	0.243	2.242	2.244	0.1306	0.043	.3085E+4	.3241E-3	.5788E-3	.2931E+0
1.00	0.68	.844	0.177	0.156	0.325	2.288	2.512	0.1497	0.051	.2012E+4	.4970E-3	.7308E-3	.3759E+0
1.00	0.82	.859	0.192	0.141	0.435	2.324	2.859	0.1652	0.061	.1387E+4	.7209E-3	.8791E-3	.4723E+0
1.00	1.00	.870	0.203	0.130	0.593	2.351	3.345	0.1769	0.077	.9739E+3	.1027E-2	.1027E-2	.5918E+0
1.00	1.20	.878	0.211	0.122	0.777	2.370	3.907	0.1856	0.095	.7260E+3	.1377E-2	.1148E-2	.7249E+0
1.00	1.50	.884	0.217	0.116	1.065	2.385	4.797	0.1913	0.124	.5228E+3	.1913E-2	.1275E-2	.9179E+0
1.00	1.80	.887	0.220	0.113	1.359	2.392	5.663	0.1954	0.154	.4077E+3	.2453E-2	.1363E-2	.1107E+1
1.00	2.20	.890	0.223	0.110	1.753	2.399	6.848	0.1988	0.193	.3149E+3	.3176E-2	.1443E-2	.1361E+1
1.00	2.70	.892	0.225	0.108	2.249	2.404	8.339	0.2010	0.243	.2458E+3	.4082E-2	.1512E-2	.1676E+1
1.00	3.30	.893	0.226	0.107	2.847	2.407	10.134	0.2021	0.305	.1934E+3	.5171E-2	.1567E-2	.2048E+1
1.00	3.90	.894	0.227	0.106	3.445	2.409	11.929	0.2032	0.365	.1597E+3	.6261E-2	.1605E-2	.2424E+1
1.00	4.70	.895	0.228	0.105	4.243	2.412	14.324	0.2044	0.446	.1296E+3	.7716E-2	.1642E-2	.2927E+1
1.00	5.60	.896	0.229	0.104	5.141	2.414	17.019	0.2055	0.535	.1069E+3	.9352E-2	.1670E-2	.3497E+1
1.00	6.80	.896	0.229	0.104	6.341	2.414	20.619	0.2055	0.659	.8669E+2	.1154E-1	.1696E-2	.4237E+1
1.00	8.20	.897	0.230	0.103	7.739	2.416	24.814	0.2066	0.797	.7101E+2	.1408E-1	.1717E-2	.5127E+1
1.00	10.00	.897	0.230	0.103	9.539	2.416	30.214	0.2066	0.983	.5761E+2	.1736E-1	.1736E-2	.6243E+1
1.00	12.00	.897	0.230	0.103	11.539	2.416	36.214	0.2066	1.189	.4763E+2	.2100E-1	.1750E-2	.7482E+1
1.00	15.00	.897	0.230	0.103	14.539	2.416	45.214	0.2066	1.498	.3780E+2	.2645E-1	.1764E-2	.9342E+1
1.00	18.00	.898	0.231	0.102	17.537	2.419	54.209	0.2077	1.789	.3133E+2	.3191E-1	.1773E-2	.1126E+2
1.00	22.00	.898	0.231	0.102	21.537	2.419	66.209	0.2077	2.197	.2551E+2	.3919E-1	.1781E-2	.1375E+2
1.00	27.00	.898	0.231	0.102	26.537	2.419	81.209	0.2077	2.707	.2071E+2	.4829E-1	.1789E-2	.1687E+2
1.00	33.00	.898	0.231	0.102	32.537	2.419	99.209	0.2077	3.319	.1689E+2	.5921E-1	.1794E-2	.2061E+2
1.20	0.33	.706	0.106	0.294	0.118	2.086	2.075	0.0748	0.035	.1074E+5	.9315E-4	.2352E-3	.1553E+0
1.20	0.39	.724	0.124	0.276	0.142	2.130	2.168	0.0898	0.039	.6742E+4	.1483E-3	.3169E-3	.1946E+0
1.20	0.47	.746	0.146	0.254	0.178	2.183	2.301	0.1089	0.045	.4070E+4	.2457E-3	.4356E-3	.2505E+0
1.20	0.56	.769	0.169	0.231	0.222	2.229	2.459	0.1100	0.051	.2585E+4	.3869E-3	.5757E-3	.3196E+0
1.20	0.68	.795	0.195	0.205	0.290	2.302	2.693	0.1550	0.059	.1611E+4	.6207E-3	.7606E-3	.4174E+0

CASE # 4   COPYRIGHT 1982 STEVE SMITH   DATA DATE 1/ 3/82   CHOKE/RESONANT CHARGE REACTOR DESIGN TABLE

$$f(D)=U/(F*G*(D*E)^2)$$

P	Q	D	E	F	G	U	Um	DE	FG	f(D)	1/f(D)	1/(PQf(D))	DEUm
1.20	0.82	.817	0.217	0.183	0.386	2.355	3.006	0.1773	0.071	.1061E+4	.9426E-3	.9579E-3	.5329E+0
1.20	1.00	.836	0.236	0.164	0.528	2.402	3.453	0.1973	0.087	.7125E+3	.1403E-2	.1170E-2	.6813E+0
1.20	1.20	.848	0.248	0.152	0.704	2.431	3.995	0.2103	0.107	.5136E+3	.1947E-2	.1352E-2	.8402E+0
1.20	1.50	.858	0.258	0.142	0.984	2.455	4.847	0.2214	0.140	.3586E+3	.2789E-2	.1549E-2	.1073E+1
1.20	1.80	.864	0.264	0.136	1.272	2.470	5.717	0.2281	0.173	.2744E+3	.3644E-2	.1687E-2	.1304E+1
1.20	2.20	.868	0.268	0.132	1.664	2.479	6.898	0.2326	0.220	.2086E+3	.4794E-2	.1816E-2	.1605E+1
1.20	2.70	.871	0.271	0.129	2.158	2.487	8.383	0.2360	0.278	.1603E+3	.6237E-2	.1925E-2	.1979E+1
1.20	3.30	.873	0.273	0.127	2.754	2.491	10.174	0.2383	0.350	.1254E+3	.7974E-2	.2014E-2	.2425E+1
1.20	3.90	.875	0.275	0.125	3.350	2.496	11.964	0.2406	0.419	.1030E+3	.9712E-2	.2075E-2	.2879E+1
1.20	4.70	.876	0.276	0.124	4.148	2.499	14.359	0.2418	0.514	.8311E+2	.1203E-1	.2133E-2	.3472E+1
1.20	5.60	.877	0.277	0.123	5.046	2.501	17.054	0.2429	0.621	.6829E+2	.1464E-1	.2179E-2	.4143E+1
1.20	6.80	.878	0.278	0.122	6.244	2.504	20.649	0.2441	0.762	.5517E+2	.1813E-1	.2221E-2	.5040E+1
1.20	8.20	.878	0.278	0.122	7.644	2.504	24.849	0.2441	0.933	.4506E+2	.2219E-1	.2255E-2	.6065E+1
1.20	10.00	.879	0.279	0.121	9.442	2.506	30.245	0.2452	1.142	.3647E+2	.2742E-1	.2285E-2	.7417E+1
1.20	12.00	.879	0.279	0.121	11.442	2.506	36.245	0.2452	1.384	.3010E+2	.3323E-1	.2307E-2	.8889E+1
1.20	15.00	.880	0.280	0.120	14.440	2.508	45.240	0.2464	1.733	.2384E+2	.4194E-1	.2330E-2	.1115E+2
1.20	18.00	.880	0.280	0.120	17.440	2.508	54.240	0.2464	2.093	.1974E+2	.5065E-1	.2345E-2	.1336E+2
1.20	22.00	.880	0.280	0.120	21.440	2.508	66.240	0.2464	2.573	.1606E+2	.6227E-1	.2359E-2	.1632E+2
1.20	27.00	.880	0.280	0.120	26.440	2.508	81.240	0.2464	3.173	.1302E+2	.7679E-1	.2370E-2	.2002E+2
1.20	33.00	.881	0.281	0.120	32.438	2.511	99.235	0.2476	3.860	.1061E+2	.9422E-1	.2379E-2	.2457E+2
1.50	0.33	.609	0.109	0.391	0.112	2.050	2.460	0.0664	0.044	.1062E+5	.9412E-4	.1901E-3	.1633E+0
1.50	0.39	.623	0.128	0.372	0.134	2.096	2.548	0.0804	0.050	.6508E+4	.1530E-3	.2626E-3	.2048E+0
1.50	0.47	.653	0.153	0.347	0.164	2.157	2.667	0.0999	0.057	.3797E+4	.2633E-3	.3735E-3	.2664E+0
1.50	0.56	.679	0.179	0.321	0.202	2.222	2.810	0.1215	0.065	.2318E+4	.4314E-3	.5136E-3	.3416E+0
1.50	0.68	.711	0.211	0.289	0.258	2.298	3.015	0.1500	0.075	.1369E+4	.7303E-3	.7159E-3	.4523E+0
1.50	0.82	.743	0.243	0.257	0.334	2.376	3.279	0.1805	0.086	.8490E+3	.1178E-2	.9576E-3	.5921E+0
1.50	1.00	.775	0.275	0.225	0.450	2.453	3.664	0.2131	0.101	.5335E+3	.1875E-2	.1250E-2	.7809E+0
1.50	1.20	.798	0.298	0.202	0.604	2.509	4.152	0.2378	0.122	.3637E+3	.2750E-2	.1528E-2	.9874E+0
1.50	1.50	.818	0.318	0.182	0.864	2.558	4.955	0.2601	0.157	.2404E+3	.4160E-2	.1849E-2	.1289E+1
1.50	1.80	.828	0.328	0.172	1.144	2.582	5.806	0.2716	0.197	.1779E+3	.5621E-2	.2082E-2	.1577E+1
1.50	2.20	.836	0.336	0.164	1.528	2.602	6.968	0.2809	0.251	.1316E+3	.7603E-2	.2303E-2	.1957E+1
1.50	2.70	.842	0.342	0.158	2.016	2.616	8.438	0.2880	0.319	.9905E+2	.1010E-1	.2493E-2	.2430E+1
1.50	3.30	.845	0.345	0.155	2.610	2.623	10.224	0.2915	0.405	.7631E+2	.1311E-1	.2648E-2	.2981E+1
1.50	3.90	.848	0.348	0.152	3.204	2.631	12.009	0.2951	0.487	.6203E+2	.1612E-1	.2756E-2	.3544E+1
1.50	4.70	.850	0.350	0.150	4.000	2.636	14.400	0.2975	0.600	.4963E+2	.2015E-1	.2858E-2	.4284E+1
1.50	5.60	.851	0.351	0.149	4.898	2.638	17.095	0.2987	0.730	.4051E+2	.2468E-1	.2938E-2	.5106E+1

CHOKE/RESONANT CHARGE REACTOR DESIGN TABLE

CASE # 4   COPYRIGHT 1982 STEVE SMITH   DATA DATE 1/ 3/82

$$f(D) = U/(F \cdot G \cdot (D \cdot E)^2)$$

P	Q	D	E	F	G	U	Um	DE	FG	f(D)	1/f(D)	1/(PQf(D))	DEUm
1.50	6.80	.852	0.352	0.148	6.096	2.640	20.690	0.2999	0.902	.3254E+2	.3073E-1	.3013E-2	.6205E+1
1.50	8.20	.853	0.353	0.147	7.494	2.643	24.885	0.3011	1.102	.2646E+2	.3779E-1	.3072E-2	.7493E+1
1.50	10.00	.854	0.354	0.146	9.292	2.645	30.280	0.3023	1.357	.2134E+2	.4687E-1	.3125E-2	.9154E+1
1.50	12.00	.855	0.355	0.145	11.290	2.648	36.275	0.3035	1.637	.1756E+2	.5690E-1	.3164E-2	.1101E+2
1.50	15.00	.856	0.355	0.145	14.290	2.648	45.275	0.3047	2.072	.1387E+2	.7210E-1	.3204E-2	.1374E+2
1.50	18.00	.856	0.356	0.144	17.288	2.650	54.270	0.3047	2.489	.1146E+2	.8723E-1	.3231E-2	.1654E+2
1.50	22.00	.856	0.356	0.144	21.288	2.650	66.270	0.3047	3.065	.9310E+1	.1074E+0	.3255E-2	.2019E+2
1.50	27.00	.856	0.356	0.144	26.288	2.650	81.270	0.3047	3.785	.7539E+1	.1326E+0	.3275E-2	.2477E+2
1.50	33.00	.856	0.356	0.144	32.288	2.650	99.270	0.3047	4.649	.6138E+1	.1629E+0	.3291E-2	.3025E+2
1.80	0.33	.511	0.111	0.489	0.108	2.012	2.851	0.0567	0.053	.1184E+5	.8444E-4	.1422E-3	.1617E+0
1.80	0.39	.531	0.131	0.469	0.128	2.061	2.934	0.0696	0.060	.7094E+4	.1410E-3	.2008E-3	.2041E+0
1.80	0.47	.557	0.157	0.443	0.156	2.124	3.047	0.0874	0.069	.4019E+4	.2488E-3	.2941E-3	.2665E+0
1.80	0.56	.586	0.186	0.414	0.188	2.194	3.176	0.1090	0.078	.2373E+4	.4214E-3	.4180E-3	.3462E+0
1.80	0.68	.622	0.221	0.378	0.236	2.282	3.361	0.1381	0.089	.1341E+4	.7455E-3	.6090E-3	.4642E+0
1.80	0.82	.661	0.262	0.339	0.298	2.376	3.592	0.1725	0.101	.7904E+3	.1265E-2	.8572E-3	.6197E+0
1.80	1.00	.703	0.303	0.297	0.394	2.479	3.928	0.2130	0.117	.4668E+3	.2142E-2	.1190E-2	.8367E+0
1.80	1.20	.739	0.339	0.261	0.522	2.566	4.353	0.2505	0.136	.3001E+3	.3332E-2	.1543E-2	.1091E+1
1.80	1.50	.772	0.372	0.228	0.756	2.646	5.093	0.2872	0.172	.1861E+3	.5372E-2	.1990E-2	.1463E+1
1.80	1.80	.791	0.391	0.209	1.018	2.692	5.900	0.3093	0.213	.1323E+3	.7559E-2	.2333E-2	.1825E+1
1.80	2.20	.804	0.404	0.196	1.392	2.724	7.037	0.3248	0.273	.9463E+2	.1057E-1	.2669E-2	.2286E+1
1.80	2.70	.813	0.413	0.187	1.874	2.746	8.493	0.3358	0.350	.6950E+2	.1439E-1	.2961E-2	.2852E+1
1.80	3.30	.819	0.419	0.181	2.462	2.760	10.264	0.3432	0.446	.5260E+2	.1901E-1	.3200E-2	.3522E+1
1.80	3.90	.822	0.422	0.178	3.056	2.768	12.050	0.3469	0.544	.4228E+2	.2365E-1	.3369E-2	.4180E+1
1.80	4.70	.825	0.425	0.175	3.850	2.775	14.435	0.3506	0.674	.3330E+2	.2985E-1	.3528E-2	.5061E+1
1.80	5.60	.828	0.428	0.172	4.744	2.782	17.121	0.3544	0.816	.2715E+2	.3683E-1	.3654E-2	.6067E+1
1.80	6.80	.829	0.429	0.171	5.942	2.785	20.716	0.3556	1.016	.2167E+2	.4615E-1	.3771E-2	.7367E+1
1.80	8.20	.831	0.431	0.168	7.338	2.789	24.906	0.3582	1.240	.1753E+2	.5703E-1	.3864E-2	.8920E+1
1.80	10.00	.832	0.432	0.167	9.136	2.792	30.301	0.3594	1.535	.1408E+2	.7102E-1	.3946E-2	.1089E+2
1.80	12.00	.833	0.433	0.167	11.134	2.794	36.296	0.3607	1.859	.1155E+2	.8657E-1	.4008E-2	.1309E+2
1.80	15.00	.833	0.433	0.166	14.134	2.794	45.296	0.3607	2.360	.9100E+1	.1099E+0	.4070E-2	.1634E+2
1.80	18.00	.834	0.434	0.166	17.132	2.797	54.291	0.3620	2.844	.7506E+1	.1332E+0	.4112E-2	.1965E+2
1.80	22.00	.834	0.434	0.165	21.132	2.797	66.291	0.3632	3.548	.6085E+1	.1643E+0	.4150E-2	.2399E+2
1.80	27.00	.835	0.435	0.165	26.130	2.799	81.287	0.3632	4.311	.4921E+1	.2032E+0	.4181E-2	.2953E+2
1.80	33.00	.835	0.435	0.165	32.130	2.799	99.287	0.3632	5.301	.4002E+1	.2499E+0	.4207E-2	.3606E+2
2.20	0.33	.381	0.114	0.615	0.101	1.963	3.368	0.0436	0.063	.1649E+5	.6063E-4	.8352E-4	.1467E+0
2.20	0.39	.402	0.135	0.598	0.119	2.014	3.446	0.0544	0.071	.9535E+4	.1049E-3	.1222E-3	.1875E+0

CASE # 4   COPYRIGHT 1982 STEVE SMITH   DATA DATE 1/ 3/82

CHOKE/RESONANT CHARGE REACTOR DESIGN TABLE

$$f(D) = U/(F \cdot G \cdot (D \cdot E)^2)$$

P	Q	D	E	F	G	U	Um	DE	FG	f(D)	1/f(D)	1/(PQf(D))	DEUm
2.20	0.47	.430	.163	.570	0.143	2.082	3.550	0.0702	0.082	.5166E+4	.1936E-3	.1872E-3	.2493E+0
2.20	0.56	.461	.194	.539	0.171	2.157	3.669	0.0896	0.092	.2911E+4	.3436E-3	.2789E-3	.3287E+0
2.20	0.68	.501	.234	.499	0.211	2.254	3.835	0.1174	0.105	.1551E+4	.6447E-3	.4310E-3	.4502E+0
2.20	0.82	.545	.278	.455	0.263	2.361	4.041	0.1517	0.120	.8565E+3	.1168E-2	.6472E-3	.6130E+0
2.20	1.00	.598	.331	.402	0.337	2.490	4.324	0.1981	0.136	.4677E+3	.2138E-2	.9718E-3	.8567E+0
2.20	1.20	.648	.381	.352	0.437	2.612	4.681	0.2471	0.154	.2778E+3	.3599E-2	.1363E-2	.1157E+1
2.20	1.50	.703	.436	.297	0.627	2.745	5.313	0.3067	0.186	.1566E+3	.6386E-2	.1935E-2	.1630E+1
2.20	1.80	.736	.469	.264	0.861	2.825	6.053	0.3454	0.227	.1041E+3	.9603E-2	.2425E-2	.2091E+1
2.20	2.20	.761	.494	.239	1.211	2.886	7.132	0.3762	0.290	.7044E+2	.1420E-1	.2933E-2	.2683E+1
2.20	2.70	.777	.510	.223	1.679	2.925	8.554	0.3965	0.374	.4967E+2	.2013E-1	.3389E-2	.3392E+1
2.20	3.00	.786	.519	.214	2.261	2.947	10.310	0.4084	0.484	.3655E+2	.2736E-1	.3769E-2	.4200E+1
2.20	3.90	.792	.525	.208	2.849	2.961	12.081	0.4161	0.593	.2886E+2	.3464E-1	.4038E-2	.5026E+1
2.20	4.70	.797	.530	.203	3.639	2.974	14.457	0.4227	0.739	.2253E+2	.4439E-1	.4293E-2	.6111E+1
2.20	5.60	.800	.533	.200	4.533	2.981	17.142	0.4267	0.907	.1806E+2	.5537E-1	.4494E-2	.7314E+1
2.20	6.80	.802	.535	.198	5.729	2.986	20.732	0.4293	1.134	.1428E+2	.7004E-1	.4682E-2	.8901E+1
2.20	8.20	.804	.537	.196	7.125	2.991	24.923	0.4320	1.397	.1147E+2	.8716E-1	.4831E-2	.1077E+2
2.20	10.00	.806	.539	.194	8.921	2.995	30.313	0.4347	1.731	.9159E+1	.1092E+0	.4963E-2	.1318E+2
2.20	12.00	.807	.540	.193	10.919	2.998	36.308	0.4360	2.107	.7461E+1	.1337E+0	.5063E-2	.1583E+2
2.20	15.00	.808	.541	.192	13.917	3.000	45.303	0.4374	2.672	.5869E+1	.1704E+0	.5163E-2	.1982E+2
2.20	18.00	.810	.543	.191	16.915	3.003	54.298	0.4387	3.231	.4828E+1	.2071E+0	.5230E-2	.2382E+2
2.20	22.00	.810	.543	.190	20.913	3.005	66.294	0.4401	3.974	.3905E+1	.2561E+0	.5291E-2	.2918E+2
2.20	27.00	.810	.543	.190	25.913	3.005	81.294	0.4401	4.924	.3151E+1	.3173E+0	.5342E-2	.3578E+2
2.20	33.00	.810	.543	.190	31.913	3.005	99.294	0.4401	6.064	.2559E+1	.3908E+0	.5383E-2	.4376E+2
2.70	0.33	.221	.121	.779	0.088	1.908	4.002	0.0267	0.069	.3892E+5	.2570E-4	.2884E-4	.1070E+0
2.70	0.39	.244	.144	.756	0.102	1.964	4.074	0.0351	0.077	.2063E+5	.4848E-4	.4604E-4	.1430E+0
2.70	0.47	.274	.174	.726	0.122	2.030	4.165	0.0477	0.089	.1012E+5	.9886E-4	.7791E-4	.1986E+0
2.70	0.56	.307	.207	.693	0.146	2.117	4.274	0.0635	0.101	.5180E+4	.1931E-3	.1277E-3	.2716E+0
2.70	0.68	.350	.250	.650	0.180	2.221	4.425	0.0875	0.117	.2479E+4	.4033E-3	.2197E-3	.3872E+0
2.70	0.82	.399	.299	.601	0.222	2.340	4.607	0.1194	0.133	.1232E+4	.8115E-3	.3665E-3	.5497E+0
2.70	1.00	.459	.359	.541	0.282	2.486	4.856	0.1648	0.153	.6001E+3	.1666E-2	.6172E-3	.8001E+0
2.70	1.20	.521	.421	.479	0.358	2.636	5.155	0.2193	0.171	.3196E+3	.3129E-2	.9658E-3	.1131E+1
2.70	1.50	.599	.499	.401	0.502	2.826	5.676	0.2989	0.201	.1571E+3	.6364E-2	.1571E-2	.1696E+1
2.70	1.80	.656	.556	.344	0.688	2.966	6.299	0.3647	0.237	.9415E+2	.1062E-1	.2185E-2	.2299E+1
2.70	2.20	.703	.603	.297	0.994	3.079	7.270	0.4239	0.295	.5803E+2	.1723E-1	.2901E-2	.3082E+1
2.70	2.70	.732	.632	.268	1.436	3.149	8.629	0.4626	0.385	.3823E+2	.2616E-1	.3588E-2	.3992E+1
2.70	3.30	.749	.649	.251	2.002	3.190	10.347	0.4861	0.503	.2687E+2	.3722E-1	.4177E-2	.5030E+1

CASE # 4   COPYRIGHT 1982 STEVE SMITH   DATA DATE 1/ 3/82   CHOKE/RESONANT CHARGE REACTOR DESIGN TABLE

$$f(D)=U/(F*G*(D*E)^2)$$

P	Q	D	E	F	G	U	Um	DE	FG	f(D)	1/f(D)	1/(PQf(D))	DEUm
2.70	3.90	.759	0.659	0.241	2.582	3.215	12.098	0.5002	0.622	.2065E+2	.4843E-1	.4599E-2	.6051E+1
2.70	4.70	.766	0.666	0.234	3.368	3.232	14.464	0.5102	0.788	.1576E+2	.6347E-1	.5002E-2	.7379E-1
2.70	5.60	.771	0.671	0.229	4.258	3.244	17.140	0.5173	0.975	.1243E+2	.8045E-1	.5321E-2	.8667E+1
2.70	6.80	.774	0.674	0.226	5.452	3.251	20.725	0.5217	1.232	.9695E+1	.1031E+0	.5618E-2	.1081E+2
2.70	8.20	.777	0.677	0.223	6.846	3.258	24.911	0.5260	1.527	.7713E+1	.1296E+0	.5856E-2	.1310E+2
2.70	10.00	.779	0.679	0.221	8.642	3.263	30.301	0.5289	1.910	.6107E+1	.1638E+0	.6065E-2	.1603E+2
2.70	12.00	.781	0.681	0.219	10.638	3.268	36.291	0.5319	2.330	.4959E+1	.2017E+0	.6224E-2	.1930E+2
2.70	15.00	.782	0.682	0.218	13.636	3.270	45.287	0.5333	2.973	.3868E+1	.2585E+0	.6384E-2	.2415E+2
2.70	18.00	.783	0.683	0.217	16.634	3.273	54.282	0.5348	3.610	.3170E+1	.3154E+0	.6490E-2	.2903E+2
2.70	22.00	.784	0.684	0.216	20.632	3.275	66.277	0.5363	4.457	.2556E+1	.3913E+0	.6587E-2	.3554E+2
2.70	27.00	.785	0.685	0.215	25.630	3.278	81.272	0.5377	5.510	.2051E+1	.4861E+0	.6668E-2	.4370E+2
2.70	33.00	.785	0.685	0.215	31.630	3.278	99.272	0.5377	6.800	.1667E+1	.5999E+0	.6733E-2	.5338E+2
3.30	0.33	.047	0.147	0.953	0.036	1.885	4.676	0.0069	0.034	.1151E+7	.8688E-6	.7978E-6	.3231E-1
3.30	0.39	.069	0.169	0.931	0.052	1.938	4.754	0.0117	0.048	.2945E+6	.3396E-5	.2639E-5	.5538E-1
3.30	0.47	.099	0.199	0.901	0.072	2.011	4.843	0.0197	0.065	.7988E+5	.1252E-4	.8071E-5	.9542E-1
3.30	0.56	.133	0.233	0.867	0.094	2.094	4.948	0.0310	0.081	.2675E+5	.3738E-4	.2023E-4	.1533E+0
3.30	0.69	.178	0.278	0.822	0.124	2.203	5.089	0.0495	0.102	.8827E+4	.1133E-3	.5048E-4	.2518E+0
3.30	0.82	.230	0.330	0.770	0.160	2.330	5.257	0.0759	0.123	.3282E+4	.3047E-3	.1126E-3	.3990E+0
3.30	1.00	.294	0.394	0.706	0.212	2.485	5.486	0.1158	0.150	.1237E+4	.8082E-3	.2449E-3	.6355E+0
3.30	1.20	.363	0.463	0.637	0.274	2.653	5.751	0.1681	0.175	.5388E+3	.1859E-2	.4694E-3	.9665E+0
3.30	1.50	.459	0.559	0.541	0.382	2.886	6.184	0.2566	0.207	.2121E+3	.4715E-2	.9524E-3	.1587E+1
3.30	1.80	.541	0.641	0.459	0.518	3.085	6.686	0.3468	0.238	.1079E+3	.9268E-2	.1560E-2	.2318E+1
3.30	2.20	.620	0.720	0.380	0.760	3.277	7.502	0.4464	0.289	.5694E+2	.1756E-1	.2419E-2	.3449E+1
3.30	2.70	.676	0.776	0.324	1.148	3.413	8.730	0.5246	0.372	.3334E+2	.2999E-1	.3366E-2	.4579E+1
3.30	3.30	.708	0.808	0.292	1.684	3.491	10.374	0.5721	0.492	.2169E+2	.4610E-1	.4233E-2	.5935E+1
3.30	3.90	.723	0.823	0.277	2.254	3.527	12.102	0.5950	0.624	.1596E+2	.6267E-1	.4870E-2	.7201E+1
3.30	4.70	.735	0.835	0.265	3.030	3.556	14.443	0.6137	0.803	.1176E+2	.8504E-1	.5483E-2	.8664E+1
3.30	5.60	.742	0.842	0.258	3.916	3.573	17.109	0.6248	1.010	.9061E+1	.1104E+0	.5972E-2	.1069E+2
3.30	6.80	.748	0.848	0.252	5.104	3.588	20.680	0.6333	1.286	.6933E+1	.1442E+0	.6428E-2	.1312E+2
3.30	8.20	.752	0.852	0.248	6.496	3.598	24.861	0.6407	1.611	.5448E+1	.1838E+0	.6793E-2	.1593E+2
3.30	10.00	.755	0.855	0.245	8.290	3.605	30.246	0.6455	2.031	.4259E+1	.2348E+0	.7115E-2	.1952E+2
3.30	12.00	.757	0.857	0.243	10.286	3.610	36.236	0.6487	2.499	.3431E+1	.2914E+0	.7359E-2	.2351E+2
3.30	15.00	.759	0.859	0.241	13.282	3.615	45.227	0.6520	3.201	.2656E+1	.3764E+0	.7605E-2	.2949E+2
3.30	18.00	.760	0.860	0.240	16.280	3.617	54.222	0.6536	3.907	.2167E+1	.4615E+0	.7769E-2	.3544E+2
3.30	22.00	.761	0.861	0.239	20.278	3.619	66.217	0.6552	4.846	.1748E+1	.5749E+0	.7918E-2	.4339E+2
3.30	27.00	.762	0.862	0.238	25.276	3.622	81.212	0.6568	6.016	.1395E+1	.7166E+0	.8043E-2	.5334E+2

CASE # 4   COPYRIGHT 1982 STEVE SMITH   DATA DATE 1/ 3/82   CHOKE/RESONANT CHARGE REACTOR DESIGN TABLE

$$f(D)=U/(F*G*(D*E)^2)$$

P	Q	D	E	F	G	U	Um	DE	FG	f(D)	1/f(D)	1/(PQf(D))	DEUm
3.30	33.00	.762	0.862	0.238	31.276	3.622	99.212	0.6568	7.444	.1128E+1	.8867E+0	.8142E-2	.6517E+2
3.90	0.82	.076	0.376	0.924	0.068	2.355	5.833	0.0286	0.063	.4591E+5	.2178E-4	.6811E-5	.1667E+0
3.90	1.00	.140	0.440	0.860	0.120	2.511	6.062	0.0616	0.103	.6412E+4	.1560E-3	.3999E-4	.3734E+0
3.90	1.20	.211	0.511	0.789	0.178	2.683	6.317	0.1078	0.140	.1644E+4	.6084E-3	.1300E-3	.6811E+0
3.90	1.50	.314	0.614	0.686	0.272	2.934	6.717	0.1928	0.187	.4230E+3	.2364E-2	.4041E-3	.1295E+1
3.90	1.80	.409	0.709	0.591	0.382	3.164	7.155	0.2900	0.226	.1667E+3	.5999E-2	.8546E-3	.2075E+1
3.90	2.20	.517	0.817	0.483	0.566	3.427	7.831	0.4224	0.273	.7026E+2	.1423E-1	.1659E-2	.3308E+1
3.90	2.70	.609	0.909	0.391	0.882	3.650	8.884	0.5536	0.345	.3454E+2	.2895E-1	.2750E-2	.4918E+1
3.90	3.00	.665	0.965	0.335	1.370	3.786	10.412	0.6417	0.459	.2003E+2	.4992E-1	.3879E-2	.6681E+1
3.90	3.90	.691	0.991	0.309	1.918	3.849	12.085	0.6848	0.593	.1385E+2	.7220E-1	.4747E-2	.8276E+1
3.90	4.70	.709	1.009	0.291	2.682	3.893	14.398	0.7154	0.780	.9747E+1	.1026E+0	.5597E-2	.1030E+2
3.90	5.60	.719	1.019	0.281	3.562	3.917	17.049	0.7327	1.001	.7291E+1	.1372E+0	.6288E-2	.1249E+2
3.90	6.20	.727	1.027	0.273	4.746	3.937	20.610	0.7466	1.296	.5451E+1	.1835E+0	.6918E-2	.1539E+2
3.90	8.20	.732	1.032	0.268	6.136	3.949	24.786	0.7554	1.644	.4208E+1	.2376E+0	.7431E-2	.1872E+2
3.90	10.00	.736	1.036	0.264	7.928	3.959	30.167	0.7625	2.093	.3253E+1	.3074E+0	.7882E-2	.2300E+2
3.90	12.00	.738	1.038	0.262	9.924	3.964	36.157	0.7660	2.600	.2598E+1	.3850E+0	.8226E-2	.2770E+2
3.90	15.00	.741	1.041	0.259	12.918	3.968	45.142	0.7714	3.346	.1995E+1	.5014E+0	.8570E-2	.3482E+2
3.90	18.00	.742	1.042	0.258	15.916	3.973	54.138	0.7732	4.106	.1619E+1	.6178E+0	.8801E-2	.4186E+2
3.90	22.00	.743	1.043	0.257	19.914	3.976	66.133	0.7749	5.118	.1294E+1	.7731E+0	.9010E-2	.5125E+2
3.90	27.00	.744	1.044	0.256	24.912	3.978	81.128	0.7767	6.382	.1034E+1	.9672E+0	.9185E-2	.6301E+2
3.90	33.00	.745	1.045	0.255	30.910	3.981	99.123	0.7785	7.882	.8333E+0	.1200E+1	.9325E-2	.7717E+2
4.70	1.50	.124	0.691	0.876	0.119	3.005	7.411	0.0856	0.104	.3942E+4	.2537E-3	.3599E-4	.6347E+0
4.70	1.80	.226	0.793	0.774	0.215	3.253	7.816	0.1791	0.166	.6101E+3	.1639E-2	.1937E-3	.1400E+1
4.70	2.00	.354	0.921	0.646	0.359	3.564	8.394	0.3259	0.232	.1448E+3	.6905E-2	.6673E-3	.2736E+1
4.70	2.70	.489	1.056	0.511	0.589	3.892	9.238	0.5162	0.301	.4855E+2	.2060E-1	.1623E-2	.4769E+1
4.70	3.30	.594	1.161	0.406	0.979	4.147	10.528	0.6894	0.397	.2196E+2	.4554E-1	.2936E-2	.7258E+1
4.70	3.90	.646	1.213	0.354	1.475	4.273	12.075	0.7834	0.522	.1334E+2	.7497E-1	.4090E-2	.9459E+1
4.70	4.70	.677	1.244	0.323	2.213	4.349	14.324	0.8420	0.715	.8583E+1	.1165E+0	.5274E-2	.1206E+2
4.70	5.60	.694	1.261	0.306	3.079	4.390	16.942	0.8749	0.942	.6088E+1	.1643E+0	.6241E-2	.1482E+2
4.70	6.80	.705	1.272	0.295	4.257	4.417	20.488	0.8965	1.256	.4376E+1	.2285E+0	.7150E-2	.1837E+2
4.70	8.20	.712	1.279	0.288	5.643	4.434	24.654	0.9104	1.625	.3292E+1	.3038E+0	.7883E-2	.2245E+2
4.70	10.00	.717	1.284	0.283	7.433	4.446	30.030	0.9204	2.103	.2495E+1	.4008E+0	.8527E-2	.2764E+2
4.70	12.00	.720	1.287	0.280	9.427	4.453	36.016	0.9264	2.639	.1966E+1	.5087E+0	.9019E-2	.3336E+2
4.70	15.00	.723	1.292	0.277	12.421	4.460	45.001	0.9324	3.441	.1491E+1	.6706E+0	.9512E-2	.4196E+2
4.70	18.00	.725	1.292	0.275	15.417	4.465	53.991	0.9365	4.240	.1201E+1	.8326E+0	.9842E-2	.5056E+2
4.70	22.00	.726	1.293	0.274	19.415	4.468	65.986	0.9385	5.320	.9536E+0	.1049E+1	.1014E-1	.6193E+2

CASE # 4   COPYRIGHT 1982 STEVE SMITH   DATA DATE 1/ 3/82   CHOKE/RESONANT CHARGE REACTOR DESIGN TABLE

$f(D) = U/(F*G*(D*E)^2)$

P	Q	D	E	F	G	U	Um	DE	FG	f(D)	1/f(D)	1/(PQf(D))	DEUm
4.70	27.00	.727	1.294	0.273	24.413	4.470	80.982	0.9405	6.665	.7583E+0	.1319E+1	.1039E-1	.7616E+2
4.70	33.00	.728	1.295	0.272	30.411	4.473	98.977	0.9425	8.272	.6087E+0	.1643E+1	.1059E-1	.9329E+2
5.60	2.20	.156	1.023	0.844	0.155	3.683	9.098	0.1595	0.131	.1109E+4	.9021E-3	.7322E-4	.1451E+1
5.60	2.70	.316	1.183	0.684	0.335	4.072	9.821	0.3737	0.229	.1274E+3	.7852E-2	.5193E-3	.3670E+1
5.60	3.30	.475	1.342	0.525	0.617	4.458	10.848	0.6373	0.324	.3390E+2	.2949E-1	.1596E-2	.6910E+1
5.60	3.90	.577	1.444	0.423	1.013	4.706	12.153	0.8330	0.428	.1583E+2	.6316E-1	.2892E-2	.1012E+2
5.60	4.70	.639	1.506	0.361	1.689	4.856	14.252	0.9621	0.610	.8606E+1	.1162E+0	.4415E-2	.1371E+2
5.60	5.60	.668	1.535	0.332	2.531	4.927	16.811	1.0252	0.840	.5579E+1	.1792E+0	.5715E-2	.1723E+2
5.60	6.80	.685	1.552	0.315	3.697	4.968	20.328	1.0629	1.164	.3777E+1	.2648E+0	.6954E-2	.2161E+2
5.60	8.20	.695	1.562	0.305	5.077	4.992	24.479	1.0854	1.548	.2737E+1	.3654E+0	.7956E-2	.2657E+2
5.60	10.00	.701	1.568	0.299	6.865	5.007	29.850	1.0989	2.053	.2020E+1	.4951E+0	.8840E-2	.3280E+2
5.60	12.00	.706	1.573	0.294	8.855	5.019	35.826	1.1103	2.603	.1564E+1	.6394E+0	.9515E-2	.3978E+2
5.60	15.00	.709	1.576	0.291	11.849	5.026	44.811	1.1171	3.448	.1168E+1	.8561E+0	.1019E-1	.5006E+2
5.60	18.00	.711	1.578	0.289	14.845	5.031	53.802	1.1217	4.290	.9321E+0	.1073E+1	.1064E-1	.6035E+2
5.60	22.00	.713	1.580	0.287	18.841	5.036	65.792	1.1263	5.407	.7342E+0	.1362E+1	.1106E-1	.7411E+2
5.60	27.00	.714	1.581	0.286	23.839	5.039	80.787	1.1286	6.818	.5802E+0	.1724E+1	.1140E-1	.9118E+2
5.60	33.00	.715	1.582	0.285	29.837	5.041	98.782	1.1309	8.503	.4635E+0	.2157E+1	.1167E-1	.1117E+3
6.80	2.70	.055	1.322	0.945	0.057	4.208	10.745	0.0727	0.054	.1498E+5	.6677E-4	.3637E-5	.7811E+0
6.80	3.30	.250	1.517	0.750	0.267	4.711	11.598	0.3792	0.200	.1639E+3	.6103E-2	.2720E-3	.4398E+1
6.80	3.90	.418	1.685	0.582	0.531	5.120	12.582	0.7042	0.309	.3343E+2	.2992E-1	.1128E-2	.8860E+1
6.80	4.70	.562	1.829	0.438	1.043	5.469	14.282	1.0277	0.457	.1134E+2	.8819E-1	.2759E-2	.1468E+2
6.80	5.60	.629	1.896	0.371	1.809	5.632	16.657	1.1924	0.671	.5904E+1	.1694E+0	.4448E-2	.1986E+2
6.80	6.80	.662	1.929	0.338	2.943	5.712	20.096	1.2768	0.995	.3523E+1	.2838E+0	.6138E-2	.2566E+2
6.80	8.20	.677	1.944	0.323	4.313	5.749	24.224	1.3159	1.393	.2383E+1	.4196E+0	.7524E-2	.3187E+2
6.80	10.00	.686	1.953	0.314	6.095	5.771	29.580	1.3395	1.914	.1680E+1	.5951E+0	.8751E-2	.3963E+2
6.80	12.00	.692	1.959	0.308	8.803	5.785	35.551	1.3554	2.489	.1265E+1	.7905E+0	.9688E-2	.4819E+2
6.80	15.00	.696	1.963	0.304	11.075	5.795	44.531	1.3660	3.367	.9224E+0	.1084E+1	.1063E-1	.6083E+2
6.80	18.00	.699	1.966	0.301	14.069	5.802	53.517	1.3740	4.235	.7258E+0	.1378E+1	.1126E-1	.7353E+2
6.80	22.00	.701	1.968	0.299	18.665	5.807	65.507	1.3793	5.401	.5651E+0	.1770E+1	.1183E-1	.9036E+2
6.80	27.00	.702	1.969	0.298	23.063	5.808	80.502	1.3820	6.873	.4426E+0	.2259E+1	.1231E-1	.1113E+3
6.80	33.00	.704	1.971	0.296	29.059	5.814	98.492	1.3873	8.601	.3512E+0	.2847E+1	.1269E-1	.1366E+3
8.20	3.90	.142	1.875	0.858	0.149	5.382	13.522	0.2663	0.128	.5924E+3	.1688E-2	.5279E-4	.3601E+1
8.20	4.70	.379	2.112	0.621	0.475	5.958	14.771	0.8006	0.295	.3175E+2	.3175E-1	.8239E-3	.1183E+2
8.20	5.60	.548	2.281	0.452	1.037	6.369	16.650	1.2502	0.469	.8691E+1	.1151E+0	.2506E-2	.2081E+2
8.20	6.80	.629	2.362	0.371	2.075	6.565	19.856	1.4859	0.770	.3862E+1	.2589E+0	.4644E-2	.2950E+2
8.20	8.20	.658	2.391	0.342	3.417	6.636	23.915	1.5735	1.169	.2293E+1	.4361E+0	.6485E-2	.3763E+2

CASE # 4   COPYRIGHT 1982 STEVE SMITH   DATA DATE 1/ 3/82   CHOKE/RESONANT CHARGE REACTOR DESIGN TABLE

$$f(D)=U/(F*G*(D*E)^2)$$

P	Q	D	E	F	G	U	Um	DE	FG	f(D)	1/f(D)	1/(PQf(D))	DEUm
8.20	10.00	.673	2.406	0.327	5.187	6.672	29.242	1.6195	1.696	.1500E+1	.6667E+0	.8131E-2	.4736E+2
8.20	12.00	.680	2.413	0.320	7.173	6.689	35.208	1.6411	2.295	.1082E+1	.9241E+0	.9392E-2	.5778E+2
8.20	15.00	.686	2.419	0.314	10.161	6.704	44.179	1.6597	3.191	.7628E+0	.1311E+1	.1066E-1	.7332E+2
8.20	18.00	.689	2.422	0.311	13.155	6.711	53.165	1.6690	4.091	.5889E+0	.1698E+1	.1150E-1	.8873E+2
8.20	22.00	.691	2.424	0.309	17.151	6.716	65.155	1.6752	5.300	.4516E+0	.2215E+1	.1228E-1	.1091E+3
8.20	27.00	.693	2.426	0.307	22.147	6.721	80.145	1.6814	6.799	.3496E+0	.2860E+1	.1292E-1	.1348E+3
8.20	33.00	.694	2.427	0.306	28.145	6.723	98.140	1.6846	8.612	.2751E+0	.3635E+1	.1343E-1	.1653E+3
10.00	5.60	.294	2.627	0.706	0.345	6.952	17.369	0.7724	0.244	.4779E+2	.2093E-1	.3737E-3	.1342E+2
10.00	6.80	.540	2.873	0.460	1.053	7.549	19.774	1.5516	0.485	.6472E+1	.1545E+0	.2272E-2	.3068E+2
10.00	8.20	.627	2.960	0.373	2.279	7.761	23.551	1.8561	0.850	.2649E+1	.3774E+0	.4603E-2	.4371E+2
10.00	10.00	.656	2.989	0.344	4.021	7.831	28.810	1.9610	1.383	.1472E+1	.6793E+0	.6793E-2	.5650E+2
10.00	12.00	.668	3.001	0.332	5.997	7.860	34.752	2.0049	1.991	.9821E+0	.1018E+1	.8485E-2	.6967E+2
10.00	15.00	.676	3.009	0.324	8.981	7.880	43.713	2.0343	2.910	.6543E+0	.1528E+1	.1019E-1	.8893E+2
10.00	18.00	.680	3.013	0.320	11.973	7.889	52.693	2.0491	3.631	.4904E+0	.2039E+1	.1133E-1	.1080E+3
10.00	22.00	.683	3.016	0.317	15.967	7.897	64.679	2.0602	5.062	.3676E+0	.2721E+1	.1237E-1	.1332E+3
10.00	27.00	.685	3.018	0.315	20.963	7.901	79.669	2.0676	6.603	.2799E+0	.3573E+1	.1323E-1	.1647E+3
10.00	33.00	.687	3.020	0.313	26.959	7.906	97.659	2.0750	8.438	.2176E+0	.4595E+1	.1392E-1	.2026E+3

APPENDIX B
TRANSFORMER DESIGN TABLES

TRANSFORMER DESIGN TABLES

OPTMIZATION PARAMETER: POWER LOSS = $[(U/A)^5 * (1/B)^4]^{(1/13)}$

$$A = (D^2 * E^2 * F * G)$$
$$B = (D * E * Um)$$

CASE # 1

E=(P-1+D)/2   F=(1-D)/2   G=Q-P+1-D
U=3*D+P-1+1.5708*(1-D)
Um=3*(1-D)+2*(Q-P)+1.5708*(P-1+D)

CASE # 2

E=(P-1+D)/2   F=(1-D)/2   G=(2*Q-P+1-D)/2
U=3*D+P-1+1.5708*(1-D)
Um=2*Q-P+2-2*D+0.7854*(P-1+D)

CASE # 3

E=(P-2+2*D)/2   F=1-D   G=Q-P+2-2*D
U=4*D+P-2+1.5708*(1-D)
Um=6-6*D+2*Q-2*P+1.5708*(P-2+2*D)

CASE # 4

E=(P-3+3*D)/3   F=1-D   G=(3*Q-2*P+6-6*D)/3
U=2*(P-3+6*D)/3+1.5708*(1-D)
Um=3*Q-2*P+10-10*D+1.713864*(P-3+3*D)

279

TRANSFORMER DESIGN TABLE

$f(D) = [(U/(FG(DE)^2))^5 * (1/(DEUm))^4]^{(1/13)}$

CASE # 1   COPYRIGHT 1982 STEVE SMITH   DATA DATE 1/ 3/82

SCALE BY 1.000E-04   FUNCTION 1/f(D)

Q/P	0.33	0.39	0.47	0.56	0.68	0.82	1.00	1.20	1.50	1.80	2.20	2.70	3.30	3.90	4.70	5.60	6.80	8.20	10.00
0.33	77	85	92	95	92	80	54												
0.39	94	107	118	124	123	110	80												
0.47	115	134	152	164	168	157	124	69											
0.56	138	162	189	211	223	218	184	120											
0.68	165	197	235	270	299	306	279	210	66										
0.82	194	234	284	333	383	413	405	341	172										
1.00	230	279	343	408	482	544	574	537	364	121									
1.20	266	325	402	485	583	677	753	766	633	359									
1.50	317	389	485	590	721	855	992	1084	1072	851	330								
1.80	364	449	562	687	847	1018	1206	1364	1474	1390	926	92							
2.20	423	523	657	807	1002	1217	1466	1697	1939	2027	1828	967							
2.70	492	609	768	947	1183	1447	1762	2072	2446	2694	2790	2388	938						
3.30	570	707	893	1104	1384	1702	2089	2482	2990	3387	3727	3785	3072	1178					
3.90	643	798	1010	1251	1573	1940	2393	2863	3488	4012	4540	4900	4789	3764	565				
4.70	735	914	1157	1436	1810	2239	2774	3336	4105	4776	5515	6176	6550	6364	4694	261			
5.60	833	1036	1314	1633	2060	2556	3176	3835	4750	5571	6516	7453	8214	8545	8153	5747			
6.80	956	1190	1510	1879	2375	2952	3679	4457	5553	6555	7744	8998	10169	10974	10935	11436	7157		
8.20	1092	1359	1726	2149	2721	3387	4230	5137	6427	7623	9070	10650	12225	13463	14591	15103	14236	8809	
10.00	1255	1563	1987	2476	3138	3911	4894	5956	7478	8903	10654	12225	14591	16346	18140	19515	20231	19071	10939
12.00	1427	1777	2261	2819	3574	4459	5587	6811	8573	10235	12296	14640	17118	19279	21696	23818	25685	26397	24361
15.00	1668	2079	2645	3300	4188	5230	6561	8011	10107	12099	14590	17453	20552	23321	26549	29602	32793	35291	36487
18.00	1895	2362	3006	3751	4763	5953	7474	9135	11544	13842	16731	20077	23740	27060	31010	34874	39166	43029	46252
22.00	2179	2717	3460	4319	5486	6860	8620	10545	13344	16024	19409	23355	27713	31707	36535	41368	46946	52337	57655
27.00	2513	3134	3992	4984	6334	7924	9963	12197	15453	18579	22541	27183	32345	37115	42946	48876	55886	62936	70427
33.00	2890	3604	4591	5735	7290	9123	11476	14057	17826	21451	26060	31481	37539	43171	50110	57243	65809	74634	84389

TRANSFORMER DESIGN TABLE

$$f(D) = [(U/([FG(DE)^2))^5 * (1/(DEUm))^4]^{(1/13)}$$

CASE # 1   COPYRIGHT 1982 STEVE SMITH   DATA DATE 1/ 3/82

FUNCTION   1/[P*Q*f(D)]

SCALE BY 1.000E-04

Q/P	0.33	0.39	0.47	0.56	0.68	0.82	1.00	1.20	1.50	1.80	2.20	2.70	3.30	3.90	4.70	5.60	6.80	8.20	10.00
0.33	707	664	596	514	409	295	164												
0.39	730	701	643	567	462	344	206	123											
0.47	743	729	689	625	526	408	263	179											
0.56	735	742	719	672	587	474	328	257	65										
0.68	718	743	736	708	646	550	410	346	140										
0.82	696	733	738	726	687	615	494	448	243	67									
1.00	672	716	729	729	709	663	574	532	352	166									
1.20	640	695	713	721	715	688	627	602	476	315	100								
1.50	613	665	688	702	706	696	662	632	546	429	234								
1.80	583	639	664	681	692	690	670	643	588	512	378	19							
2.20	553	609	635	655	670	675	666	639	604	554	470	163							
2.70	523	579	605	627	644	653	652	627	604	570	513	328	105						
3.30	500	549	576	597	617	629	633	612	596	572	529	425	282	92					
3.90	474	525	551	573	593	607	614	593	582	565	533	487	372	247	31				
4.70	451	498	524	546	566	581	590	571	565	553	529	493	422	347	212	10			
5.60	426	474	499	521	541	557	567	546	544	536	518	490	453	391	310	183	155		
6.80	403	449	473	493	514	529	541	522	523	516	503	481	452	414	358	287	255		
8.00	380	425	448	468	488	504	516	496	499	495	484	467	444	419	379	348	298	131	
10.00	360	401	423	442	461	477	489	473	476	474	466	452	432	412	386	354	315	233	109
12.00	337	380	401	419	438	453	466	445	449	448	442	431	415	399	377	352	322	268	203
15.00	319	355	375	393	411	425	437	423	428	427	422	413	400	385	367	346	320	287	243
18.00	300	336	355	372	389	403	415	399	404	405	401	393	382	370	353	336	314	292	257
22.00	282	317	335	351	367	380	392	376	382	382	379	373	363	352	338	323	304	290	262
27.00	265	298	315	330	345	358	369	376	382	382	379	373	363	352	338	323	304	284	261
33.00		280	296	310	325	337	348	355	360	361	359	353	345	335	323	310	293	276	256

TRANSFORMER DESIGN TABLE

$$f(D)=[(U/(FG(DE)^2))^5 * (1/(DEUm))^4]^{1/13}$$

CASE # 1   COPYRIGHT 1982 STEVE SMITH   DATA DATE 1/ 3/82

P	Q	D	E	F	G	U	Um	DE	FG	f(D)	1/f(D)	1/(PQf(D))	DEUm	FG(DE)^2/U
0.33	0.33	.868	0.099	0.066	0.132	2.141	0.707	0.0859	0.009	.1299E+3	.7698E-2	.7069E-1	.6076E-1	.3004E-4
0.33	0.39	.883	0.107	0.058	0.177	2.163	0.806	0.0940	0.010	.1064E+3	.9401E-2	.7305E-1	.7576E-1	.4234E-4
0.33	0.47	.896	0.113	0.052	0.244	2.181	0.947	0.1012	0.013	.8672E+2	.1153E-1	.7434E-1	.9588E-1	.5963E-4
0.33	0.56	.903	0.117	0.049	0.327	2.191	1.117	0.1052	0.016	.7272E+2	.1375E-1	.7441E-1	.1175E+0	.8009E-4
0.33	0.68	.909	0.119	0.046	0.441	2.200	1.348	0.1086	0.020	.6066E+2	.1649E-1	.7347E-1	.1465E+0	.1076E-3
0.33	0.82	.913	0.122	0.044	0.577	2.206	1.623	0.1109	0.025	.5145E+2	.1944E-1	.7183E-1	.1800E+0	.1400E-3
0.33	1.00	.916	0.123	0.042	0.754	2.213	1.978	0.1127	0.032	.4356E+2	.2296E-1	.6956E-1	.2229E+0	.1819E-3
0.33	1.20	.918	0.124	0.041	0.952	2.213	2.376	0.1138	0.039	.3760E+2	.2660E-1	.6717E-1	.2704E+0	.2286E-3
0.33	1.50	.920	0.125	0.040	1.250	2.216	2.973	0.1150	0.050	.3157E+2	.3168E-1	.6399E-1	.3419E+0	.2984E-3
0.33	1.80	.921	0.126	0.039	1.549	2.217	3.571	0.1156	0.061	.2746E+2	.3641E-1	.6130E-1	.4128E+0	.3687E-3
0.33	2.20	.922	0.126	0.039	1.948	2.219	4.370	0.1162	0.076	.2362E+2	.4233E-1	.5830E-1	.5077E+0	.4622E-3
0.33	2.70	.922	0.126	0.039	2.448	2.219	5.370	0.1162	0.095	.2031E+2	.4924E-1	.5527E-1	.6238E+0	.5008E-3
0.33	3.30	.923	0.126	0.039	3.047	2.220	6.568	0.1168	0.117	.1754E+2	.5701E-1	.5235E-1	.7669E+0	.7204E-3
0.33	3.90	.923	0.127	0.039	3.647	2.221	7.768	0.1168	0.140	.1555E+2	.6432E-1	.4998E-1	.9070E+0	.8623E-3
0.33	4.70	.924	0.127	0.038	4.446	2.221	9.367	0.1173	0.169	.1360E+2	.7354E-1	.4741E-1	.1099E+1	.1047E-2
0.33	5.60	.924	0.127	0.038	5.346	2.221	11.167	0.1173	0.203	.1200E+2	.8333E-1	.4509E-1	.1310E+1	.1259E-2
0.33	6.80	.924	0.128	0.038	6.546	2.221	13.567	0.1173	0.249	.1046E+2	.9564E-1	.4262E-1	.1592E+1	.1542E-2
0.33	8.20	.924	0.128	0.038	7.946	2.221	16.367	0.1173	0.302	.9160E+1	.1092E+0	.4034E-1	.1921E+1	.1872E-2
0.33	10.00	.925	0.128	0.037	9.745	2.223	19.966	0.1179	0.365	.7966E+1	.1255E+0	.3804E-1	.2355E+1	.2287E-2
0.33	12.00	.925	0.128	0.037	11.745	2.223	23.966	0.1179	0.440	.7009E+1	.1427E+0	.3603E-1	.2826E+1	.2756E-2
0.33	15.00	.925	0.128	0.037	14.745	2.223	29.966	0.1179	0.553	.5995E+1	.1668E+0	.3370E-1	.3534E+1	.3460E-2
0.33	18.00	.925	0.128	0.037	17.745	2.223	35.966	0.1179	0.665	.5278E+1	.1895E+0	.3190E-1	.4242E+1	.4164E-2
0.33	22.00	.925	0.128	0.037	21.745	2.223	43.966	0.1179	0.815	.4589E+1	.2179E+0	.3002E-1	.5185E+1	.5103E-2
0.33	27.00	.925	0.128	0.037	26.745	2.223	53.966	0.1179	1.003	.3979E+1	.2513E+0	.2821E-1	.6365E+1	.6276E-2
0.33	33.00	.925	0.128	0.037	32.745	2.223	65.966	0.1179	1.228	.3460E+1	.2890E+0	.2654E-1	.7780E+1	.7684E-2
0.39	0.33	.824	0.107	0.088	0.116	2.138	0.744	0.0882	0.010	.1178E+3	.8549E-2	.6643E-1	.6561E-1	.3711E-4
0.39	0.39	.847	0.118	0.077	0.153	2.171	0.831	0.1004	0.012	.9383E+2	.1066E-1	.7007E-1	.8344E-1	.5430E-4
0.39	0.47	.868	0.129	0.066	0.212	2.201	0.961	0.1120	0.014	.7486E+2	.1336E-1	.7288E-1	.1076E+0	.7969E-4
0.39	0.56	.881	0.136	0.055	0.289	2.220	1.123	0.1194	0.017	.6173E+2	.1620E-1	.7417E-1	.1340E+0	.1104E-3
0.39	0.68	.890	0.140	0.051	0.400	2.233	1.350	0.1246	0.022	.5078E+2	.1969E-1	.7426E-1	.1682E+0	.1530E-3
0.39	0.82	.897	0.144	0.049	0.533	2.243	1.620	0.1287	0.027	.4265E+2	.2345E-1	.7332E-1	.2085E+0	.2028E-3
0.39	1.00	.901	0.146	0.048	0.709	2.249	1.974	0.1311	0.035	.3583E+2	.2791E-1	.7156E-1	.2588E+0	.2682E-3
0.39	1.20	.903	0.147	0.047	0.906	2.253	2.370	0.1329	0.043	.3076E+2	.3251E-1	.6946E-1	.3149E+0	.3499E-3
0.39	1.50	.906	0.148	0.047	1.204	2.256	2.967	0.1341	0.057	.2577E+2	.3890E-1	.6650E-1	.3978E+0	.4511E-3
0.39	1.80	.908	0.149	0.046	1.502	2.259	3.564	0.1353	0.069	.2229E+2	.4486E-1	.6390E-1	.4822E+0	.5600E-3
0.39	2.20	.909	0.149	0.046	1.901	2.260	4.363	0.1359	0.086	.1913E+2	.5228E-1	.6091E-1	.5923E+0	.7068E-3

CASE # 1   COPYRIGHT 1982 STEVE SMITH   DATA DATE 1/ 3/82   TRANSFORMER DESIGN TABLE

$$f(D) = \left[\left(\frac{U}{FG(DE)^2}\right)^5 * \left(\frac{1}{DEUm}\right)^4\right]^{1/13}$$

P	Q	D	E	F	G	U	Um	DE	FG	f(D)	1/f(D)	1/(PQf(D))	DEUm	FG(DE)^2/U
0.39	2.70	.910	0.150	0.045	2.400	2.261	5.361	0.1365	0.108	.1641E+2	.6094E-1	.5788E-1	.7318E+0	.8899E-3
0.39	3.30	.911	0.151	0.044	2.999	2.263	6.560	0.1371	0.133	.1415E+2	.7067E-1	.5491E-1	.8994E+0	.1199E-2
0.39	3.90	.912	0.151	0.044	3.598	2.264	7.758	0.1377	0.158	.1253E+2	.7792E-1	.5248E-1	.1068E+1	.1326E-2
0.39	4.70	.912	0.151	0.044	4.398	2.264	9.358	0.1377	0.194	.1095E+2	.9135E-1	.4984E-1	.1289E+1	.1621E-2
0.39	5.60	.912	0.151	0.044	5.298	2.264	11.158	0.1377	0.233	.9653E+1	.1036E+0	.4743E-1	.1537E+1	.1952E-2
0.39	6.80	.913	0.151	0.044	6.497	2.266	13.557	0.1383	0.283	.8405E+1	.1190E+0	.4486E-1	.1875E+1	.2387E-2
0.39	8.20	.913	0.151	0.044	7.897	2.266	16.357	0.1383	0.344	.7360E+1	.1359E+0	.4249E-1	.2262E+1	.2901E-2
0.39	10.00	.913	0.151	0.044	9.697	2.266	19.957	0.1383	0.422	.6397E+1	.1563E+0	.4008E-1	.2760E+1	.3562E-2
0.39	12.00	.913	0.151	0.044	11.697	2.266	23.957	0.1383	0.509	.5626E+1	.1777E+0	.3798E-1	.3314E+1	.4297E-2
0.39	15.00	.913	0.151	0.044	14.697	2.266	29.957	0.1383	0.639	.4811E+1	.2079E+0	.3553E-1	.4144E+1	.5399E-2
0.39	18.00	.914	0.152	0.043	17.696	2.267	35.956	0.1389	0.761	.4235E+1	.2362E+0	.3364E-1	.4995E+1	.6478E-2
0.39	22.00	.914	0.152	0.043	21.696	2.267	43.956	0.1389	0.933	.3661E+1	.2717E+0	.3167E-1	.6107E+1	.7943E-2
0.39	27.00	.914	0.152	0.043	26.696	2.267	53.956	0.1389	1.148	.3191E+1	.3134E+0	.2976E-1	.7496E+1	.9773E-2
0.39	33.00	.914	0.154	0.043	32.696	2.267	65.956	0.1389	1.406	.2775E+1	.3604E+0	.2800E-1	.9163E+1	.1197E-1
0.47	0.33	.758	0.114	0.121	0.102	2.168	0.804	0.0864	0.012	.1082E+3	.9241E-2	.5958E-1	.6949E-1	.4339E-4
0.47	0.39	.789	0.129	0.106	0.131	2.216	0.880	0.1022	0.014	.8481E+2	.1179E-1	.6433E-1	.8990E-1	.6654E-4
0.47	0.47	.822	0.146	0.089	0.178	2.248	0.993	0.1200	0.016	.6574E+2	.1521E-1	.6886E-1	.1191E+0	.1030E-3
0.47	0.56	.845	0.158	0.077	0.245	2.274	1.140	0.1331	0.019	.5287E+2	.1891E-1	.7186E-1	.1517E+0	.1496E-3
0.47	0.68	.863	0.166	0.069	0.347	2.274	1.354	0.1437	0.024	.4253E+2	.2351E-1	.7356E-1	.1946E+0	.2158E-3
0.47	0.82	.874	0.172	0.063	0.476	2.290	1.618	0.1503	0.030	.3517E+2	.2844E-1	.7379E-1	.2433E+0	.2959E-3
0.47	1.00	.881	0.175	0.060	0.649	2.300	1.968	0.1546	0.039	.2919E+2	.3425E-1	.7288E-1	.3043E+0	.4014E-3
0.47	1.20	.886	0.178	0.057	0.844	2.307	2.361	0.1577	0.048	.2486E+2	.4022E-1	.7132E-1	.3724E+0	.5186E-3
0.47	1.50	.890	0.180	0.055	1.140	2.313	2.955	0.1602	0.063	.2062E+2	.4849E-1	.6878E-1	.4735E+0	.6938E-3
0.47	1.80	.892	0.181	0.054	1.438	2.316	3.553	0.1615	0.078	.1781E+2	.5616E-1	.6638E-1	.5736E+0	.8741E-3
0.47	2.20	.894	0.182	0.053	1.836	2.319	4.350	0.1627	0.097	.1522E+2	.6570E-1	.6354E-1	.7077E+0	.1111E-2
0.47	2.70	.895	0.182	0.053	2.335	2.320	5.348	0.1633	0.123	.1302E+2	.7682E-1	.6053E-1	.8736E+0	.1410E-2
0.47	3.30	.896	0.183	0.052	2.934	2.321	6.547	0.1640	0.153	.1120E+2	.8927E-1	.5756E-1	.1073E+1	.1767E-2
0.47	3.90	.897	0.183	0.051	3.533	2.323	7.745	0.1646	0.182	.9901E+1	.1010E+0	.5510E-1	.1275E+1	.2122E-2
0.47	4.70	.898	0.184	0.051	4.332	2.324	9.344	0.1652	0.221	.8640E+1	.1157E+0	.5240E-1	.1544E+1	.2595E-2
0.47	5.60	.898	0.184	0.051	5.232	2.324	11.144	0.1652	0.267	.7611E+1	.1314E+0	.4992E-1	.1841E+1	.3134E-2
0.47	6.80	.899	0.184	0.051	6.431	2.326	13.543	0.1659	0.325	.6620E+1	.1510E+0	.4726E-1	.2246E+1	.3842E-2
0.47	8.00	.899	0.184	0.051	7.831	2.326	16.343	0.1659	0.395	.5793E+1	.1726E+0	.4479E-1	.2711E+1	.4678E-2
0.47	10.00	.899	0.184	0.051	9.631	2.326	19.943	0.1659	0.486	.5032E+1	.1987E+0	.4229E-1	.3308E+1	.5753E-2
0.47	12.00	.899	0.184	0.050	11.631	2.326	23.943	0.1659	0.587	.4423E+1	.2261E+0	.4008E-1	.3971E+1	.6948E-2
0.47	15.00	.900	0.185	0.050	14.630	2.327	29.941	0.1665	0.732	.3780E+1	.2645E+0	.3752E-1	.4985E+1	.8714E-2
0.47	18.00	.900	0.185	0.050	17.630	2.327	35.941	0.1665	0.882	.3326E+1	.3006E+0	.3553E-1	.5984E+1	.1050E-1

TRANSFORMER DESIGN TABLE

CASE # 1   COPYRIGHT 1982 STEVE SMITH   DATA DATE 1/ 3/82

$$f(D)=[(U/(FG(DE)^2))^5 * (1/(DEUm))^4]^{(1/13)}$$

P	Q	D	E	F	G	U	Um	DE	FG	f(D)	1/f(D)	1/(PQf(D))	DEUm	FG(DE)^2/U
0.47	22.00	.900	0.185	0.050	21.630	2.327	43.941	0.1665	1.082	.2890E+1	.3460E+0	.3346E-1	.7316E+1	.1288E-1
0.47	27.00	.900	0.185	0.050	26.630	2.327	53.941	0.1665	1.332	.2505E+1	.3992E+0	.3146E-1	.8981E+1	.1586E-1
0.47	33.00	.900	0.185	0.050	32.630	2.327	65.941	0.1665	1.632	.2178E+1	.4591E+0	.2960E-1	.1098E+2	.1944E-1
0.56	0.33	.678	0.119	0.161	0.092	2.100	0.880	0.0807	0.015	.1052E+3	.9507E-2	.5145E-1	.7099E-1	.4592E-4
0.56	0.39	.715	0.137	0.143	0.115	2.153	0.947	0.0983	0.016	.8072E+2	.1239E-1	.5673E-1	.9102E-1	.7358E-4
0.56	0.47	.758	0.159	0.121	0.152	2.214	1.046	0.1205	0.018	.6080E+2	.1645E-1	.6249E-1	.1260E+0	.1207E-3
0.56	0.56	.796	0.178	0.102	0.204	2.268	1.171	0.1417	0.021	.4748E+2	.2106E-1	.6716E-1	.1659E+0	.1841E-3
0.56	0.68	.827	0.194	0.086	0.293	2.313	1.367	0.1600	0.025	.3709E+2	.2696E-1	.7081E-1	.2187E+0	.2806E-3
0.56	0.82	.846	0.203	0.077	0.414	2.340	1.620	0.1717	0.032	.3000E+2	.3333E-1	.7259E-1	.2782E+0	.4018E-3
0.56	1.00	.858	0.209	0.071	0.582	2.357	1.963	0.1793	0.041	.2450E+2	.4082E-1	.7289E-1	.3519E+0	.5637E-3
0.56	1.20	.865	0.213	0.067	0.775	2.367	2.153	0.1838	0.052	.2064E+2	.4846E-1	.7211E-1	.4334E+0	.7467E-3
0.56	1.50	.871	0.216	0.065	1.069	2.376	2.944	0.1877	0.069	.1696E+2	.5897E-1	.7020E-1	.5526E+0	.1023E-2
0.56	1.80	.875	0.218	0.063	1.365	2.381	3.538	0.1903	0.085	.1456E+2	.6868E-1	.6813E-1	.6734E+0	.1298E-2
0.56	2.20	.877	0.219	0.060	1.761	2.384	4.335	0.1916	0.108	.1239E+2	.8073E-1	.6553E-1	.8308E+0	.1670E-2
0.56	2.70	.879	0.220	0.060	2.261	2.387	5.333	0.1929	0.137	.1056E+2	.9473E-1	.6266E-1	.1029E+1	.2133E-2
0.56	3.30	.881	0.221	0.060	2.859	2.390	6.530	0.1943	0.170	.9058E+1	.1104E+0	.5974E-1	.1268E+1	.2686E-2
0.56	3.90	.882	0.221	0.059	3.458	2.391	7.728	0.1949	0.209	.7992E+1	.1251E+0	.5729E-1	.1506E+1	.3242E-2
0.56	4.70	.883	0.222	0.058	4.257	2.393	9.327	0.1956	0.249	.6962E+1	.1436E+0	.5457E-1	.1824E+1	.3981E-2
0.56	5.60	.884	0.222	0.058	5.156	2.394	11.125	0.1962	0.299	.6125E+1	.1633E+0	.5206E-1	.2183E+1	.4810E-2
0.56	6.80	.884	0.222	0.058	6.356	2.394	13.525	0.1962	0.369	.5322E+1	.1879E+0	.4934E-1	.2654E+1	.5930E-2
0.56	8.20	.885	0.223	0.058	7.755	2.396	16.324	0.1969	0.446	.4652E+1	.2149E+0	.4681E-1	.3214E+1	.7217E-2
0.56	10.00	.885	0.223	0.058	9.555	2.396	19.924	0.1969	0.549	.4038E+1	.2476E+0	.4422E-1	.3923E+1	.8892E-2
0.56	12.00	.885	0.223	0.058	11.555	2.396	23.924	0.1969	0.664	.3548E+1	.2819E+0	.4194E-1	.4711E+1	.1075E-1
0.56	15.00	.885	0.223	0.058	14.555	2.396	29.924	0.1969	0.837	.3031E+1	.3300E+0	.3928E-1	.5892E+1	.1355E-1
0.56	18.00	.886	0.223	0.057	17.554	2.397	35.923	0.1976	1.001	.2666E+1	.3751E+0	.3722E-1	.7098E+1	.1629E-1
0.56	22.00	.886	0.223	0.057	21.554	2.397	43.923	0.1976	1.229	.2315E+1	.4319E+0	.3506E-1	.8678E+1	.2001E-1
0.56	27.00	.886	0.223	0.057	26.554	2.397	53.923	0.1976	1.514	.2006E+1	.4984E+0	.3297E-1	.1065E+2	.2465E-1
0.56	33.00	.886	0.223	0.057	32.554	2.397	65.923	0.1976	1.856	.1744E+1	.5735E+0	.3103E-1	.1302E+2	.3022E-1
0.68	0.33	.567	0.123	0.217	0.083	2.061	0.987	0.0700	0.018	.1090E+3	.9173E-2	.4088E-1	.6911E-1	.4275E-4
0.68	0.39	.607	0.145	0.196	0.101	2.121	1.047	0.0880	0.020	.8162E+2	.1225E-1	.4620E-1	.9213E-1	.7209E-4
0.68	0.47	.662	0.171	0.169	0.128	2.197	1.131	0.1132	0.022	.5947E+2	.1682E-1	.5262E-1	.1281E+0	.1262E-3
0.68	0.56	.713	0.197	0.144	0.167	2.270	1.238	0.1401	0.024	.4475E+2	.2235E-1	.5868E-1	.1735E+0	.2072E-3
0.68	0.68	.765	0.223	0.118	0.235	2.344	1.407	0.1702	0.028	.3347E+2	.2987E-1	.6461E-1	.2390E+0	.3413E-3
0.68	0.82	.802	0.241	0.099	0.338	2.397	1.631	0.1933	0.033	.2611E+2	.3830E-1	.6868E-1	.3153E+0	.5215E-3
0.68	1.00	.826	0.253	0.087	0.494	2.431	1.957	0.2090	0.043	.2073E+2	.4824E-1	.7095E-1	.4089E+0	.7720E-3
0.68	1.20	.839	0.260	0.081	0.681	2.450	2.338	0.2177	0.055	.1715E+2	.5831E-1	.7146E-1	.5091E+0	.1061E-2

CASE # 1   COPYRIGHT 1982 STEVE SMITH   DATA DATE 1/ 3/82   TRANSFORMER DESIGN TABLE

$$f(D) = \left[\left(\frac{U}{FG(DE)^2}\right)^5 * \left(\frac{1}{DEUm}\right)^4\right]^{1/13}$$

P	Q	D	E	F	G	U	Um	DE	FG	f(D)	1/f(D)	1/(PQf(D))	DEUm	FG(DE)^2/U
0.68	1.50	.848	.264	0.076	0.972	2.463	2.925	0.2239	0.074	.1388E+2	.7206E-1	.7065E-1	.6549E+0	.1503E-2
0.68	1.80	.854	.267	0.073	1.266	2.471	3.517	0.2280	0.092	.1181E+2	.8467E-1	.6917E-1	.8019E+0	.1944E-2
0.68	2.20	.858	.269	0.071	1.662	2.477	4.311	0.2308	0.118	.9977E+1	.1002E+0	.6700E-1	.9950E+0	.2538E-2
0.68	2.70	.861	.271	0.070	2.159	2.481	5.307	0.2329	0.150	.8455E+1	.1183E+0	.6442E-1	.1236E+1	.3280E-2
0.68	3.30	.863	.272	0.069	2.757	2.484	6.504	0.2343	0.189	.7226E+1	.1384E+0	.6167E-1	.1524E+1	.4174E-2
0.68	3.90	.864	.272	0.068	3.356	2.486	7.703	0.2350	0.228	.6359E+1	.1573E+0	.5930E-1	.1810E+1	.5071E-2
0.68	4.70	.865	.273	0.067	4.155	2.487	9.301	0.2357	0.280	.5526E+1	.1810E+0	.5662E-1	.2192E+1	.6265E-2
0.68	5.60	.866	.273	0.067	5.054	2.488	11.100	0.2364	0.339	.4855E+1	.2060E+0	.5411E-1	.2624E+1	.7606E-2
0.68	6.80	.867	.274	0.067	6.253	2.490	13.498	0.2371	0.416	.4210E+1	.2377E+0	.5137E-1	.3201E+1	.9390E-2
0.68	8.20	.868	.274	0.066	7.652	2.491	16.297	0.2378	0.505	.3676E+1	.2721E+0	.4879E-1	.3876E+1	.1147E-1
0.68	10.00	.868	.274	0.066	9.452	2.491	19.897	0.2378	0.624	.3187E+1	.3138E+0	.4614E-1	.4732E+1	.1416E-1
0.68	12.00	.869	.275	0.065	11.451	2.493	23.895	0.2385	0.750	.2798E+1	.3574E+0	.4380E-1	.5700E+1	.1712E-1
0.68	15.00	.869	.275	0.065	14.451	2.493	29.895	0.2385	0.947	.2388E+1	.4188E+0	.4106E-1	.7131E+1	.2161E-1
0.68	18.00	.869	.275	0.065	17.450	2.493	35.895	0.2385	1.143	.2099E+1	.4763E+0	.3922E-1	.8563E+1	.2609E-1
0.68	22.00	.870	.275	0.065	21.450	2.494	43.894	0.2393	1.394	.1823E+1	.5486E+0	.3667E-1	.1050E+2	.3200E-1
0.68	27.00	.870	.275	0.065	26.450	2.494	53.894	0.2393	1.719	.1579E+1	.6334E+0	.3450E-1	.1289E+2	.3946E-1
0.68	33.00	.870	.275	0.065	32.450	2.494	65.894	0.2393	2.109	.1372E+1	.7290E+0	.3249E-1	.1577E+2	.4881E-1
0.82	0.33	.437	.129	0.282	0.073	2.015	1.113	0.0562	0.021	.1255E+3	.7970E-2	.2945E-1	.6248E-1	.3215E-4
0.82	0.39	.483	.152	0.259	0.087	2.081	1.167	0.0732	0.022	.9092E+2	.1100E-1	.3439E-1	.8539E-1	.5786E-4
0.82	0.47	.541	.181	0.230	0.138	2.164	1.244	0.0977	0.025	.6366E+2	.1571E-1	.4076E-1	.1215E+0	.1102E-3
0.82	0.56	.602	.211	0.199	0.188	2.251	1.337	0.1270	0.027	.4595E+2	.2176E-1	.4739E-1	.1698E+0	.1968E-3
0.82	0.68	.674	.247	0.163	0.186	2.354	1.474	0.1665	0.030	.3263E+2	.3064E-1	.5496E-1	.2454E+0	.3569E-3
0.82	0.82	.736	.278	0.132	0.264	2.484	1.665	0.2046	0.035	.2420E+2	.4132E-1	.6146E-1	.3407E+0	.5972E-3
0.82	1.00	.781	.301	0.109	0.399	2.507	1.961	0.2347	0.044	.1838E+2	.5440E-1	.6634E-1	.4602E+0	.9599E-3
0.82	1.20	.806	.313	0.097	0.574	2.543	2.325	0.2523	0.056	.1478E+2	.6766E-1	.6876E-1	.5866E+0	.1394E-2
0.82	1.50	.823	.322	0.088	0.857	2.567	2.901	0.2646	0.076	.1169E+2	.8555E-1	.6955E-1	.7676E+0	.2068E-2
0.82	1.80	.831	.326	0.085	1.149	2.578	3.490	0.2711	0.097	.9824E+1	.1018E+0	.6897E-1	.9439E+0	.2755E-2
0.82	2.20	.837	.329	0.082	1.543	2.587	4.281	0.2750	0.126	.8216E+1	.1217E+0	.6747E-1	.1177E+1	.3675E-2
0.82	2.70	.841	.331	0.079	2.039	2.593	5.275	0.2780	0.162	.6913E+1	.1447E+0	.6534E-1	.1466E+1	.4830E-2
0.82	3.30	.844	.332	0.078	2.636	2.597	6.471	0.2802	0.206	.5871E+1	.1702E+0	.6289E-1	.1813E+1	.6216E-2
0.82	3.90	.846	.333	0.077	3.234	2.600	7.668	0.2817	0.249	.5154E+1	.1940E+0	.6067E-1	.2160E+1	.7602E-2
0.82	4.70	.848	.334	0.076	4.031	2.603	9.265	0.2832	0.306	.4465E+1	.2239E+0	.5811E-1	.2624E+1	.9445E-2
0.82	5.60	.849	.334	0.076	4.931	2.604	11.064	0.2840	0.372	.3913E+1	.2556E+0	.5566E-1	.3142E+1	.1153E-1
0.82	6.80	.850	.335	0.075	6.130	2.606	13.462	0.2848	0.460	.3387E+1	.2952E+0	.5295E-1	.3833E+1	.1431E-1
0.82	8.20	.851	.336	0.074	7.529	2.607	16.261	0.2855	0.561	.2953E+1	.3387E+0	.5037E-1	.4643E+1	.1754E-1
0.82	10.00	.852	.336	0.074	9.328	2.608	19.860	0.2863	0.690	.2557E+1	.3911E+0	.4770E-1	.5685E+1	.2169E-1

CASE # 1   COPYRIGHT 1982 STEVE SMITH   DATA DATE 1/ 3/82   TRANSFORMER DESIGN TABLE

$$f(D)=[(U/(FG(DE)^2))^5 * (1/(DEUm))^4]^{(1/13)}$$

P	Q	D	E	F	G	U	Um	DE	FG	f(D)	1/f(D)	1/(PQf(D))	DEUm	FG(DE)^2/U
0.82	12.00	.852	0.336	0.074	11.328	2.608	23.860	0.2863	0.838	.2242E+1	.4459E+0	.4532E-1	.6830E+1	.2634E-1
0.82	15.00	.853	0.337	0.074	14.327	2.610	29.858	0.2870	1.053	.1912E+1	.5230E+0	.4252E-1	.8570E+1	.3324E-1
0.82	18.00	.853	0.337	0.074	17.327	2.610	35.858	0.2870	1.274	.1688E+1	.5953E+0	.4033E-1	.1029E+2	.4020E-1
0.82	22.00	.853	0.337	0.074	21.327	2.610	43.858	0.2870	1.568	.1458E+1	.6860E+0	.3803E-1	.1259E+2	.4948E-1
0.82	27.00	.854	0.337	0.073	26.326	2.611	53.857	0.2878	1.922	.1262E+1	.7924E+0	.3579E-1	.1550E+2	.6096E-1
0.82	33.00	.854	0.337	0.073	32.326	2.611	65.857	0.2878	2.360	.1096E+1	.9123E+0	.3371E-1	.1895E+2	.7485E-1
1.00	0.33	.273	0.137	0.364	0.057	1.961	1.270	0.0373	0.021	.1848E+3	.5411E-2	.1640E-1	.4732E-1	.1667E-4
1.00	0.41	.321	0.161	0.340	0.069	2.069	1.321	0.0515	0.023	.1245E+3	.8032E-2	.2060E-1	.6807E-1	.3064E-4
1.00	0.47	.383	0.192	0.309	0.087	2.118	1.393	0.0733	0.027	.8079E+2	.1238E-1	.2634E-1	.1021E+0	.6816E-4
1.00	0.56	.451	0.226	0.275	0.109	2.215	1.475	0.1017	0.030	.5446E+2	.1836E-1	.3272E-1	.1501E+0	.1397E-3
1.00	0.68	.536	0.268	0.232	0.144	2.337	1.594	0.1436	0.033	.3582E+2	.2788E-1	.4108E-1	.2290E+0	.2950E-3
1.00	0.82	.624	0.312	0.188	0.196	2.463	1.748	0.1947	0.037	.2469E+2	.4049E-1	.4938E-1	.3403E+0	.5671E-3
1.00	1.00	.706	0.353	0.147	0.294	2.580	1.991	0.2492	0.043	.1741E+2	.5743E-1	.5743E-1	.4962E+0	.1040E-2
1.00	1.20	.757	0.379	0.122	0.443	2.653	2.318	0.2865	0.054	.1328E+2	.7529E-1	.6274E-1	.6642E+0	.1666E-2
1.00	1.50	.790	0.395	0.105	0.710	2.700	2.871	0.3121	0.075	.1008E+2	.9924E-1	.6616E-1	.8959E+0	.2689E-2
1.00	1.80	.804	0.407	0.098	0.996	2.720	3.451	0.3232	0.098	.8289E+1	.1206E+0	.6702E-1	.1115E+1	.3749E-2
1.00	2.20	.814	0.407	0.093	1.386	2.734	4.237	0.3313	0.129	.6823E+1	.1466E+0	.6662E-1	.1404E+1	.5174E-2
1.00	2.70	.820	0.410	0.090	1.880	2.743	5.228	0.3362	0.169	.5677E+1	.1762E+0	.6524E-1	.1758E+1	.6973E-2
1.00	3.30	.824	0.412	0.088	2.476	2.748	6.422	0.3395	0.218	.4788E+1	.2089E+0	.6329E-1	.2180E+1	.9137E-2
1.00	3.90	.827	0.414	0.086	3.093	2.753	7.618	0.3420	0.266	.4178E+1	.2393E+0	.6137E-1	.2605E+1	.1129E-1
1.00	4.70	.829	0.415	0.085	3.894	2.756	9.215	0.3436	0.331	.3605E+1	.2774E+0	.5903E-1	.3167E+1	.1418E-1
1.00	5.60	.831	0.416	0.084	4.798	2.758	11.012	0.3453	0.403	.3148E+1	.3176E+0	.5672E-1	.3802E+1	.1742E-1
1.00	6.80	.832	0.416	0.083	6.036	2.760	13.411	0.3461	0.501	.2718E+1	.3679E+0	.5411E-1	.4642E+1	.2176E-1
1.00	8.20	.833	0.417	0.083	7.410	2.761	16.209	0.3469	0.615	.2364E+1	.4230E+0	.5158E-1	.5624E+1	.2682E-1
1.00	10.00	.834	0.417	0.082	9.280	2.763	19.808	0.3478	0.761	.2043E+1	.4894E+0	.4894E-1	.6889E+1	.3331E-1
1.00	12.00	.835	0.417	0.082	11.232	2.764	23.807	0.3486	0.921	.1790E+1	.5587E+0	.4656E-1	.8299E+1	.4049E-1
1.00	15.00	.835	0.417	0.082	14.256	2.764	29.807	0.3486	1.169	.1524E+1	.6561E+0	.4374E-1	.1039E+2	.5138E-1
1.00	18.00	.836	0.418	0.082	17.159	2.766	35.805	0.3494	1.407	.1338E+1	.7474E+0	.4152E-1	.1251E+2	.6215E-1
1.00	22.00	.836	0.418	0.082	21.159	2.766	43.805	0.3494	1.735	.1160E+1	.8620E+0	.3918E-1	.1531E+2	.7663E-1
1.00	27.00	.836	0.418	0.082	26.159	2.766	53.805	0.3494	2.145	.1004E+1	.9963E+0	.3690E-1	.1880E+2	.9473E-1
1.00	33.00	.837	0.419	0.082	31.963	2.767	65.804	0.3503	2.621	.8714E+0	.1148E+1	.3478E-1	.2305E+2	.1162E+0
1.20	0.47	.212	0.206	0.394	0.058	2.074	1.551	0.0437	0.023	.1441E+3	.6938E-2	.1230E-1	.6774E-1	.2102E-4
1.20	0.56	.283	0.241	0.359	0.078	2.175	1.630	0.0683	0.028	.8301E+2	.1205E-1	.1793E-1	.1114E+0	.5928E-4
1.20	0.68	.374	0.287	0.313	0.105	2.305	1.740	0.1073	0.033	.4767E+2	.2098E-1	.2571E-1	.1867E+0	.1658E-3
1.20	0.82	.476	0.338	0.262	0.145	2.451	1.874	0.1609	0.038	.2936E+2	.3406E-1	.3461E-1	.3015E+0	.3984E-3
1.20	1.00	.590	0.395	0.205	0.210	2.614	2.071	0.2331	0.043	.1861E+2	.5372E-1	.4477E-1	.4826E+0	.8945E-3

CASE # 1 COPYRIGHT 1982 STEVE SMITH DATA DATE 1/ 3/82 TRANSFORMER DESIGN TABLE

$$f(D) = [(U/(FG(DE)^2))^5 * (1/(DEUm))^4]^{1/13}$$

P	Q	D	E	F	G	U	Um	DE	FG	f(D)	1/f(D)	1/(PQf(D))	DEUm	$FG(DE)^2/U$
1.20	1.20	.681	0.441	0.160	0.319	2.744	2.341	0.3000	0.051	.1305E+2	.7664E-1	.5322E-1	.7022E+0	.1669E-2
1.20	1.50	.748	0.474	0.126	0.552	2.840	2.845	0.3546	0.070	.9222E+1	.1084E+0	.6024E-1	.1009E+1	.3079E-2
1.20	1.80	.775	0.488	0.113	0.825	2.878	3.407	0.3778	0.093	.7330E+1	.1364E+0	.6316E-1	.1287E+1	.4663E-2
1.20	2.20	.791	0.496	0.105	1.209	2.901	4.184	0.3919	0.126	.5893E+1	.1697E+0	.6428E-1	.1640E+1	.6689E-2
1.20	2.70	.800	0.500	0.100	1.700	2.914	5.171	0.4000	0.170	.4827E+1	.2072E+0	.6395E-1	.2068E+1	.9334E-2
1.20	3.30	.806	0.503	0.097	2.294	2.923	6.362	0.4054	0.223	.4029E+1	.2482E+0	.6268E-1	.2579E+1	.1251E-1
1.20	3.90	.809	0.505	0.095	2.891	2.927	7.558	0.4081	0.276	.3493E+1	.2863E+0	.6117E-1	.3085E+1	.1571E-1
1.20	4.70	.812	0.506	0.093	3.688	2.931	9.154	0.4109	0.347	.2997E+1	.3336E+0	.5916E-1	.3761E+1	.1997E-1
1.20	5.60	.814	0.507	0.093	4.586	2.934	10.951	0.4127	0.426	.2608E+1	.3835E+0	.5707E-1	.4519E+1	.2476E-1
1.20	6.80	.816	0.508	0.092	5.784	2.937	13.348	0.4145	0.532	.2244E+1	.4457E+0	.5462E-1	.5533E+1	.3113E-1
1.20	8.20	.818	0.509	0.091	7.182	2.940	16.145	0.4164	0.654	.1947E+1	.5137E+0	.5221E-1	.6722E+1	.3854E-1
1.20	10.00	.819	0.510	0.091	8.981	2.941	19.744	0.4173	0.813	.1679E+1	.5956E+0	.4964E-1	.8239E+1	.4812E-1
1.20	12.00	.819	0.510	0.091	10.981	2.941	23.744	0.4173	0.994	.1468E+1	.6811E+0	.4730E-1	.9908E+1	.5883E-1
1.20	15.00	.820	0.510	0.090	13.980	2.943	29.742	0.4182	1.258	.1248E+1	.8011E+0	.4451E-1	.1244E+2	.7478E-1
1.20	18.00	.821	0.511	0.090	16.979	2.944	35.741	0.4191	1.520	.1095E+1	.9135E+0	.4229E-1	.1498E+2	.9067E-1
1.20	22.00	.821	0.511	0.090	20.979	2.944	43.741	0.4191	1.878	.9483E+0	.1054E+1	.3994E-1	.1833E+2	.1120E+0
1.20	27.00	.822	0.511	0.089	25.978	2.946	53.739	0.4200	2.312	.8199E+0	.1220E+1	.3765E-1	.2257E+2	.1385E+0
1.20	33.00	.822	0.511	0.089	31.978	2.946	65.739	0.4200	2.846	.7114E+0	.1406E+1	.3550E-1	.2761E+2	.1705E+0
1.20	0.68	.136	0.318	0.432	0.044	2.265	1.951	0.0432	0.019	.1507E+3	.6635E-2	.6508E-2	.8438E-1	.1570E-4
1.20	0.82	.242	0.371	0.379	0.078	2.417	2.080	0.0898	0.030	.5822E+2	.1718E-1	.1396E-1	.1867E+0	.9860E-4
1.20	1.00	.375	0.438	0.313	0.125	2.607	2.249	0.1641	0.039	.2746E+2	.3642E-1	.2428E-1	.3691E+0	.4033E-3
1.50	1.20	.510	0.505	0.245	0.190	2.800	2.457	0.2576	0.047	.1580E+2	.6329E-1	.3516E-1	.6327E+0	.1103E-2
1.50	1.50	.656	0.578	0.172	0.344	2.808	2.848	0.3792	0.059	.9332E+1	.1072E+0	.4763E-1	.1080E+1	.2828E-2
1.50	1.80	.724	0.612	0.138	0.576	3.106	3.351	0.4431	0.079	.6782E+1	.1474E+0	.5461E-1	.1485E+1	.5025E-2
1.50	2.20	.758	0.629	0.121	0.942	3.154	4.102	0.4768	0.114	.5158E+1	.1939E+0	.5876E-1	.1956E+1	.8215E-2
1.50	2.70	.774	0.637	0.113	1.426	3.177	5.079	0.4930	0.161	.4089E+1	.2446E+0	.6039E-1	.2504E+1	.1233E-1
1.50	3.30	.784	0.642	0.108	2.016	3.191	6.265	0.5035	0.218	.3345E+1	.2990E+0	.6040E-1	.3153E+1	.1728E-1
1.50	3.90	.789	0.645	0.106	2.611	3.198	7.458	0.5085	0.275	.2867E+1	.3488E+0	.5963E-1	.3792E+1	.2227E-1
1.50	4.70	.793	0.647	0.104	3.407	3.204	9.052	0.5127	0.353	.2436E+1	.4105E+0	.5822E-1	.4641E+1	.2893E-1
1.50	5.60	.796	0.648	0.102	4.304	3.208	10.848	0.5158	0.439	.2105E+1	.4750E+0	.5655E-1	.5595E+1	.3640E-1
1.50	6.80	.798	0.649	0.101	5.502	3.211	13.245	0.5179	0.556	.1801E+1	.5553E+0	.5444E-1	.6860E+1	.4641E-1
1.50	8.20	.800	0.650	0.100	6.900	3.214	16.042	0.5200	0.690	.1556E+1	.6427E+0	.5225E-1	.8342E+1	.5805E-1
1.50	10.00	.801	0.651	0.100	8.699	3.216	19.641	0.5211	0.866	.1337E+1	.7478E+0	.4985E-1	.1023E+2	.7308E-1
1.50	12.00	.802	0.651	0.099	10.698	3.217	23.639	0.5221	1.059	.1166E+1	.8573E+0	.4763E-1	.1234E+2	.8974E-1
1.50	15.00	.803	0.652	0.099	13.697	3.218	29.638	0.5232	1.349	.9994E+0	.1011E+1	.4492E-1	.1551E+2	.1147E+0
1.50	18.00	.804	0.652	0.098	16.696	3.220	35.636	0.5242	1.636	.8663E+0	.1154E+1	.4276E-1	.1868E+2	.1396E+0

CASE # 1   COPYRIGHT 1982 STEVE SMITH   DATA DATE 1/ 3/82   TRANSFORMER DESIGN TABLE

$$f(D)=\left[\left(\frac{U}{FG(DE)^2}\right)^5 \cdot \left(\frac{1}{DEUm}\right)^4\right]^{1/13}$$

P	Q	D	E	F	G	U	Um	DE	FG	f(D)	1/f(D)	1/(PQf(D))	DEUm	FG(DE)^2/U
1.50	22.00	.804	0.652	0.098	20.696	3.220	43.636	0.5242	2.028	.7494E+0	.1334E+1	.4044E-1	.2287E+2	.1731E+0
1.50	27.00	.805	0.653	0.098	25.695	3.221	53.635	0.5253	2.505	.6471E+0	.1545E+1	.3816E-1	.2817E+2	.2146E+0
1.50	33.00	.805	0.653	0.098	31.695	3.221	65.635	0.5253	3.090	.5610E+0	.1783E+1	.3601E-1	.3448E+2	.2647E+0
1.80	1.00	.149	0.475	0.426	0.051	2.584	2.444	0.0707	0.022	.8281E+2	.1208E-1	.6709E-2	.1728E+0	.4198E-4
1.80	1.20	.296	0.548	0.352	0.104	2.794	2.634	0.1622	0.037	.2789E+2	.3586E-1	.1660E-1	.4272E+0	.3448E-3
1.80	1.50	.498	0.649	0.251	0.222	3.083	2.945	0.3232	0.057	.1175E+2	.8511E-1	.3152E-1	.9518E+0	.1718E-2
1.80	1.80	.639	0.720	0.181	0.361	3.284	3.343	0.4598	0.065	.7193E+1	.1390E+0	.4291E-1	.1537E+1	.4194E-2
1.80	2.20	.719	0.760	0.141	0.681	3.398	4.029	0.5461	0.096	.4932E+1	.2027E+0	.5120E-1	.2200E+1	.8396E-2
1.80	2.70	.751	0.776	0.125	1.149	3.466	4.983	0.5824	0.143	.3712E+1	.2694E+0	.5543E-1	.2902E+1	.1409E-1
1.80	3.30	.766	0.783	0.117	1.734	3.466	6.162	0.5998	0.203	.2952E+1	.3387E+0	.5702E-1	.3696E+1	.2106E-1
1.80	3.90	.773	0.787	0.114	2.327	3.476	7.352	0.6080	0.264	.2493E+1	.4012E+0	.5715E-1	.4470E+1	.2809E-1
1.80	4.70	.778	0.789	0.111	3.122	3.483	8.945	0.6138	0.347	.2094E+1	.4776E+0	.5646E-1	.5491E+1	.3749E-1
1.80	5.60	.782	0.791	0.109	4.018	3.488	10.739	0.6186	0.438	.1795E+1	.5571E+0	.5527E-1	.6643E+1	.4804E-1
1.80	6.80	.785	0.794	0.107	5.215	3.493	13.135	0.6221	0.561	.1526E+1	.6555E+0	.5355E-1	.8171E+1	.6212E-1
1.80	8.20	.787	0.794	0.107	6.613	3.496	15.932	0.6245	0.704	.1312E+1	.7623E+0	.5165E-1	.9949E+1	.7857E-1
1.80	10.00	.788	0.794	0.106	8.412	3.497	19.530	0.6257	0.892	.1123E+1	.8903E+0	.4946E-1	.1222E+2	.9982E-1
1.80	12.00	.790	0.796	0.105	10.410	3.501	23.528	0.6281	1.093	.9777E+0	.1023E+1	.4738E-1	.1478E+2	.1232E+0
1.80	15.00	.791	0.796	0.105	13.409	3.501	29.526	0.6292	1.401	.8265E+0	.1210E+1	.4481E-1	.1858E+2	.1585E+0
1.80	18.00	.791	0.796	0.105	16.409	3.501	35.526	0.6292	1.715	.7225E+0	.1384E+1	.4272E-1	.2235E+2	.1939E+0
1.80	22.00	.792	0.796	0.104	20.408	3.503	43.525	0.6304	2.122	.6241E+0	.1602E+1	.4046E-1	.2744E+2	.2408E+0
1.80	27.00	.793	0.796	0.104	25.407	3.504	53.523	0.6316	2.630	.5383E+0	.1858E+1	.3823E-1	.3381E+2	.2994E+0
1.80	33.00	.793	0.796	0.104	31.407	3.504	65.523	0.6316	3.251	.4662E+0	.2145E+1	.3611E-1	.4139E+2	.3701E+0
2.20	1.50	.220	0.710	0.390	0.080	3.085	3.171	0.1562	0.031	.3030E+2	.3300E-1	.1000E-1	.4952E+0	.2467E-3
2.20	1.80	.427	0.814	0.287	0.173	3.381	3.475	0.3474	0.050	.1080E+2	.9259E-1	.2338E-1	.1207E+1	.1769E-2
2.20	2.20	.625	0.913	0.188	0.375	3.664	3.992	0.5703	0.070	.5471E+1	.1828E+0	.3777E-1	.2277E+1	.6242E-2
2.20	2.70	.714	0.957	0.143	0.786	3.791	4.865	0.6833	0.112	.3584E+1	.2790E+0	.4697E-1	.3324E+1	.1384E-1
2.20	3.30	.744	0.972	0.128	1.356	3.834	6.022	0.7232	0.174	.2683E+1	.3727E+0	.5133E-1	.4355E+1	.2367E-1
2.20	3.90	.756	0.978	0.122	1.944	3.851	7.204	0.7394	0.237	.2203E+1	.4540E+0	.5289E-1	.5327E+1	.3366E-1
2.20	4.70	.763	0.984	0.118	2.737	3.861	8.794	0.7449	0.324	.1813E+1	.5513E+0	.5334E-1	.6586E+1	.4711E-1
2.20	5.60	.768	0.988	0.116	3.632	3.868	10.587	0.7557	0.421	.1535E+1	.6516E+0	.5288E-1	.8001E+1	.6220E-1
2.20	6.80	.772	0.986	0.114	4.828	3.874	12.982	0.7612	0.550	.1291E+1	.7744E+0	.5177E-1	.9882E+1	.8232E-1
2.20	8.20	.776	0.987	0.113	6.226	3.877	15.779	0.7639	0.704	.1103E+1	.9070E+0	.5028E-1	.1205E+2	.1059E+0
2.20	10.00	.778	0.988	0.112	8.024	3.880	19.376	0.7667	0.899	.9386E+0	.1065E+1	.4843E-1	.1486E+2	.1362E+0
2.20	12.00	.778	0.989	0.111	10.022	3.883	23.373	0.7694	1.112	.8133E+0	.1230E+1	.4658E-1	.1798E+2	.1696E+0
2.20	15.00	.779	0.990	0.111	13.021	3.884	29.372	0.7708	1.439	.6854E+0	.1459E+1	.4421E-1	.2264E+2	.2201E+0
2.20	18.00	.780	0.990	0.110	16.020	3.886	35.370	0.7722	1.762	.5977E+0	.1673E+1	.4225E-1	.2731E+2	.2704E+0

CASE # 1 COPYRIGHT 1982 STEVE SMITH DATA DATE 1/ 3/82 TRANSFORMER DESIGN TABLE

$$f(D) = [(U/(FG(DE)^2))^5 * (1/(DEUm))^4]^{(1/13)}$$

P	Q	D	E	F	G	U	Um	DE	FG	f(D)	1/f(D)	1/(PQf(D))	DEUm	FG(DE)^2/U
2.20	22.00	.780	.990	0.110	20.020	3.886	43.370	0.7722	2.202	.5152E+0	.1941E+1	.4010E-1	.3349E+2	.3380E+0
2.20	27.00	.781	.991	0.109	25.019	3.887	53.369	0.7736	2.740	.4436E+0	.2254E+1	.3795E-1	.4129E+2	.4218E+0
2.20	33.00	.782	.991	0.109	31.018	3.888	65.367	0.7750	3.381	.3837E+0	.2606E+1	.3590E-1	.5666E+2	.5222E+0
2.70	1.80	.074	.887	0.463	0.026	3.377	3.765	0.0656	0.012	.1092E+3	.9158E-2	.1884E-2	.2471E+0	.1536E-4
2.70	2.20	.357	1.029	0.322	0.143	3.781	4.160	0.3672	0.046	.1034E+2	.9668E-1	.1628E-1	.1527E+1	.1639E-2
2.70	2.70	.614	1.157	0.193	0.386	4.148	4.793	0.7104	0.074	.4187E+1	.2388E+0	.3276E-1	.3405E+1	.9063E-2
2.70	3.30	.710	1.205	0.145	0.890	4.286	5.856	0.8556	0.129	.2642E+1	.3785E+0	.4249E-1	.5010E+1	.2204E-1
2.70	3.90	.735	1.218	0.132	1.465	4.321	7.020	0.8949	0.194	.2041E+1	.4900E+0	.4653E-1	.6282E+1	.3597E-1
2.70	4.70	.749	1.225	0.126	2.251	4.341	8.600	0.9172	0.283	.1619E+1	.6176E+0	.4867E-1	.7887E+1	.5474E-1
2.70	5.60	.755	1.228	0.123	3.145	4.350	10.391	0.9268	0.385	.1342E+1	.7453E+0	.4930E-1	.9630E+1	.7607E-1
2.70	6.80	.760	1.230	0.120	4.340	4.357	12.784	0.9348	0.521	.1111E+1	.8998E+0	.4901E-1	.1195E+2	.1045E+0
2.70	8.20	.763	1.232	0.118	5.737	4.361	15.580	0.9396	0.680	.9390E+0	.1065E+1	.4810E-1	.1464E+2	.1370E+0
2.70	10.00	.766	1.233	0.117	7.534	4.366	19.176	0.9445	0.881	.7930E+0	.1261E+1	.4671E-1	.1811E+2	.1801E+0
2.70	12.00	.767	1.234	0.117	9.533	4.367	23.174	0.9461	1.111	.6333E+0	.1463E+1	.4517E-1	.2192E+2	.2276E+0
2.70	15.00	.769	1.235	0.116	12.531	4.370	29.171	0.9493	1.447	.5730E+0	.1745E+1	.4309E-1	.2769E+2	.2985E+0
2.70	18.00	.770	1.235	0.115	15.530	4.371	35.170	0.9510	1.786	.4981E+0	.2008E+1	.4131E-1	.3344E+2	.3695E+0
2.70	22.00	.771	1.236	0.114	19.529	4.373	43.168	0.9526	2.236	.4282E+0	.2335E+1	.3992E-1	.4112E+2	.4648E+0
2.70	27.00	.771	1.236	0.114	24.529	4.373	53.168	0.9526	2.809	.3679E+0	.2718E+1	.3729E-1	.5065E+2	.5828E+0
2.70	33.00	.772	1.236	0.114	30.528	4.374	65.167	0.9542	3.480	.3177E+0	.3148E+1	.3533E-1	.6218E+2	.7244E+0
3.30	2.70	.287	1.293	0.357	0.113	4.281	5.003	0.3712	0.040	.1066E+2	.9382E-1	.1053E-1	.1857E+1	.1297E-2
3.30	3.30	.606	1.453	0.197	0.394	4.737	5.747	0.8805	0.078	.3255E+1	.3072E+0	.2821E-1	.5060E+1	.1270E-1
3.30	3.90	.702	1.501	0.149	0.898	4.874	6.810	1.0537	0.134	.2088E+1	.4789E+0	.3721E-1	.7175E+1	.3048E-1
3.30	4.70	.731	1.516	0.135	1.669	4.916	8.368	1.1078	0.224	.1527E+1	.6551E+0	.4223E-1	.9270E+1	.5605E-1
3.30	5.60	.742	1.522	0.129	2.557	4.933	10.151	1.1305	0.329	.1217E+1	.8214E+0	.4445E-1	.1148E+2	.8513E-1
3.30	6.80	.750	1.525	0.125	3.750	4.943	12.541	1.1438	0.469	.9833E+0	.1017E+1	.4532E-1	.1434E+2	.1241E+0
3.30	8.20	.754	1.527	0.123	5.146	4.948	15.335	1.1514	0.633	.8180E+0	.1223E+1	.4518E-1	.1766E+2	.1696E+0
3.30	10.00	.757	1.529	0.122	6.943	4.953	18.931	1.1571	0.844	.6830E+0	.1464E+1	.4436E-1	.2190E+2	.2280E+0
3.30	12.00	.759	1.530	0.121	8.941	4.956	22.928	1.1609	1.077	.5842E+0	.1712E+1	.4323E-1	.2662E+2	.2930E+0
3.30	15.00	.761	1.530	0.119	11.939	4.958	28.925	1.1647	1.427	.4866E+0	.2055E+1	.4152E-1	.3369E+2	.3902E+0
3.30	18.00	.762	1.531	0.119	14.938	4.960	34.924	1.1666	1.778	.4212E+0	.2374E+1	.3997E-1	.4074E+2	.4878E+0
3.30	22.00	.763	1.532	0.118	18.937	4.961	42.922	1.1685	2.244	.3608E+0	.2771E+1	.3817E-1	.5016E+2	.6178E+0
3.30	27.00	.764	1.532	0.118	23.936	4.963	52.921	1.1704	2.824	.3092E+0	.3235E+1	.3636E-1	.6194E+2	.7797E+0
3.30	33.00	.764	1.532	0.118	29.936	4.963	64.921	1.1704	3.532	.2664E+0	.3754E+1	.3447E-1	.7599E+2	.9751E+0
3.90	3.30	.286	1.593	0.357	0.114	4.880	5.947	0.4556	0.041	.8492E+1	.1178E+0	.9150E-2	.2709E+1	.1731E-2
3.90	3.90	.601	1.751	0.200	0.399	5.330	6.696	1.0521	0.080	.2657E+1	.3764E+0	.2474E-1	.7045E+1	.1653E-1
3.90	4.70	.707	1.804	0.147	1.093	5.481	8.145	1.2751	0.160	.1571E+1	.6364E+0	.3472E-1	.1039E+2	.4750E-1

TRANSFORMER DESIGN TABLE

CASE # 1   COPYRIGHT 1982 STEVE SMITH   DATA DATE 1/ 3/82

$$f(D) = \left[\left(\frac{U}{FG(DE)^2}\right)^5 \cdot \left(\frac{1}{DE\,Um}\right)^4\right]^{1/13}$$

P	Q	D	E	F	G	U	Um	DE	FG	f(D)	1/f(D)	1/(PQf(D))	DEUm	FG(DE)²/U
3.90	5.60	.731	1.816	0.135	1.969	5.516	9.911	1.3271	0.265	.1170E+1	.8545E+0	.3912E-1	.1315E+2	.8457E-1
3.90	6.80	.742	1.821	0.129	3.158	5.531	12.295	1.3512	0.407	.9112E+0	.1097E+1	.4138E-1	.1661E+2	.1345E+0
3.90	8.20	.747	1.824	0.127	4.553	5.538	15.088	1.3622	0.576	.7428E+0	.1346E+1	.4210E-1	.2055E+2	.1930E+0
3.90	10.00	.751	1.826	0.125	6.349	5.544	18.682	1.3710	0.790	.6118E+0	.1635E+1	.4191E-1	.2561E+2	.2680E+0
3.90	12.00	.753	1.827	0.123	8.347	5.547	22.679	1.3754	1.031	.5187E+0	.1928E+1	.4119E-1	.3119E+2	.3515E+0
3.90	15.00	.755	1.828	0.122	11.345	5.550	28.676	1.3798	1.390	.4288E+0	.2332E+1	.3986E-1	.3957E+2	.4767E+0
3.90	18.00	.756	1.828	0.122	14.344	5.551	34.675	1.3820	1.750	.3696E+0	.2706E+1	.3855E-1	.4792E+2	.6021E+0
3.90	22.00	.757	1.829	0.122	18.343	5.553	42.673	1.3842	2.229	.3154E+0	.3171E+1	.3695E-1	.5907E+2	.7690E+0
3.90	27.00	.758	1.829	0.121	23.342	5.554	52.672	1.3864	2.824	.2694E+0	.3712E+1	.3525E-1	.7302E+2	.9774E+0
3.90	33.00	.759	1.830	0.120	29.341	5.556	64.671	1.3886	3.536	.2316E+0	.4317E+1	.3354E-1	.8980E+2	.1227E+1
4.70	3.90	.146	1.923	0.427	0.054	6.124	7.303	0.2808	0.023	.1769E+2	.5652E-1	.3003E-2	.2051E+1	.2962E-3
4.70	4.70	.597	2.149	0.202	0.403	6.281	7.959	1.2827	0.081	.2131E+1	.4694E+0	.2125E-1	.1021E+2	.2182E-1
4.70	5.60	.707	2.204	0.147	1.193	6.316	9.602	1.5579	0.175	.1227E+1	.8153E+0	.3098E-1	.1496E+2	.6753E-1
4.70	6.80	.731	2.216	0.130	2.454	6.328	11.967	1.6195	0.319	.8744E+0	.1144E+1	.3708E-1	.1938E+2	.1323E+0
4.70	8.20	.740	2.220	0.128	3.760	6.336	14.754	1.6428	0.489	.6854E+0	.1459E+1	.3786E-1	.2424E+2	.2083E+0
4.70	10.00	.745	2.222	0.128	5.555	6.340	18.347	1.6558	0.708	.5513E+0	.1814E+1	.3860E-1	.3038E+2	.3065E+0
4.70	12.00	.748	2.224	0.126	7.552	6.343	22.343	1.6636	0.952	.4609E+0	.2170E+1	.3847E-1	.3717E+2	.4154E+0
4.70	15.00	.750	2.225	0.125	10.550	6.344	28.343	1.6688	1.319	.3767E+0	.2655E+1	.3766E-1	.4729E+2	.5790E+0
4.70	18.00	.751	2.226	0.125	13.549	6.346	34.139	1.6714	1.687	.3225E+0	.3101E+1	.3666E-1	.5739E+2	.7427E+0
4.70	22.00	.752	2.226	0.124	17.548	6.347	42.337	1.6740	2.176	.2737E+0	.3653E+1	.3533E-1	.7087E+2	.9699E+0
4.70	27.00	.753	2.227	0.123	22.547	6.348	52.336	1.6766	2.785	.2329E+0	.4295E+1	.3384E-1	.8774E+2	.1233E+1
4.70	33.00	.754	2.227	0.123	28.546	6.351	64.334	1.6792	3.511	.1996E+0	.5011E+1	.3231E-1	.1080E+3	.1559E+1
5.60	4.70	.073	2.337	0.464	0.028	6.275	7.321	0.1706	0.013	.3825E+2	.2615E-1	.9934E-3	.1249E+1	.6029E-4
5.60	5.60	.594	2.597	0.203	0.406	7.020	9.377	1.5426	0.082	.1740E+1	.5747E+0	.1832E-1	.1446E+2	.2794E-1
5.60	6.80	.713	2.657	0.144	1.487	7.190	11.607	1.8941	0.213	.9145E+0	.1094E+1	.2872E-1	.2198E+2	.1065E+0
5.60	8.00	.732	2.666	0.134	2.868	7.217	14.380	1.9515	0.384	.6621E+0	.1510E+1	.3289E-1	.2806E+2	.2028E+0
5.60	10.00	.739	2.670	0.131	4.661	7.227	17.970	1.9728	0.608	.5124E+0	.1952E+1	.3485E-1	.3545E+2	.3276E+0
5.60	12.00	.743	2.672	0.129	6.657	7.233	21.964	1.9849	0.855	.4199E+0	.2382E+1	.3544E-1	.4360E+2	.4660E+0
5.60	15.00	.746	2.673	0.127	9.654	7.237	27.959	1.9941	1.226	.3378E+0	.2960E+1	.3524E-1	.5575E+2	.6736E+0
5.60	18.00	.747	2.674	0.127	12.653	7.238	33.958	1.9971	1.601	.2867E+0	.3487E+1	.3460E-1	.6782E+2	.8819E+0
5.60	22.00	.749	2.675	0.126	16.651	7.241	41.955	2.0032	2.090	.2417E+0	.4137E+1	.3358E-1	.8404E+2	.1158E+1
5.60	27.00	.750	2.675	0.125	21.650	7.243	51.954	2.0063	2.706	.2046E+0	.4888E+1	.3233E-1	.1042E+3	.1584E+1
5.60	33.00	.750	2.675	0.125	27.650	7.243	63.954	2.0063	3.456	.1747E+0	.5724E+1	.3098E-1	.1283E+3	.1921E+1
6.80	6.80	.592	3.196	0.204	0.408	8.217	11.265	1.8920	0.083	.1397E+1	.7157E+0	.1548E-1	.2131E+2	.3626E-1
6.80	8.20	.715	3.258	0.143	1.685	8.393	13.889	2.3291	0.240	.7025E+0	.1424E+1	.2553E-1	.3235E+2	.1552E+0
6.80	10.00	.732	3.266	0.134	3.468	8.417	17.464	2.3907	0.465	.4943E+0	.2023E+1	.2975E-1	.4175E+2	.3156E+0

CASE # 1 COPYRIGHT 1982 STEVE SMITH DATA DATE 1/ 3/82 TRANSFORMER DESIGN TABLE

$$f(D)=[(U/(FG(DE)^2))^5 * (1/(DEUm))^4]^{(1/13)}$$

P	Q	D	E	F	G	U	Um	DE	FG	f(D)	1/f(D)	1/(PQf(D))	DEUm	FG(DE)^2/U
6.80	12.00	.738	3.269	0.131	5.462	8.426	21.456	2.4125	0.716	.3893E+0	.2569E+1	.3148E-1	.5176E+2	.4943E+0
6.80	15.00	.742	3.271	0.129	8.458	8.431	27.450	2.4271	1.091	.3049E+0	.3279E+1	.3215E-1	.6662E+2	.7623E+0
6.80	18.00	.744	3.272	0.128	11.456	8.434	33.447	2.4344	1.466	.2553E+0	.3917E+1	.3200E-1	.8142E+2	.1030E+1
6.80	22.00	.745	3.273	0.128	15.455	8.436	41.446	2.4380	1.971	.2130E+0	.4695E+1	.3138E-1	.1010E+3	.1388E+1
6.80	27.00	.746	3.273	0.127	20.454	8.437	51.444	2.4417	2.598	.1789E+0	.5589E+1	.3044E-1	.1256E+3	.1836E+1
6.80	33.00	.747	3.274	0.127	26.453	8.438	63.443	2.4453	3.346	.1520E+0	.6581E+1	.2933E-1	.1551E+3	.2371E+1
8.20	8.00	.590	3.895	0.205	0.410	9.614	13.467	2.2981	0.084	.1135E+1	.8009E+0	.1310E-1	.3095E+2	.4617E-1
8.20	10.00	.719	3.959	0.141	2.081	9.798	16.882	2.8469	0.292	.5244E+0	.1907E+1	.2326E-1	.4806E+2	.2418E+0
8.20	12.00	.732	3.966	0.134	4.068	9.831	20.864	2.9031	0.545	.3788E+0	.2640E+1	.2683E-1	.6057E+2	.4680E+0
8.20	15.00	.738	3.969	0.131	7.062	9.826	26.855	2.9291	0.925	.2834E+0	.3529E+1	.2869E-1	.7866E+2	.8078E+0
8.20	18.00	.741	3.971	0.130	10.059	9.830	32.851	2.9421	1.303	.2324E+0	.4303E+1	.2915E-1	.9665E+2	.1147E+1
8.20	22.00	.742	3.971	0.129	14.058	9.831	40.849	2.9465	1.813	.1911E+0	.5234E+1	.2901E-1	.1204E+3	.1601E+1
8.20	27.00	.743	3.972	0.129	19.057	9.833	50.848	2.9508	2.449	.1589E+0	.6294E+1	.2843E-1	.1500E+3	.2169E+1
8.20	33.00	.744	3.972	0.128	25.056	9.834	62.846	2.9552	3.207	.1340E+0	.7463E+1	.2758E-1	.1857E+3	.2848E+1
10.00	10.00	.588	4.794	0.206	0.412	11.411	16.297	2.8189	0.085	.9142E+0	.1094E+1	.1094E-1	.4594E+2	.5910E-1
10.00	12.00	.719	4.860	0.141	2.281	11.598	20.110	3.4940	0.320	.4105E+0	.2436E+1	.2030E-1	.7026E+2	.3773E+0
10.00	15.00	.733	4.867	0.134	5.267	11.618	26.084	3.5671	0.703	.2741E+0	.3649E+1	.2432E-1	.9307E+2	.7701E+0
10.00	18.00	.737	4.869	0.132	8.263	11.624	32.084	3.5581	1.087	.2162E+0	.4625E+1	.2570E-1	.1151E+3	.1203E+1
10.00	22.00	.740	4.870	0.130	12.260	11.628	40.080	3.6038	1.594	.1734E+0	.5766E+1	.2621E-1	.1444E+3	.1780E+1
10.00	27.00	.741	4.871	0.130	17.259	11.630	50.078	3.6090	2.235	.1420E+0	.7043E+1	.2608E-1	.1807E+3	.2503E+1
10.00	33.00	.742	4.871	0.129	23.258	11.631	62.077	3.6143	3.000	.1185E+0	.8439E+1	.2557E-1	.2244E+3	.3370E+1

CASE # 2   COPYRIGHT 1982 STEVE SMITH   DATA DATE 1/ 3/82   TRANSFORMER DESIGN TABLE

$$f(D) = [(U/(FG(DE)^2))^5 * (1/(DEUm))^4]^{1/13}$$

SCALE BY 1.000E-04   FUNCTION 1/f(D)

Q/P	0.33	0.39	0.47	0.56	0.68	0.82	1.00	1.20	1.50	1.80	2.20	2.70	3.30	3.90	4.70	5.60	6.80	8.20	10.00
0.33	101	121	145	167	186	192	176	132											
0.39	117	142	172	201	231	249	243	202	94										
0.47	137	166	204	243	287	322	336	309	197	48									
0.56	158	193	238	287	344	398	438	437	343	169									
0.68	183	225	280	340	415	489	561	601	561	399	86								
0.82	212	261	326	398	490	587	691	772	807	711	385								
1.00	245	303	381	467	580	702	841	967	1084	1094	886	302							
1.20	281	347	438	539	672	820	994	1162	1351	1454	1418	985	82						
1.50	330	410	517	640	801	984	1205	1427	1706	1913	2060	1977	1280	104					
1.80	377	468	592	733	921	1136	1399	1670	2027	2318	2595	2737	2512	1578					
2.20	435	541	685	850	1071	1325	1640	1970	2419	2806	3222	3573	3712	3448					
2.70	504	626	794	987	1246	1546	1921	2319	2870	3363	3926	4477	4906	5055	1978				
3.30	581	722	917	1141	1443	1794	2235	2708	3372	3979	4696	5448	6140	6600	4689	2839	3540		
3.90	653	813	1033	1286	1628	2027	2531	3073	3842	4553	5410	6338	7252	7951	6819	6384	7852	10185	
4.70	745	927	1179	1469	1862	2320	2902	3532	4430	5271	6298	7440	8612	9582	8549	8722	11464	14716	2300
5.60	842	1049	1334	1664	2110	2632	3296	4017	5053	6028	7234	8594	10029	11262	11267	11445	14716	18458	12515
6.80	965	1202	1530	1908	2421	3023	3791	4627	5834	6977	8402	10031	11782	13331	13762	14664	18373	22205	19432
8.20	1099	1370	1744	2177	2764	3453	4335	5296	6689	8015	9678	11597	13687	15569	16750	18373	22237	24205	25584
10.00	1263	1574	2004	2502	3178	3974	4992	6105	7722	9267	11216	13481	15973	17787	19925	22237	26754	29644	32330
12.00	1434	1788	2277	2843	3612	4518	5680	6951	8802	10576	12821	15444	18350	20974	24272	26754	31375	35146	39010
15.00	1674	2088	2660	3322	4223	5285	6647	8141	10320	12414	15073	18196	21023	24906	28534	32926	37765	42700	48067
18.00	1901	2371	3020	3773	4797	6004	7555	9257	11744	14138	17184	21678	24790	28689	33158	37930	43690	49671	56361
22.00	2185	2725	3473	4339	5518	6908	8696	10659	13531	16301	19832	24003	28787	33073	38516	44173	51067	58323	66606
27.00	2519	3142	4004	5003	6364	7970	10035	12304	15628	18837	22936	27787	33252	38383	44779	51461	59662	68381	78476
33.00	2895	3612	4603	5752	7318	9165	11543	14158	17990	21694	26430	32045	38385	44352	51813	59639	69295	79634	91723

CASE # 2   COPYRIGHT 1982 STEVE SMITH   DATA DATE 1/ 3/82   TRANSFORMER DESIGN TABLE

SCALE BY 1.000E-04   FUNCTION   1/[P*Q*f(D)]

$$f(D)=[(U/[FG(DE)^2))^5 * (1/(DEUm))^4]^{(1/13)}$$

Q/P	0.33	0.39	0.47	0.56	0.68	0.82	1.00	1.20	1.50	1.80	2.20	2.70	3.30	3.90	4.70	5.60	6.80	8.20	10.00
0.33	931	944	938	906	830	711	533	333	161										
0.39	911	931	937	922	872	779	623	431	280	56									
0.47	882	908	925	924	898	835	716	548	408	168	57								
0.56	853	882	905	914	904	866	782	651	556	326	213	112							
0.68	817	849	877	894	897	878	826	737	656	482	403	304	21						
0.82	782	815	846	867	879	873	843	784	722	608	537	488	259	18					
1.00	744	777	810	835	852	856	841	806	751	673	624	563	423	225	191				
1.20	709	742	776	802	824	833	828	807	758	709	655	601	511	402	369	188			
1.50	668	700	734	761	785	800	803	793	751	715	661	614	551	480	440	345	158		
1.80	635	666	700	727	753	769	777	773	733	709	661	616	564	513	466	399	296	97	
2.20	600	630	663	690	716	735	746	746	709	692	661	611	566	523	479	428	358	264	49
2.70	565	595	626	653	679	698	712	716	681	670	647	602	563	526	482	439	385	320	223
3.30	533	561	591	618	643	663	677	684	657	649	631	586	555	523	479	442	397	349	286
3.90	507	535	564	589	614	634	649	657	628	623	609	568	543	516	473	440	399	360	312
4.70	480	506	534	558	582	602	618	626	598	598	598	546	525	503	462	434	393	362	323
5.60	456	480	507	531	554	573	589	598	572	570	587	524	506	487	449	428	397	362	325
6.80	430	453	479	501	524	542	558	567	543	544	562	499	484	468	446	426	393	360	325
8.20	406	429	453	474	496	514	529	538	515	515	536	477	463	449	438	423	385	357	323
10.00	383	404	426	447	467	485	499	509	489	490	510	449	449	438	430	410	384	349	320
12.00	362	382	404	423	443	459	473	483	459	460	486	438	438	426	417	399	372	337	318
15.00	338	357	377	396	414	430	443	452	435	436	457	427	427	406	410	376	370	323	315
18.00	320	338	357	374	392	407	420	429	410	412	434	404	417	392	392	359	357	309	313
22.00	301	318	336	352	369	383	395	404	386	388	410	381	395	385	372	340	341	294	303
27.00	283	298	316	331	347	360	372	380	363	365	386	360	373	365	353	323	325	309	291
33.00	266	281	297	311	326	339	350	358	363	365	364	360	352	345	334	323	309	294	278

CASE # 2   COPYRIGHT 1982 STEVE SMITH   DATA DATE 1/ 3/82   TRANSFORMER DESIGN TABLE

$$f(D) = \left[\left(\frac{U}{FG(DE)^2}\right)^5 * \left(\frac{1}{DEUm}\right)^4\right]^{1/13}$$

P	Q	D	E	F	G	U	Um	DE	FG	f(D)	1/f(D)	1/(PQf(D))	DEUm	FG(DE)²/U
0.33	0.33	.906	0.118	0.047	0.212	2.196	0.703	0.1069	0.010	.9861E+2	.1014E-1	.9312E-1	.7519E-1	.5187E-4
0.33	0.39	.910	0.120	0.045	0.270	2.201	0.818	0.1092	0.012	.8532E+2	.1172E-1	.9107E-1	.8938E-1	.6582E-4
0.33	0.47	.913	0.122	0.044	0.349	2.206	0.975	0.1109	0.015	.7306E+2	.1369E-1	.8825E-1	.1081E+0	.8458E-4
0.33	0.56	.915	0.123	0.042	0.438	2.209	1.152	0.1121	0.019	.6347E+2	.1576E-1	.8526E-1	.1292E+0	.1058E-3
0.33	0.68	.917	0.123	0.042	0.557	2.211	1.390	0.1132	0.023	.5454E+2	.1834E-1	.8171E-1	.1574E+0	.1339E-3
0.33	0.82	.919	0.125	0.040	0.696	2.214	1.668	0.1144	0.028	.4728E+2	.2115E-1	.7817E-1	.1908E+0	.1665E-3
0.33	1.00	.920	0.125	0.040	0.875	2.216	2.026	0.1150	0.035	.4074E+2	.2454E-1	.7438E-1	.2330E+0	.2089E-3
0.33	1.20	.921	0.126	0.040	1.075	2.217	2.425	0.1156	0.044	.3561E+2	.2808E-1	.7092E-1	.2803E+0	.2558E-3
0.33	1.50	.922	0.126	0.039	1.374	2.219	3.024	0.1162	0.054	.3026E+2	.3305E-1	.6676E-1	.3513E+0	.3360E-3
0.33	1.80	.923	0.126	0.039	1.674	2.220	3.623	0.1168	0.064	.2653E+2	.3770E-1	.6346E-1	.4230E+0	.3957E-3
0.33	2.20	.923	0.126	0.039	2.074	2.220	4.423	0.1168	0.080	.2297E+2	.4353E-1	.5996E-1	.5164E+0	.4902E-3
0.33	2.70	.923	0.126	0.039	2.574	2.220	5.423	0.1168	0.099	.1986E+2	.5036E-1	.5652E-1	.6332E+0	.6085E-3
0.33	3.30	.924	0.127	0.038	3.173	2.221	6.621	0.1173	0.121	.1723E+2	.5805E-1	.5331E-1	.7770E+0	.7474E-3
0.33	3.90	.924	0.127	0.038	3.773	2.221	7.821	0.1173	0.143	.1531E+2	.6531E-1	.5075E-1	.9178E+0	.8888E-3
0.33	4.70	.924	0.127	0.038	4.573	2.221	9.421	0.1173	0.174	.1343E+2	.7447E-1	.4802E-1	.1108E+1	.1077E-2
0.33	5.60	.924	0.127	0.038	5.473	2.221	11.221	0.1173	0.208	.1187E+2	.8421E-1	.4557E-1	.1317E+1	.1289E-2
0.33	6.80	.925	0.128	0.037	6.673	2.223	13.620	0.1179	0.250	.1037E+2	.9647E-1	.4299E-1	.1606E+1	.1566E-2
0.33	8.20	.925	0.128	0.037	8.073	2.223	16.420	0.1179	0.303	.9095E+1	.1099E+0	.4063E-1	.1937E+1	.1894E-2
0.33	10.00	.925	0.128	0.037	9.873	2.223	20.020	0.1179	0.370	.7920E+1	.1263E+0	.3826E-1	.2361E+1	.2317E-2
0.33	12.00	.925	0.128	0.037	11.873	2.223	24.020	0.1179	0.445	.6975E+1	.1434E+0	.3621E-1	.2833E+1	.2786E-2
0.33	15.00	.925	0.128	0.037	14.873	2.223	30.020	0.1179	0.558	.5972E+1	.1674E+0	.3383E-1	.3541E+1	.3490E-2
0.33	18.00	.925	0.128	0.037	17.872	2.223	36.020	0.1179	0.670	.5261E+1	.1901E+0	.3200E-1	.4248E+1	.4194E-2
0.33	22.00	.925	0.128	0.037	21.873	2.223	44.020	0.1179	0.820	.4577E+1	.2185E+0	.3010E-1	.5192E+1	.5133E-2
0.33	27.00	.925	0.128	0.037	26.873	2.223	54.020	0.1179	1.008	.3970E+1	.2519E+0	.2827E-1	.6371E+1	.6306E-2
0.33	33.00	.925	0.128	0.037	32.872	2.223	66.020	0.1179	1.233	.3454E+1	.2895E+0	.2659E-1	.7786E+1	.7714E-2
0.39	0.33	.885	0.137	0.058	0.193	2.226	0.716	0.1217	0.011	.8235E+2	.1214E-1	.9435E-1	.8713E-1	.7364E-4
0.39	0.39	.892	0.141	0.054	0.249	2.236	0.827	0.1258	0.013	.7065E+2	.1415E-1	.9305E-1	.1041E+0	.9514E-4
0.39	0.47	.897	0.144	0.051	0.326	2.243	0.981	0.1287	0.017	.6000E+2	.1665E-1	.9081E-1	.1263E+0	.1242E-3
0.39	0.56	.901	0.146	0.049	0.415	2.251	1.157	0.1311	0.021	.5193E+2	.1926E-1	.8818E-1	.1516E+0	.1568E-3
0.39	0.68	.903	0.147	0.049	0.534	2.251	1.394	0.1323	0.026	.4443E+2	.2251E-1	.8488E-1	.1844E+0	.2011E-3
0.39	0.82	.906	0.148	0.047	0.672	2.256	1.670	0.1341	0.032	.3839E+2	.2605E-1	.8146E-1	.2240E+0	.2518E-3
0.39	1.00	.907	0.149	0.046	0.852	2.257	2.029	0.1347	0.040	.3299E+2	.3031E-1	.7772E-1	.2733E+0	.3182E-3
0.39	1.20	.909	0.149	0.046	1.051	2.260	2.427	0.1359	0.048	.2878E+2	.3474E-1	.7424E-1	.3298E+0	.3906E-3
0.39	1.50	.910	0.150	0.045	1.350	2.261	3.026	0.1365	0.061	.2441E+2	.4096E-1	.7002E-1	.4130E+0	.5095E-3
0.39	1.80	.911	0.151	0.044	1.650	2.263	3.624	0.1371	0.073	.2138E+2	.4678E-1	.6664E-1	.4969E+0	.5698E-3
0.39	2.20	.911	0.151	0.044	2.050	2.263	4.424	0.1371	0.090	.1849E+2	.5408E-1	.6302E-1	.6066E+0	.7577E-3

CASE # 2   COPYRIGHT 1982 STEVE SMITH   DATA DATE 1/ 3/82   TRANSFORMER DESIGN TABLE

$$f(D) = [(U/(FG(DE)^2))^5 * (1/(DEUm))^4]^{(1/13)}$$

P	Q	D	E	F	G	U	Um	DE	FG	f(D)	1/f(D)	1/(PQf(D))	DEUm	FG(DE)^2/U
0.39	2.70	.912	0.151	0.044	2.549	2.264	5.423	0.1377	0.112	.1597E+2	.6262E-1	.5947E-1	.7468E+0	.9394E-3
0.39	3.30	.912	0.151	0.044	3.149	2.264	6.623	0.1377	0.139	.1384E+2	.7223E-1	.5612E-1	.9121E+0	.1161E-2
0.39	3.90	.913	0.151	0.044	3.749	2.266	7.822	0.1383	0.163	.1230E+2	.8130E-1	.5345E-1	.1082E+1	.1377E-2
0.39	4.70	.913	0.151	0.044	4.549	2.266	9.422	0.1383	0.198	.1078E+2	.9274E-1	.5060E-1	.1303E+1	.1671E-2
0.39	5.60	.913	0.151	0.044	5.449	2.266	11.222	0.1383	0.237	.9532E+1	.1049E+0	.4803E-1	.1552E+1	.2001E-2
0.39	6.80	.913	0.151	0.044	6.649	2.266	13.622	0.1383	0.289	.8318E+1	.1202E+0	.4533E-1	.1884E+1	.2442E-2
0.39	8.20	.913	0.151	0.044	8.049	2.266	16.422	0.1383	0.350	.7297E+1	.1370E+0	.4285E-1	.2271E+1	.2956E-2
0.39	10.00	.913	0.152	0.044	9.849	2.267	20.022	0.1383	0.428	.6353E+1	.1574E+0	.4036E-1	.2769E+1	.3618E-2
0.39	12.00	.914	0.152	0.043	11.848	2.267	24.021	0.1389	0.509	.5594E+1	.1788E+0	.3820E-1	.3337E+1	.4337E-2
0.39	15.00	.914	0.152	0.043	14.848	2.267	30.021	0.1389	0.638	.4789E+1	.2088E+0	.3570E-1	.4171E+1	.5436E-2
0.39	18.00	.914	0.152	0.043	17.848	2.267	36.021	0.1389	0.767	.4218E+1	.2371E+0	.3377E-1	.5004E+1	.6534E-2
0.39	22.00	.914	0.152	0.043	21.848	2.267	44.021	0.1389	0.939	.3669E+1	.2725E+0	.3177E-1	.6116E+1	.7998E-2
0.39	27.00	.914	0.152	0.043	26.848	2.267	54.021	0.1389	1.154	.3183E+1	.3142E+0	.2984E-1	.7505E+1	.9829E-2
0.39	33.00	.914	0.152	0.043	32.848	2.267	66.021	0.1389	1.412	.2769E+1	.3612E+0	.2806E-1	.9172E+1	.1203E-1
0.47	0.33	.856	0.163	0.072	0.167	2.264	0.734	0.1395	0.012	.6877E+2	.1454E-1	.9375E-1	.1024E+0	.1034E-3
0.47	0.39	.867	0.168	0.067	0.222	2.280	0.841	0.1461	0.015	.5821E+2	.1718E-1	.9371E-1	.1228E+0	.1379E-3
0.47	0.47	.873	0.173	0.063	0.298	2.291	0.991	0.1509	0.023	.4895E+2	.2043E-1	.9249E-1	.1496E+0	.1849E-3
0.47	0.56	.881	0.175	0.060	0.385	2.300	1.164	0.1546	0.023	.4197E+2	.2382E-1	.9052E-1	.1799E+0	.2378E-3
0.47	0.68	.885	0.177	0.058	0.503	2.306	1.399	0.1571	0.029	.3568E+2	.2803E-1	.8770E-1	.2197E+0	.3092E-3
0.47	0.82	.889	0.179	0.056	0.641	2.311	1.674	0.1596	0.036	.3068E+2	.3260E-1	.8458E-1	.2671E+0	.3916E-3
0.47	1.00	.891	0.180	0.055	0.820	2.314	2.032	0.1608	0.045	.2627E+2	.3807E-1	.8101E-1	.3267E+0	.4992E-3
0.47	1.20	.893	0.182	0.054	1.019	2.317	2.429	0.1621	0.054	.2285E+2	.4377E-1	.7760E-1	.3937E+0	.6178E-3
0.47	1.50	.895	0.182	0.053	1.318	2.320	3.027	0.1633	0.069	.1933E+2	.5174E-1	.7338E-1	.4944E+0	.7954E-3
0.47	1.80	.896	0.183	0.051	1.617	2.321	3.625	0.1640	0.084	.1690E+2	.5916E-1	.6996E-1	.5945E+0	.9736E-3
0.47	2.20	.897	0.183	0.051	2.017	2.323	4.424	0.1646	0.104	.1460E+2	.6852E-1	.6626E-1	.7282E+0	.1221E-2
0.47	2.70	.897	0.183	0.051	2.517	2.323	5.424	0.1646	0.130	.1259E+2	.7944E-1	.6260E-1	.8928E+0	.1512E-2
0.47	3.30	.898	0.184	0.051	3.116	2.324	6.623	0.1652	0.159	.1090E+2	.9172E-1	.5914E-1	.1094E+1	.1867E-2
0.47	3.90	.898	0.184	0.051	3.716	2.324	7.823	0.1652	0.190	.9680E+1	.1033E+0	.5636E-1	.1293E+1	.2226E-2
0.47	4.70	.899	0.184	0.051	4.515	2.326	9.422	0.1659	0.228	.8451E+1	.1179E+0	.5338E-1	.1563E+1	.2698E-2
0.47	5.60	.899	0.184	0.051	5.416	2.326	11.222	0.1659	0.273	.7494E+1	.1334E+0	.5070E-1	.1861E+1	.3235E-2
0.47	6.80	.899	0.184	0.051	6.616	2.326	13.622	0.1659	0.334	.6537E+1	.1530E+0	.4786E-1	.2259E+1	.3952E-2
0.47	8.20	.900	0.185	0.051	8.015	2.327	16.421	0.1665	0.401	.5732E+1	.1744E+0	.4526E-1	.2734E+1	.4774E-2
0.47	10.00	.900	0.185	0.050	9.815	2.327	20.021	0.1665	0.491	.4989E+1	.2004E+0	.4265E-1	.3333E+1	.5846E-2
0.47	12.00	.900	0.185	0.050	11.815	2.327	24.021	0.1665	0.591	.4392E+1	.2277E+0	.4037E-1	.3999E+1	.7038E-2
0.47	15.00	.900	0.185	0.050	14.815	2.327	30.021	0.1665	0.741	.3759E+1	.2660E+0	.3773E-1	.4998E+1	.8824E-2
0.47	18.00	.900	0.185	0.050	17.815	2.327	36.021	0.1665	0.891	.3311E+1	.3020E+0	.3570E-1	.5997E+1	.1061E-1

CASE # 2   COPYRIGHT 1982 STEVE SMITH   DATA DATE 1/ 3/82   TRANSFORMER DESIGN TABLE

$f(D) = [(U/(FG(DE)^2))^5 * (1/(DEUm)^4)]^{1/13}$

P	Q	D	E	F	G	U	Um	DE	FG	f(D)	1/f(D)	1/(PQf(D))	DEUm	FG(DE)^2/U
0.47	22.00	.900	0.185	0.050	21.815	2.327	44.021	0.1665	1.091	.2879E+1	.3473E+0	.3359E-1	.7329E+1	.1299E-1
0.47	27.00	.900	0.185	0.050	26.815	2.327	54.021	0.1665	1.341	.2497E+1	.4004E+0	.3155E-1	.8994E+1	.1597E-1
0.47	33.00	.900	0.185	0.050	32.815	2.327	66.021	0.1665	1.641	.2172E+1	.4603E+0	.2968E-1	.1099E+2	.1955E-1
0.56	0.33	.816	0.188	0.092	0.142	2.297	0.763	0.1534	0.013	.5975E+2	.1674E-1	.9056E-1	.1171E+0	.1338E-3
0.56	0.39	.835	0.198	0.083	0.193	2.324	0.860	0.1649	0.016	.4965E+2	.2014E-1	.9222E-1	.1419E+0	.1858E-3
0.56	0.47	.850	0.205	0.075	0.265	2.346	1.002	0.1743	0.020	.4110E+2	.2433E-1	.9244E-1	.1746E+0	.2573E-3
0.56	0.56	.859	0.210	0.070	0.351	2.358	1.171	0.1800	0.025	.3487E+2	.2868E-1	.9145E-1	.2107E+0	.3393E-3
0.56	0.66	.866	0.213	0.067	0.467	2.368	1.403	0.1845	0.031	.2938E+2	.3403E-1	.8938E-1	.2587E+0	.4495E-3
0.56	0.82	.870	0.215	0.065	0.605	2.374	1.678	0.1871	0.039	.2511E+2	.3983E-1	.8673E-1	.3138E+0	.5795E-3
0.56	1.00	.874	0.217	0.063	0.783	2.380	2.033	0.1897	0.049	.2139E+2	.4675E-1	.8348E-1	.3855E+0	.7456E-3
0.56	1.20	.877	0.219	0.062	0.982	2.384	2.429	0.1916	0.060	.1854E+2	.5393E-1	.8025E-1	.4655E+0	.9297E-3
0.56	1.50	.879	0.220	0.060	1.281	2.387	3.027	0.1929	0.077	.1564E+2	.6395E-1	.7613E-1	.5840E+0	.1208E-2
0.56	1.80	.880	0.220	0.060	1.580	2.388	3.626	0.1936	0.095	.1364E+2	.7332E-1	.7273E-1	.7019E+0	.1488E-2
0.56	2.20	.882	0.221	0.059	1.979	2.391	4.423	0.1949	0.117	.1176E+2	.8503E-1	.6902E-1	.8622E+0	.1855E-2
0.56	2.70	.883	0.222	0.059	2.479	2.393	5.422	0.1956	0.145	.1013E+2	.9872E-1	.6509E-1	.1060E+1	.2318E-2
0.56	3.30	.883	0.222	0.058	3.079	2.393	6.622	0.1956	0.180	.8763E+1	.1141E+0	.6175E-1	.1295E+1	.2879E-2
0.56	3.90	.884	0.222	0.058	3.678	2.394	7.821	0.1962	0.213	.7774E+1	.1286E+0	.5890E-1	.1535E+1	.3432E-2
0.56	4.70	.884	0.222	0.058	4.478	2.394	9.421	0.1962	0.260	.6806E+1	.1469E+0	.5582E-1	.1849E+1	.4178E-2
0.56	5.60	.885	0.223	0.058	5.378	2.396	11.220	0.1969	0.309	.6011E+1	.1664E+0	.5305E-1	.2209E+1	.5005E-2
0.56	6.80	.885	0.223	0.058	6.578	2.396	13.620	0.1969	0.378	.5241E+1	.1908E+0	.5011E-1	.2682E+1	.6121E-2
0.56	8.20	.885	0.223	0.058	7.978	2.396	16.420	0.1969	0.459	.4594E+1	.2177E+0	.4741E-1	.3233E+1	.7424E-2
0.56	10.00	.886	0.223	0.057	9.777	2.397	20.018	0.1976	0.557	.3997E+1	.2502E+0	.4468E-1	.3955E+1	.9076E-2
0.56	12.00	.886	0.223	0.057	11.777	2.397	24.018	0.1976	0.671	.3518E+1	.2843E+0	.4230E-1	.4745E+1	.1093E-1
0.56	15.00	.886	0.223	0.057	14.777	2.397	30.018	0.1976	0.842	.3010E+1	.3322E+0	.3955E-1	.5931E+1	.1372E-1
0.56	18.00	.886	0.223	0.057	17.777	2.397	36.018	0.1976	1.013	.2651E+1	.3773E+0	.3743E-1	.7116E+1	.1650E-1
0.56	22.00	.886	0.223	0.057	21.777	2.397	44.018	0.1976	1.241	.2305E+1	.4339E+0	.3522E-1	.8697E+1	.2021E-1
0.56	27.00	.886	0.223	0.057	26.777	2.397	54.018	0.1976	1.526	.1999E+1	.5003E+0	.3309E-1	.1067E+2	.2486E-1
0.56	33.00	.886	0.223	0.057	32.777	2.397	66.018	0.1976	1.868	.1738E+1	.5752E+0	.3113E-1	.1304E+2	.3043E-1
0.68	0.33	.752	0.216	0.124	0.114	2.326	0.815	0.1624	0.014	.5367E+2	.1863E-1	.8303E-1	.1324E+0	.1644E-3
0.68	0.39	.786	0.233	0.107	0.157	2.374	0.894	0.1831	0.017	.4324E+2	.2313E-1	.8720E-1	.1637E+0	.2373E-3
0.68	0.47	.813	0.247	0.093	0.223	2.413	1.021	0.2004	0.021	.3485E+2	.2869E-1	.8978E-1	.2047E+0	.3479E-3
0.68	0.56	.829	0.255	0.086	0.306	2.436	1.182	0.2110	0.026	.2904E+2	.3444E-1	.9443E-1	.2493E+0	.4774E-3
0.68	0.68	.840	0.260	0.080	0.420	2.451	1.408	0.2184	0.034	.2412E+2	.4146E-1	.8966E-1	.3076E+0	.6538E-3
0.68	0.82	.847	0.264	0.077	0.557	2.470	1.680	0.2273	0.043	.2041E+2	.4900E-1	.8787E-1	.3749E+0	.8616E-3
0.68	1.00	.853	0.267	0.074	0.734	2.470	2.033	0.2273	0.054	.1725E+2	.5796E-1	.8523E-1	.4621E+0	.1128E-2
0.68	1.20	.857	0.269	0.072	0.932	2.476	2.428	0.2301	0.067	.1488E+2	.6721E-1	.8237E-1	.5586E+0	.1444E-2

CASE # 2   COPYRIGHT 1982 STEVE SMITH   DATA DATE 1/ 3/82

TRANSFORMER DESIGN TABLE

$$f(D) = \left[\left(\frac{U}{FG(DE)^2}\right)^5 * \left(\frac{1}{DEUm}\right)^4 \right]^{1/13}$$

P	Q	D	E	F	G	U	Um	DE	FG	f(D)	1/f(D)	1/(PQf(D))	DEUm	FG(DE)^2/U
0.68	1.50	.862	0.270	0.070	1.230	2.480	3.024	0.2322	0.086	.1248E+2	.8010E-1	.7853E-1	.7022E+0	.1872E-2
0.68	1.80	.862	0.271	0.070	1.529	2.483	3.622	0.2322	0.107	.1086E+2	.9211E-1	.7525E-1	.8466E+0	.2319E-2
0.68	2.20	.864	0.272	0.068	1.928	2.486	4.419	0.2350	0.131	.9337E+1	.1071E+0	.7159E-1	.1039E+1	.2913E-2
0.68	2.70	.865	0.273	0.067	2.428	2.487	5.418	0.2357	0.164	.8024E+1	.1246E+0	.6788E-1	.1277E+1	.3361E-2
0.68	3.30	.866	0.273	0.067	3.027	2.488	6.617	0.2364	0.203	.6931E+1	.1443E+0	.6430E-1	.1564E+1	.4555E-2
0.68	3.90	.867	0.274	0.067	3.627	2.490	7.816	0.2371	0.241	.6142E+1	.1628E+0	.6139E-1	.1853E+1	.5446E-2
0.68	4.70	.868	0.274	0.066	4.426	2.491	9.414	0.2378	0.292	.5372E+1	.1862E+0	.5825E-1	.2239E+1	.6632E-2
0.68	5.60	.869	0.274	0.066	5.326	2.491	11.214	0.2378	0.352	.4740E+1	.2110E+0	.5540E-1	.2667E+1	.7981E-2
0.68	6.80	.869	0.275	0.065	6.526	2.493	13.613	0.2385	0.427	.4130E+1	.2421E+0	.5236E-1	.3247E+1	.9757E-2
0.68	8.20	.869	0.275	0.065	7.926	2.493	16.413	0.2385	0.519	.3618E+1	.2764E+0	.4957E-1	.3915E+1	.1185E-1
0.68	10.00	.869	0.275	0.065	9.726	2.494	20.013	0.2393	0.637	.3146E+1	.3178E+0	.4674E-1	.4774E+1	.1454E-1
0.68	12.00	.870	0.275	0.065	11.725	2.494	24.012	0.2393	0.762	.2768E+1	.3612E+0	.4427E-1	.5745E+1	.1749E-1
0.68	15.00	.870	0.275	0.065	14.725	2.494	30.012	0.2393	0.957	.2368E+1	.4223E+0	.4140E-1	.7180E+1	.2197E-1
0.68	18.00	.870	0.275	0.065	17.725	2.494	36.012	0.2393	1.152	.2085E+1	.4797E+0	.3919E-1	.8616E+1	.2644E-1
0.68	22.00	.870	0.275	0.065	21.725	2.494	44.012	0.2393	1.412	.1812E+1	.5518E+0	.3688E-1	.1053E+2	.3241E-1
0.68	27.00	.870	0.275	0.065	26.725	2.494	54.012	0.2393	1.737	.1571E+1	.6364E+0	.3466E-1	.1292E+2	.3987E-1
0.68	33.00	.870	0.275	0.065	32.725	2.494	66.012	0.2393	2.127	.1367E+1	.7318E+0	.3261E-1	.1579E+2	.4882E-1
0.82	0.33	.658	0.239	0.171	0.091	2.331	0.899	0.1573	0.016	.5201E+2	.1923E-1	.7105E-1	.1414E+0	.1651E-3
0.82	0.39	.714	0.267	0.143	0.123	2.411	0.951	0.1906	0.018	.4015E+2	.2490E-1	.7787E-1	.1814E+0	.2651E-3
0.82	0.47	.762	0.291	0.119	0.179	2.480	1.053	0.2217	0.021	.3066E+2	.3220E-1	.8354E-1	.2335E+0	.4223E-3
0.82	0.56	.791	0.306	0.105	0.255	2.521	1.198	0.2417	0.027	.2514E+2	.3977E-1	.8661E-1	.2895E+0	.6160E-3
0.82	0.68	.810	0.315	0.095	0.365	2.548	1.415	0.2552	0.035	.2043E+2	.4895E-1	.8778E-1	.3610E+0	.8858E-3
0.82	0.82	.831	0.321	0.089	0.499	2.566	1.680	0.2639	0.044	.1704E+2	.5870E-1	.8730E-1	.4433E+0	.1205E-2
0.82	1.00	.831	0.326	0.085	0.675	2.578	2.029	0.2705	0.057	.1425E+2	.7019E-1	.8560E-1	.5489E+0	.1617E-2
0.82	1.20	.836	0.328	0.082	0.872	2.586	2.423	0.2744	0.072	.1220E+2	.8200E-1	.8333E-1	.6645E+0	.2079E-2
0.82	1.50	.840	0.330	0.080	1.170	2.591	3.018	0.2772	0.094	.1017E+2	.9837E-1	.7997E-1	.8367E+0	.2775E-2
0.82	1.80	.843	0.332	0.079	1.469	2.596	3.615	0.2795	0.115	.8806E+1	.1136E+0	.7694E-1	.1010E+1	.3468E-2
0.82	2.20	.846	0.333	0.077	1.867	2.600	4.411	0.2817	0.144	.7547E+1	.1325E+0	.7345E-1	.1243E+1	.4388E-2
0.82	2.70	.848	0.334	0.076	2.366	2.603	5.409	0.2832	0.180	.6469E+1	.1546E+0	.6982E-1	.1532E+1	.5328E-2
0.82	3.30	.848	0.334	0.076	2.966	2.603	6.609	0.2832	0.225	.5576E+1	.1794E+0	.6628E-1	.1872E+1	.6948E-2
0.82	3.90	.851	0.335	0.075	3.565	2.606	8.006	0.2848	0.267	.4934E+1	.2027E+0	.6337E-1	.2223E+1	.8310E-2
0.82	4.70	.851	0.336	0.074	4.365	2.607	9.405	0.2855	0.325	.4310E+1	.2328E+0	.6020E-1	.2685E+1	.1012E-1
0.82	5.60	.851	0.336	0.074	5.265	2.607	11.205	0.2863	0.392	.3800E+1	.2632E+0	.5731E-1	.3199E+1	.1226E-1
0.82	6.80	.852	0.336	0.074	6.464	2.608	13.604	0.2863	0.478	.3308E+1	.3023E+0	.5422E-1	.3894E+1	.1503E-1
0.82	8.20	.852	0.336	0.074	7.864	2.608	16.404	0.2863	0.582	.2896E+1	.3453E+0	.5136E-1	.4696E+1	.1828E-1
0.82	10.00	.853	0.337	0.074	9.664	2.610	20.003	0.2870	0.710	.2517E+1	.3974E+0	.4846E-1	.5741E+1	.2242E-1

TRANSFORMER DESIGN TABLE

CASE # 2   COPYRIGHT 1982 STEVE SMITH   DATA DATE 1/ 3/82

$$f(D)=[(U/(FG(DE)^2))^5 * (1/(DEUm))^4]^{(1/13)}$$

P	Q	D	E	F	G	U	Um	DE	FG	f(D)	1/f(D)	1/(PQf(D))	DEUm	FG(DE)^2/U
0.82	12.00	.853	0.337	0.074	11.664	2.610	24.003	0.2870	0.857	.2213E+1	.4518E+0	.4592E-1	.6889E+1	.2706E-1
0.82	15.00	.853	0.337	0.074	14.664	2.610	30.003	0.2870	1.078	.1892E+1	.5285E+0	.4296E-1	.8612E+1	.3402E-1
0.82	18.00	.854	0.337	0.073	17.663	2.611	36.001	0.2878	1.289	.1665E+1	.6004E+0	.4068E-1	.1036E+2	.4098E-1
0.82	22.00	.854	0.337	0.073	21.663	2.611	44.001	0.2878	1.581	.1447E+1	.6908E+0	.3830E-1	.1266E+2	.5016E-1
0.82	27.00	.854	0.337	0.073	26.663	2.611	54.001	0.2878	1.946	.1255E+1	.7979E+0	.3600E-1	.1554E+2	.6174E-1
0.82	33.00	.854	0.337	0.073	32.663	2.611	66.001	0.2878	2.384	.1091E+1	.9165E+0	.3387E-1	.1900E+2	.7563E-1
1.00	0.33	.520	0.260	0.240	0.070	2.314	1.028	0.1352	0.017	.5686E+2	.1759E-1	.5329E-1	.1390E+0	.1327E-3
1.00	0.39	.596	0.298	0.202	0.092	2.423	1.056	0.1776	0.019	.4116E+2	.2429E-1	.6229E-1	.1876E+0	.2420E-3
1.00	0.47	.675	0.338	0.163	0.132	2.536	1.111	0.2278	0.022	.2973E+2	.3363E-1	.7156E-1	.2552E+0	.4407E-3
1.00	0.56	.732	0.366	0.134	0.194	2.617	1.231	0.2679	0.026	.2283E+2	.4379E-1	.7820E-1	.3298E+0	.7130E-3
1.00	0.68	.770	0.385	0.115	0.295	2.671	1.425	0.2965	0.034	.1781E+2	.5615E-1	.8257E-1	.4224E+0	.1116E-2
1.00	0.82	.791	0.396	0.105	0.425	2.701	1.679	0.3128	0.044	.1447E+2	.6909E-1	.8425E-1	.5253E+0	.1607E-2
1.00	1.00	.805	0.403	0.098	0.598	2.721	2.022	0.3240	0.058	.1189E+2	.8413E-1	.8413E-1	.6552E+0	.2247E-2
1.00	1.20	.819	0.407	0.093	0.794	2.731	2.413	0.3305	0.074	.1006E+2	.9941E-1	.8284E-1	.7973E+0	.2965E-2
1.00	1.50	.819	0.410	0.091	1.091	2.741	3.005	0.3354	0.099	.8302E+1	.1205E+0	.8030E-1	.1008E+1	.4049E-2
1.00	1.80	.823	0.412	0.088	1.389	2.747	3.600	0.3387	0.123	.7148E+1	.1399E+0	.7772E-1	.1219E+1	.5131E-2
1.00	2.20	.826	0.413	0.087	1.787	2.751	4.397	0.3411	0.155	.6096E+1	.1648E+0	.7456E-1	.1500E+1	.6576E-2
1.00	2.70	.828	0.414	0.086	2.286	2.754	5.394	0.3428	0.197	.5205E+1	.1921E+0	.7115E-1	.1849E+1	.8388E-2
1.00	3.30	.831	0.416	0.085	2.885	2.758	6.591	0.3453	0.244	.4474E+1	.2235E+0	.6774E-1	.2276E+1	.1053E-1
1.00	3.90	.832	0.416	0.084	3.484	2.760	7.789	0.3461	0.293	.3951E+1	.2531E+0	.6489E-1	.2696E+1	.1270E-1
1.00	4.70	.833	0.417	0.083	4.284	2.761	9.388	0.3469	0.358	.3445E+1	.2902E+0	.6175E-1	.3257E+1	.1555E-1
1.00	5.60	.834	0.417	0.083	5.183	2.763	11.187	0.3478	0.430	.3034E+1	.3296E+0	.5887E-1	.3891E+1	.1883E-1
1.00	6.80	.835	0.417	0.083	6.383	2.764	13.586	0.3486	0.527	.2637E+1	.3791E+0	.5576E-1	.4736E+1	.2315E-1
1.00	8.20	.835	0.417	0.083	7.783	2.764	16.386	0.3486	0.642	.2307E+1	.4335E+0	.5286E-1	.5712E+1	.2823E-1
1.00	10.00	.836	0.418	0.082	9.582	2.766	19.985	0.3494	0.786	.2003E+1	.4992E+0	.4992E-1	.6984E+1	.3469E-1
1.00	12.00	.836	0.418	0.082	11.582	2.766	23.985	0.3494	0.950	.1761E+1	.5680E+0	.4733E-1	.8381E+1	.4193E-1
1.00	15.00	.836	0.418	0.082	14.582	2.766	29.985	0.3494	1.196	.1504E+1	.6647E+0	.4431E-1	.1048E+2	.5208E-1
1.00	18.00	.837	0.419	0.082	17.582	2.767	35.983	0.3503	1.433	.1324E+1	.7555E+0	.4197E-1	.1260E+2	.6354E-1
1.00	22.00	.837	0.419	0.082	21.582	2.767	43.983	0.3503	1.759	.1150E+1	.8696E+0	.3953E-1	.1541E+2	.7799E-1
1.00	27.00	.837	0.419	0.082	26.582	2.767	53.983	0.3503	2.166	.9965E+0	.1003E+1	.3717E-1	.1891E+2	.9606E-1
1.00	33.00	.837	0.419	0.082	32.582	2.767	65.983	0.3503	2.655	.8663E+0	.1154E+1	.3498E-1	.2311E+2	.1177E+0
1.20	0.33	.358	0.273	0.321	0.051	2.282	1.182	0.0999	0.016	.7583E+2	.1319E-1	.3330E-1	.1181E+0	.7156E-4
1.20	0.39	.445	0.323	0.278	0.067	2.407	1.197	0.1435	0.019	.4955E+2	.2018E-1	.4312E-1	.1717E+0	.1603E-3
1.20	0.47	.550	0.375	0.225	0.095	2.557	1.229	0.2063	0.021	.3235E+2	.3091E-1	.5480E-1	.2535E+0	.3556E-3
1.20	0.56	.642	0.421	0.179	0.139	2.688	1.297	0.2703	0.025	.2287E+2	.4373E-1	.6507E-1	.3506E+0	.6761E-3
1.20	0.68	.716	0.458	0.142	0.222	2.794	1.447	0.3279	0.032	.1664E+2	.6010E-1	.7365E-1	.4747E+0	.1213E-2

CASE # 2   COPYRIGHT 1982 STEVE SMITH   DATA DATE 1/ 3/82   TRANSFORMER DESIGN TABLE

$$f(D)=[(U/(FG(DE)^2))^5 * (1/(DEUm))^4]^{(1/13)}$$

P	Q	D	E	F	G	U	Um	DE	FG	f(D)	1/f(D)	1/(PQf(D))	DEUm	FG(DE)^2/U
1.20	0.82	.756	0.478	0.122	0.342	2.851	1.679	0.3614	0.042	.1296E+2	.7719E-1	.7844E-1	.6067E+0	.1911E-2
1.20	1.00	.778	0.489	0.111	0.511	2.883	2.012	0.3804	0.057	.1035E+2	.9666E-1	.8055E-1	.7655E+0	.2884E-2
1.20	1.20	.790	0.495	0.105	0.705	2.900	2.398	0.3991	0.074	.8699E+1	.1162E+0	.8066E-1	.9376E+0	.3904E-2
1.20	1.50	.799	0.500	0.100	1.001	2.913	2.987	0.3911	0.101	.7009E+1	.1427E+0	.7926E-1	.1192E+1	.5499E-2
1.20	1.80	.805	0.503	0.098	1.298	2.921	3.579	0.4045	0.127	.5988E+1	.1670E+0	.7731E-1	.1448E+1	.7086E-2
1.20	2.20	.809	0.505	0.095	1.696	2.927	4.374	0.4081	0.162	.5075E+1	.1970E+0	.7464E-1	.1785E+1	.9215E-2
1.20	2.70	.812	0.506	0.094	2.194	2.931	5.371	0.4109	0.206	.4312E+1	.2319E+0	.7157E-1	.2207E+1	.1188E-1
1.20	3.30	.814	0.507	0.093	2.793	2.934	6.568	0.4127	0.260	.3693E+1	.2708E+0	.6838E-1	.2711E+1	.1508E-1
1.20	3.90	.815	0.508	0.093	3.393	2.936	7.767	0.4136	0.314	.3254E+1	.3073E+0	.6566E-1	.3213E+1	.1829E-1
1.20	4.70	.817	0.509	0.092	4.191	2.938	9.365	0.4154	0.384	.2832E+1	.3532E+0	.6262E-1	.3891E+1	.2251E-1
1.20	5.60	.818	0.509	0.091	5.090	2.940	11.164	0.4164	0.463	.2489E+1	.4017E+0	.5978E-1	.4648E+1	.2732E-1
1.20	6.80	.819	0.510	0.091	6.291	2.941	13.562	0.4173	0.569	.2161E+1	.4627E+0	.5671E-1	.5659E+1	.3370E-1
1.20	8.20	.820	0.510	0.090	7.690	2.943	16.361	0.4182	0.692	.1888E+1	.5296E+0	.5382E-1	.6842E+1	.4113E-1
1.20	10.00	.821	0.511	0.090	9.490	2.944	19.960	0.4191	0.849	.1638E+1	.6105E+0	.5087E-1	.8366E+1	.5067E-1
1.20	12.00	.821	0.511	0.090	11.490	2.944	23.960	0.4191	1.028	.1439E+1	.6951E+0	.4827E-1	.1004E+2	.6135E-1
1.20	15.00	.821	0.511	0.090	14.490	2.944	29.960	0.4191	1.297	.1228E+1	.8141E+0	.4523E-1	.1256E+2	.7737E-1
1.20	18.00	.822	0.511	0.089	17.489	2.946	35.959	0.4200	1.557	.1000E+1	.9257E+0	.4286E-1	.1510E+2	.9321E-1
1.20	22.00	.822	0.511	0.089	21.489	2.946	43.959	0.4200	1.913	.9382E+0	.1066E+1	.4038E-1	.1846E+2	.1146E+0
1.20	27.00	.822	0.511	0.089	26.489	2.946	53.959	0.4200	2.358	.8127E+0	.1230E+1	.3798E-1	.2266E+2	.1412E+0
1.20	33.00	.822	0.511	0.089	32.489	2.946	65.959	0.4200	2.892	.7063E+0	.1416E+1	.3575E-1	.2771E+2	.1732E+0
1.50	0.39	.212	0.356	0.394	0.034	2.374	1.415	0.0755	0.013	.1064E+3	.9399E-2	.1607E-1	.1068E+0	.3214E-4
1.50	0.47	.330	0.415	0.335	0.055	2.542	1.432	0.1370	0.018	.5069E+2	.1973E-1	.2798E-1	.1961E+0	.1359E-3
1.50	0.56	.455	0.478	0.204	0.134	2.721	1.460	0.2173	0.022	.2915E+2	.3431E-1	.4085E-1	.3172E+0	.3900E-3
1.50	0.68	.592	0.546	0.134	0.227	2.917	1.534	0.3232	0.027	.1783E+2	.5609E-1	.5499E-1	.4957E+0	.9791E-3
1.50	0.82	.687	0.594	0.157	0.382	3.053	1.698	0.4077	0.035	.1239E+2	.8071E-1	.6552E-1	.6924E+0	.1930E-2
1.50	1.00	.737	0.619	0.132	0.570	3.124	1.998	0.4558	0.050	.9228E+1	.1084E+0	.7225E-1	.9105E+0	.3337E-2
1.50	1.20	.760	0.630	0.120	0.863	3.157	2.370	0.4788	0.068	.7408E+1	.1351E+0	.7507E-1	.1135E+1	.4967E-2
1.50	1.50	.775	0.638	0.113	1.159	3.178	2.951	0.4941	0.097	.5861E+1	.1706E+0	.7583E-1	.1458E+1	.7452E-2
1.50	1.80	.783	0.642	0.106	1.556	3.190	3.542	0.5023	0.126	.4934E+1	.2027E+0	.7506E-1	.1779E+1	.9942E-2
1.50	2.20	.789	0.645	0.106	2.054	3.198	4.334	0.5085	0.164	.4134E+1	.2419E+0	.7330E-1	.2204E+1	.1327E-1
1.50	2.70	.793	0.647	0.104	2.652	3.204	5.330	0.5127	0.213	.3484E+1	.2870E+0	.7087E-1	.2732E+1	.1743E-1
1.50	3.30	.796	0.648	0.102	3.251	3.208	6.526	0.5158	0.271	.2965E+1	.3372E+0	.6812E-1	.3360E+1	.2243E-1
1.50	3.90	.798	0.649	0.101	3.251	3.211	7.723	0.5179	0.328	.2603E+1	.3842E+0	.6567E-1	.4000E+1	.2743E-1
1.50	4.70	.799	0.650	0.100	4.051	3.213	9.322	0.5190	0.407	.2257E+1	.4430E+0	.6284E-1	.4838E+1	.3412E-1
1.50	5.60	.801	0.651	0.100	4.950	3.216	11.120	0.5211	0.492	.1979E+1	.5053E+0	.6016E-1	.5794E+1	.4158E-1
1.50	6.80	.802	0.651	0.099	6.149	3.217	13.519	0.5221	0.609	.1714E+1	.5834E+0	.5719E-1	.7058E+1	.5158E-1

CASE # 2   COPYRIGHT 1982 STEVE SMITH   DATA DATE 1/ 3/82   TRANSFORMER DESIGN TABLE

$$f(D)=\left[\left(\frac{U}{FG(DE)^2}\right)^5 \cdot \left(\frac{1}{DEU_m}\right)^4\right]^{(1/13)}$$

P	Q	D	E	F	G	U	U_m	DE	FG	$f(D)$	$1/f(D)$	$1/(FG(DE)^2)$	$1/(PQf(D))$	$DE\,U_m$	$FG(DE)^2/U$
1.50	8.20	.803	.652	0.099	7.549	3.218	16.317	0.5232	0.744	.1495E+1	.6689E+0	.4915E+1	.5438E-1	.8537E+1	.6323E-1
1.50	10.00	.803	.652	0.099	9.349	3.218	19.917	0.5232	0.921	.1295E+1	.7720E+0	.3968E+1	.5148E-1	.1042E+2	.7831E-1
1.50	12.00	.804	.652	0.098	11.348	3.220	23.916	0.5242	1.112	.1136E+1	.8802E+0	.3272E+1	.4890E-1	.1254E+2	.9491E-1
1.50	15.00	.805	.653	0.098	14.348	3.220	29.915	0.5253	1.399	.9690E+0	.1032E+1	.2592E+1	.4587E-1	.1571E+2	.1198E+0
1.50	18.00	.805	.653	0.098	17.348	3.221	35.915	0.5253	1.691	.8515E+0	.1174E+1	.2143E+1	.4350E-1	.1886E+2	.1449E+0
1.50	22.00	.805	.653	0.098	21.348	3.221	43.915	0.5253	2.081	.7390E+0	.1353E+1	.1736E+1	.4100E-1	.2307E+2	.1788E+0
1.50	27.00	.805	.653	0.098	26.348	3.221	53.915	0.5253	2.569	.6399E+0	.1563E+1	.1411E+1	.3859E-1	.2832E+2	.2200E+0
1.50	33.00	.806	.653	0.097	32.347	3.223	65.914	0.5263	3.118	.5559E+0	.1799E+1	.1150E+1	.3634E-1	.3469E+2	.2697E+0
1.80	0.47	.104	.452	0.448	0.018	2.519	1.642	0.0470	0.008	.2105E+3	.4751E-2	.5613E+5	.5616E-2	.7719E-1	.7073E-5
1.80	0.56	.237	.519	0.382	0.042	2.710	1.660	0.1229	0.016	.5912E+2	.1691E-1	.4182E+4	.1678E-1	.2040E+0	.8824E-4
1.80	0.68	.406	.603	0.297	0.077	2.951	1.695	0.2448	0.023	.3986E+2	.2509E-1	.7295E+3	.2050E-1	.4150E+0	.4645E-3
1.80	0.82	.568	.684	0.216	0.136	3.183	1.778	0.3885	0.029	.1406E+2	.7114E-1	.2255E+3	.4820E-1	.6909E+0	.1393E-2
1.80	1.00	.682	.741	0.159	0.259	3.346	2.000	0.5054	0.041	.9143E+1	.1094E+0	.9506E+2	.6076E-1	.1011E+1	.3144E-2
1.80	1.20	.729	.765	0.136	0.436	3.413	2.343	0.5573	0.059	.6877E+1	.1454E+0	.5455E+2	.6732E-1	.1306E+1	.5371E-2
1.80	1.50	.754	.777	0.123	0.723	3.448	2.913	0.5859	0.089	.5226E+1	.1913E+0	.3277E+2	.7086E-1	.1706E+1	.8851E-2
1.80	1.80	.765	.783	0.118	1.018	3.464	3.499	0.5986	0.120	.4315E+1	.2318E+0	.2334E+2	.7153E-1	.2095E+1	.1237E-1
1.80	2.20	.773	.787	0.114	1.414	3.476	4.289	0.6080	0.160	.3564E+1	.2806E+0	.1686E+2	.7086E-1	.2608E+1	.1706E-1
1.80	2.70	.778	.789	0.111	1.911	3.483	5.283	0.6138	0.212	.2973E+1	.3363E+0	.1251E+2	.6920E-1	.3243E+1	.2295E-1
1.80	3.30	.782	.791	0.108	2.509	3.488	6.479	0.6186	0.273	.2513E+1	.3978E+0	.9557E+1	.6698E-1	.4007E+1	.3000E-1
1.80	3.90	.783	.792	0.108	3.108	3.491	7.676	0.6209	0.336	.2196E+1	.4553E+0	.7727E+1	.6486E-1	.4766E+1	.3707E-1
1.80	4.70	.786	.793	0.107	3.907	3.494	9.274	0.6233	0.418	.1897E+1	.5271E+0	.6158E+1	.6230E-1	.5780E+1	.4648E-1
1.80	5.60	.788	.794	0.106	4.806	3.497	11.071	0.6257	0.509	.1659E+1	.6028E+0	.5014E+1	.5981E-1	.6927E+1	.5703E-1
1.80	6.80	.789	.795	0.106	6.006	3.498	13.470	0.6269	0.634	.1433E+1	.6977E+0	.4017E+1	.5700E-1	.8444E+1	.7117E-1
1.80	8.20	.790	.795	0.105	7.405	3.500	16.269	0.6281	0.778	.1248E+1	.8015E+0	.3260E+1	.5430E-1	.1022E+2	.8763E-1
1.80	10.00	.791	.796	0.105	9.205	3.501	19.868	0.6292	0.962	.1079E+1	.9267E+0	.2625E+1	.5149E-1	.1250E+2	.1088E+0
1.80	12.00	.792	.796	0.104	11.204	3.503	23.866	0.6304	1.165	.9456E+0	.1058E+1	.2159E+1	.4896E-1	.1505E+2	.1322E+0
1.80	15.00	.793	.796	0.104	14.204	3.503	29.866	0.6304	1.477	.8055E+0	.1241E+1	.1703E+1	.4598E-1	.1883E+2	.1676E+0
1.80	18.00	.793	.796	0.104	17.204	3.504	35.865	0.6316	1.781	.7073E+0	.1414E+1	.1408E+1	.4364E-1	.2265E+2	.2027E+0
1.80	22.00	.793	.796	0.104	21.204	3.504	43.865	0.6316	2.195	.6135E+0	.1630E+1	.1142E+1	.4116E-1	.2771E+2	.2499E+0
1.80	27.00	.793	.796	0.104	26.204	3.504	53.865	0.6316	2.712	.5308E+0	.1884E+1	.9242E+0	.3876E-1	.3402E+2	.3088E+0
1.80	33.00	.794	.797	0.103	32.203	3.506	65.864	0.6328	3.317	.4610E+0	.2169E+1	.7528E+0	.3652E-1	.4168E+2	.3789E+0
2.20	0.68	.118	.659	0.441	0.021	2.939	1.959	0.0778	0.009	.1166E+3	.8573E-2	.1786E+5	.5731E-2	.1523E+0	.1905E-4
2.20	0.82	.318	.759	0.341	0.061	3.225	1.996	0.2414	0.021	.2600E+2	.3846E-1	.8254E+3	.2132E-1	.4818E+0	.3757E-3
2.20	1.00	.537	.869	0.232	0.131	3.538	2.090	0.4664	0.030	.1129E+2	.8860E-1	.1511E+3	.4027E-1	.9749E+0	.1871E-2
2.20	1.20	.668	.934	0.166	0.266	3.726	2.331	0.6238	0.044	.7053E+1	.1418E+0	.5817E+2	.5370E-1	.1454E+1	.4614E-2
2.20	1.50	.727	.964	0.137	0.537	3.810	2.859	0.7005	0.073	.4855E+1	.2060E+0	.2783E+2	.6242E-1	.2003E+1	.9431E-2

TRANSFORMER DESIGN TABLE

$$f(D) = [(U/(FG(DE)^2))^5 * (1/(DEUm))^4]^{(1/13)}$$

P	Q	D	E	F	G	U	Um	DE	FG	f(D)	1/f(D)	1/(PQf(D))	DEUm	FG(DE)^2/U
2.20	1.80	.746	.973	.127	0.827	3.837	3.436	0.7259	0.105	.3854E+1	.2595E+0	.6552E-1	.2494E+1	.1442E-1
2.20	2.20	.757	.979	.127	1.222	3.853	4.223	0.7407	0.148	.3103E+1	.3222E+0	.6658E-1	.3128E+1	.2114E-1
2.20	2.70	.764	.982	.118	1.718	3.863	5.215	0.7502	0.203	.2547E+1	.3926E+0	.6610E-1	.3912E+1	.2954E-1
2.20	3.30	.769	.985	.116	2.316	3.870	6.408	0.7571	0.267	.2129E+1	.4696E+0	.6469E-1	.4852E+1	.3961E-1
2.20	3.90	.772	.986	.114	2.914	3.874	7.605	0.7612	0.332	.1848E+1	.5410E+0	.6305E-1	.5789E+1	.4968E-1
2.20	4.70	.774	.987	.113	3.713	3.877	9.202	0.7639	0.420	.1588E+1	.6298E+0	.6091E-1	.7030E+1	.6316E-1
2.20	5.60	.776	.988	.112	4.612	3.880	11.000	0.7667	0.517	.1382E+1	.7234E+0	.5872E-1	.8434E+1	.7826E-1
2.20	6.80	.777	.989	.112	5.812	3.881	13.399	0.7681	0.648	.1190E+1	.8402E+0	.5616E-1	.1029E+2	.9849E-1
2.20	8.20	.778	.989	.111	7.211	3.883	16.198	0.7694	0.800	.1033E+1	.9678E+0	.5365E-1	.1246E+2	.1228E+0
2.20	10.00	.779	.990	.111	9.011	3.884	19.796	0.7708	0.996	.8916E+0	.1122E+1	.5098E-1	.1526E+2	.1523E+0
2.20	12.00	.780	.990	.110	11.010	3.886	23.795	0.7722	1.211	.7800E+0	.1282E+1	.4856E-1	.1837E+2	.1855E+0
2.20	15.00	.780	.991	.110	14.010	3.887	29.794	0.7736	1.534	.6634E+0	.1507E+1	.4568E-1	.2305E+2	.2362E+0
2.20	18.00	.781	.991	.109	17.010	3.887	35.794	0.7736	1.863	.5819E+0	.1718E+1	.4340E-1	.2769E+2	.2867E+0
2.20	22.00	.782	.991	.109	21.009	3.888	43.793	0.7750	2.290	.5042E+0	.1983E+1	.4098E-1	.3394E+2	.3537E+0
2.20	27.00	.782	.991	.109	26.009	3.888	53.793	0.7750	2.835	.4360E+0	.2294E+1	.3861E-1	.4169E+2	.4379E+0
2.20	33.00	.782	.991	.109	32.009	3.888	65.793	0.7750	3.489	.3784E+0	.2643E+1	.3640E-1	.5099E+2	.5389E+0
2.70	1.00	.218	.959	.391	0.041	3.582	2.370	0.2091	0.016	.3311E+2	.3018E-1	.1118E-1	.4956E+0	.1956E-3
2.70	1.20	.477	1.089	.262	0.111	3.953	2.456	0.5192	0.029	.1015E+2	.9851E-1	.3046E-1	.1275E+1	.1989E-2
2.70	1.50	.673	1.187	.164	0.311	4.233	2.818	0.7985	0.051	.5059E+1	.1977E+0	.4881E-1	.2250E+1	.7722E-2
2.70	1.80	.720	1.210	.140	0.590	4.300	3.361	0.8712	0.083	.3653E+1	.2737E+0	.5632E-1	.2928E+1	.1458E-1
2.70	2.20	.740	1.220	.130	0.980	4.328	4.136	0.9028	0.127	.2799E+1	.3573E+0	.6015E-1	.3734E+1	.2399E-1
2.70	2.70	.751	1.226	.125	1.475	4.344	5.123	0.9204	0.184	.2234E+1	.4477E+0	.6141E-1	.4715E+1	.3579E-1
2.70	3.30	.757	1.229	.122	2.072	4.353	6.316	0.9300	0.252	.1836E+1	.5448E+0	.6113E-1	.5873E+1	.5001E-1
2.70	3.90	.763	1.230	.120	2.670	4.357	7.512	0.9348	0.320	.1578E+1	.6338E+0	.6019E-1	.7022E+1	.6426E-1
2.70	4.70	.765	1.232	.118	3.468	4.361	9.108	0.9396	0.411	.1344E+1	.7440E+0	.5863E-1	.8559E+1	.8321E-1
2.70	5.60	.767	1.233	.118	4.590	4.364	10.906	0.9429	0.513	.1164E+1	.8594E+0	.5684E-1	.1028E+2	.1045E+0
2.70	6.80	.768	1.234	.117	5.567	4.367	13.304	0.9461	0.648	.9969E+0	.1003E+1	.5464E-1	.1259E+2	.1329E+0
2.70	8.20	.769	1.235	.116	6.966	4.368	16.102	0.9477	0.811	.8623E+0	.1160E+1	.5238E-1	.1526E+2	.1661E+0
2.70	10.00	.770	1.235	.116	8.766	4.370	19.701	0.9493	1.012	.7418E+0	.1348E+1	.4993E-1	.1877E+2	.2088E+0
2.70	12.00	.771	1.235	.115	10.765	4.371	23.700	0.9510	1.238	.6475E+0	.1544E+1	.4767E-1	.2254E+2	.2561E+0
2.70	15.00	.772	1.236	.114	13.765	4.373	29.699	0.9526	1.576	.5496E+0	.1820E+1	.4493E-1	.2829E+2	.3270E+0
2.70	18.00	.772	1.236	.114	16.764	4.374	35.698	0.9542	1.911	.4814E+0	.2077E+1	.4274E-1	.3466E+2	.3978E+0
2.70	22.00	.773	1.236	.114	20.764	4.374	43.698	0.9542	2.367	.4166E+0	.2400E+1	.4041E-1	.4178E+2	.4927E+0
2.70	27.00	.773	1.236	.114	25.764	4.374	53.698	0.9542	2.937	.3599E+0	.2779E+1	.3812E-1	.5124E+2	.6114E+0
2.70	33.00	.774	1.237	.114	31.764	4.376	65.696	0.9558	3.596	.3121E+0	.3204E+1	.3596E-1	.6279E+2	.7527E+0
3.10	1.00	.073	1.187	.464	0.014	3.975	2.818	0.0866	0.006	.1213E+3	.8246E-2	.2660E-2	.2441E+0	.1181E-4

CASE # 2   COPYRIGHT 1982 STEVE SMITH   DATA DATE 1/ 3/82

TRANSFORMER DESIGN TABLE

$$f(D)=[(U/(FG(DE)^2))^5 * (1/(DEUm))^4]^{(1/13)}$$

P	Q	D	E	F	G	U	Um	DE	FG	f(D)	1/f(D)	1/(PQf(D))	DEUm	FG(DE)^2/U
3.30	1.50	.472	1.386	0.264	0.114	4.545	2.933	0.6542	0.030	.7812E+1	.1280E+0	.2586E-1	.1919E+1	.2834E-2
3.30	1.80	.665	1.483	0.168	0.317	4.821	3.299	0.9859	0.053	.3981E+1	.2512E+0	.4229E-1	.3252E+1	.1072E-1
3.30	2.20	.719	1.510	0.141	0.691	4.898	4.033	1.0853	0.097	.2694E+1	.3712E+0	.5112E-1	.4377E+1	.2333E-1
3.30	2.70	.738	1.519	0.131	1.181	4.926	5.010	1.1210	0.155	.2038E+1	.4906E+0	.5586E-1	.5616E+1	.3947E-1
3.30	3.30	.746	1.523	0.127	1.777	4.937	6.200	1.1362	0.226	.1629E+1	.6140E+0	.5638E-1	.7045E+1	.5901E-1
3.30	3.90	.751	1.525	0.125	2.375	4.944	7.394	1.1457	0.296	.1379E+1	.7252E+0	.5634E-1	.8471E+1	.7848E-1
3.30	4.70	.754	1.527	0.123	3.173	4.948	8.991	1.1514	0.390	.1161E+1	.8612E+0	.5553E-1	.1035E+2	.1046E+0
3.30	5.60	.757	1.529	0.122	4.072	4.953	10.787	1.1571	0.495	.9972E+0	.1003E+1	.5427E-1	.1248E+2	.1333E+0
3.30	6.80	.759	1.530	0.121	5.271	4.956	13.185	1.1609	0.635	.8487E+0	.1178E+1	.5251E-1	.1531E+2	.1727E+0
3.30	8.20	.760	1.530	0.120	6.670	4.957	15.983	1.1628	0.800	.7306E+0	.1369E+1	.5058E-1	.1859E+2	.2183E+0
3.30	10.00	.762	1.531	0.119	8.469	4.960	19.581	1.1666	1.008	.6261E+0	.1597E+1	.4840E-1	.2284E+2	.2765E+0
3.30	12.00	.762	1.531	0.119	10.469	4.960	23.581	1.1666	1.246	.5450E+0	.1835E+1	.4634E-1	.2751E+2	.3419E+0
3.30	15.00	.763	1.532	0.118	13.469	4.961	29.580	1.1685	1.596	.4613E+0	.2168E+1	.4379E-1	.3456E+2	.4393E+0
3.30	18.00	.764	1.532	0.118	16.468	4.963	35.578	1.1704	1.943	.4034E+0	.2479E+1	.4173E-1	.4164E+2	.5364E+0
3.30	22.00	.764	1.532	0.118	20.468	4.963	43.578	1.1704	2.415	.3486E+0	.2869E+1	.3952E-1	.5101E+2	.6667E+0
3.30	27.00	.765	1.533	0.118	25.468	4.964	53.577	1.1724	2.992	.3007E+0	.3325E+1	.3732E-1	.6281E+2	.8285E+0
3.30	33.00	.765	1.533	0.118	31.468	4.964	65.577	1.1724	3.697	.2605E+0	.3839E+1	.3525E-1	.7688E+2	.1024E+1
3.90	1.50	.073	1.487	0.464	0.013	4.575	3.289	0.1085	0.006	.9575E+2	.1044E-1	.1785E-2	.3669E+0	.1610E-4
3.90	1.80	.470	1.685	0.265	0.115	5.143	3.407	0.7920	0.030	.6338E+1	.1578E+0	.2248E-1	.2698E+1	.3717E-2
3.90	2.20	.683	1.792	0.159	0.409	5.447	3.948	1.2236	0.065	.2901E+1	.3448E+0	.4018E-1	.4831E+1	.1780E-1
3.90	2.70	.724	1.812	0.138	0.888	5.506	4.898	1.3119	0.123	.1978E+1	.5055E+0	.4800E-1	.6426E+1	.3831E-1
3.90	3.30	.738	1.819	0.131	1.481	5.526	6.081	1.3424	0.194	.1515E+1	.6600E+0	.5128E-1	.8164E+1	.6327E-1
3.90	3.90	.744	1.822	0.128	2.078	5.534	7.274	1.3556	0.266	.1258E+1	.7951E+0	.5228E-1	.9860E+1	.8832E-1
3.90	4.70	.748	1.824	0.126	2.876	5.540	8.869	1.3644	0.362	.1044E+1	.9582E+0	.5227E-1	.1210E+2	.1218E+0
3.90	5.60	.751	1.826	0.125	3.775	5.544	10.665	1.3710	0.470	.8879E+0	.1126E+1	.5157E-1	.1462E+2	.1593E+0
3.90	6.80	.753	1.827	0.123	4.974	5.547	13.063	1.3754	0.614	.7501E+0	.1333E+1	.5027E-1	.1797E+2	.2095E+0
3.90	8.20	.755	1.828	0.123	6.373	5.550	15.861	1.3798	0.781	.6423E+0	.1557E+1	.4868E-1	.2188E+2	.2678E+0
3.90	10.00	.756	1.828	0.122	8.172	5.551	19.459	1.3820	0.997	.5481E+0	.1825E+1	.4678E-1	.2689E+2	.3430E+0
3.90	12.00	.757	1.829	0.122	10.172	5.553	23.458	1.3842	1.236	.4757E+0	.2102E+1	.4492E-1	.3247E+2	.4264E+0
3.90	15.00	.758	1.829	0.121	13.171	5.554	29.457	1.3864	1.594	.4015E+0	.2491E+1	.4258E-1	.4084E+2	.5515E+0
3.90	18.00	.758	1.829	0.121	16.171	5.554	35.457	1.3864	1.957	.3505E+0	.2853E+1	.4065E-1	.4916E+2	.6771E+0
3.90	22.00	.759	1.830	0.121	20.171	5.556	43.456	1.3886	2.431	.3024E+0	.3307E+1	.3855E-1	.6034E+2	.8436E+0
3.90	27.00	.759	1.830	0.121	25.171	5.556	53.456	1.3886	3.033	.2608E+0	.3838E+1	.3645E-1	.7423E+2	.1053E+1
3.90	33.00	.760	1.830	0.121	31.170	5.557	65.455	1.3908	3.740	.2255E+0	.4435E+1	.3446E-1	.9103E+2	.1302E+1
4.70	2.20	.467	2.084	0.267	0.117	5.938	4.039	0.9730	0.031	.5056E+1	.1978E+0	.1913E-1	.3930E+1	.4950E-2
4.70	2.70	.693	2.196	0.154	0.504	6.261	4.764	1.5222	0.077	.2133E+1	.4689E+0	.3695E-1	.7252E+1	.2860E-1

CASE # 2   DATA DATE 1/ 3/82

TRANSFORMER DESIGN TABLE

$$f(D) = [(U/(FG/(DE)^2))^5 * (1/(DEUm))^4]^{1/13}$$

P	Q	D	E	F	G	U	Um	DE	FG	f(D)	1/f(D)	1/(PQf(D))	DEUm	FG(DE)^2/U
4.70	3.30	.725	2.212	0.137	1.088	6.307	5.925	1.6041	0.150	.1466E+1	.6819E+0	.4397E-1	.9505E+1	.6100E-1
4.70	3.90	.735	2.212	0.132	1.683	6.321	7.113	1.6299	0.223	.1177E+1	.8549E+0	.4664E-1	.1159E+2	.9368E-1
4.70	4.70	.741	2.221	0.130	2.479	6.330	8.706	1.6454	0.321	.9460E+0	.1057E+1	.4785E-1	.1432E+2	.1373E+0
4.70	5.60	.745	2.222	0.128	3.378	6.336	10.501	1.6558	0.431	.7926E+0	.1262E+1	.4794E-1	.1739E+2	.1863E+0
4.70	6.80	.748	2.226	0.126	4.576	6.340	12.897	1.6636	0.577	.6618E+0	.1511E+1	.4728E-1	.2146E+2	.2517E+0
4.70	8.20	.750	2.225	0.125	5.975	6.343	15.695	1.6688	0.747	.5622E+0	.1779E+1	.4615E-1	.2619E+2	.3279E+0
4.70	10.00	.751	2.226	0.125	7.775	6.344	19.294	1.6714	0.968	.4768E+0	.2097E+1	.4463E-1	.3225E+2	.4262E+0
4.70	12.00	.752	2.227	0.123	9.774	6.346	23.293	1.6740	1.212	.4120E+0	.2427E+1	.4304E-1	.3899E+2	.5352E+0
4.70	15.00	.753	2.227	0.123	12.774	6.347	29.293	1.6766	1.578	.3464E+0	.2887E+1	.4095E-1	.4911E+2	.6986E+0
4.70	18.00	.754	2.227	0.123	15.773	6.348	35.290	1.6792	1.940	.3016E+0	.3316E+1	.3919E-1	.5926E+2	.8617E+0
4.70	22.00	.754	2.227	0.123	19.773	6.348	43.290	1.6792	2.432	.2596E+0	.3852E+1	.3735E-1	.7269E+2	.1080E+1
4.70	27.00	.755	2.228	0.123	24.773	6.350	53.289	1.6818	3.035	.2233E+0	.4478E+1	.3529E-1	.8962E+2	.1352E+1
4.70	33.00	.755	2.228	0.123	30.773	6.350	65.287	1.6818	3.770	.1930E+0	.5181E+1	.3341E-1	.1098E+3	.1679E+1
5.60	2.70	.515	2.557	0.243	0.143	6.907	4.787	1.3171	0.035	.3522E+1	.2839E+0	.1878E-1	.6305E+1	.8679E-2
5.60	3.30	.703	2.652	0.149	0.649	7.176	5.759	1.8640	0.096	.1566E+1	.6384E+0	.3455E-1	.1073E+2	.4663E-1
5.60	3.90	.725	2.663	0.137	1.238	7.207	6.932	1.9303	0.170	.1147E+1	.8727E+0	.3993E-1	.1338E+2	.8797E-1
5.60	4.70	.735	2.668	0.132	2.032	7.221	8.520	1.9606	0.269	.8876E+0	.1127E+1	.4281E-1	.1670E+2	.1434E+0
5.60	5.60	.740	2.670	0.130	2.930	7.228	10.314	1.9758	0.381	.7266E+0	.1376E+1	.4388E-1	.2038E+2	.2057E+0
5.60	6.80	.743	2.672	0.128	4.129	7.233	12.710	1.9849	0.531	.5970E+0	.1675E+1	.4399E-1	.2523E+2	.2890E+0
5.60	8.00	.746	2.674	0.128	5.528	7.236	15.508	1.9910	0.705	.5019E+0	.1992E+1	.4339E-1	.3088E+2	.3861E+0
5.60	10.00	.747	2.674	0.127	7.327	7.238	19.106	1.9971	0.927	.4223E+0	.2368E+1	.4229E-1	.3816E+2	.5107E+0
5.60	12.00	.748	2.674	0.126	9.326	7.240	23.104	2.0002	1.175	.3630E+0	.2755E+1	.4100E-1	.4621E+2	.6493E+0
5.60	15.00	.749	2.675	0.126	12.325	7.241	29.103	2.0032	1.547	.3037E+0	.3293E+1	.3920E-1	.5830E+2	.8572E+0
5.60	18.00	.750	2.675	0.125	15.325	7.243	35.102	2.0063	1.916	.2636E+0	.3793E+1	.3763E-1	.7042E+2	.1065E+1
5.60	22.00	.750	2.675	0.125	19.325	7.243	43.102	2.0063	2.416	.2264E+0	.4417E+1	.3585E-1	.8647E+2	.1342E+1
5.60	27.00	.751	2.676	0.125	24.325	7.244	53.101	2.0093	3.028	.1943E+0	.5146E+1	.3403E-1	.1067E+3	.1688E+1
5.60	33.00	.751	2.676	0.125	30.325	7.244	65.101	2.0093	3.777	.1677E+0	.5964E+1	.3227E-1	.1308E+3	.2104E+1
6.80	3.30	.514	3.157	0.243	0.143	8.105	5.731	1.6227	0.035	.2825E+1	.3540E+0	.1578E-1	.9300E+1	.1129E-1
6.80	3.90	.701	3.251	0.150	0.650	8.373	6.704	2.2786	0.097	.1274E+1	.7852E+0	.2961E-1	.1528E+2	.6021E-1
6.80	4.70	.726	3.263	0.137	1.437	8.408	8.274	2.3689	0.197	.8738E+0	.1144E+1	.3581E-1	.1960E+2	.1314E+0
6.80	5.60	.734	3.267	0.133	2.333	8.427	10.064	2.3980	0.310	.6820E+0	.1466E+1	.3851E-1	.2413E+2	.2115E+0
6.80	6.80	.739	3.270	0.131	3.531	8.427	12.458	2.4162	0.461	.5443E+0	.1837E+1	.3973E-1	.3010E+2	.3194E+0
6.80	8.20	.742	3.271	0.129	4.929	8.431	15.254	2.4271	0.636	.4497E+0	.2224E+1	.3988E-1	.3702E+2	.4442E+0
6.80	10.00	.743	3.272	0.129	6.729	8.433	18.853	2.4307	0.865	.3738E+0	.2675E+1	.3934E-1	.4583E+2	.6058E+0
6.80	12.00	.746	3.273	0.128	8.728	8.436	22.850	2.4380	1.113	.3187E+0	.3137E+1	.3845E-1	.5571E+2	.7841E+0
6.80	15.00	.746	3.273	0.127	11.727	8.437	28.849	2.4417	1.489	.2648E+0	.3776E+1	.3702E-1	.7044E+2	.1052E+1

TRANSFORMER DESIGN TABLE

CASE # 2   COPYRIGHT 1982 STEVE SMITH   DATA DATE 1/ 3/82

$$f(D)=[(U/(FG(DE)^2))^5 * (1/(DEUm))^4]^{1/13}$$

P	Q	D	E	F	G	U	Um	DE	FG	f(D)	1/f(D)	1/(FQf(D))	DEUm	FG(DE)^2/U
6.80	18.00	.747	3.274	0.127	14.727	8.438	34.848	2.4453	1.863	.2289E+0	.4369E+1	.3569E-1	.8521E+2	.1320E+1
6.80	22.00	.747	3.274	0.127	18.727	8.438	42.848	2.4453	2.369	.1958E+0	.5107E+1	.3414E-1	.1048E+3	.1679E+1
6.80	27.00	.748	3.274	0.126	23.726	8.440	52.847	2.4490	2.989	.1676E+0	.5966E+1	.3250E-1	.1294E+3	.2124E+1
8.20	33.00	.748	3.274	0.126	29.726	8.440	64.847	2.4490	3.745	.1443E+0	.6930E+1	.3088E-1	.1588E+3	.2662E+1
8.20	3.90	.409	3.804	0.296	0.096	9.355	6.758	1.5560	0.028	.3216E+1	.3110E+0	.9723E-2	.1052E+2	.7304E-2
8.20	4.70	.705	3.952	0.148	0.747	9.778	7.999	2.7865	0.110	.9818E+0	.1019E+1	.2643E-1	.2229E+2	.8755E-1
8.20	5.60	.727	3.964	0.137	1.637	9.810	9.772	2.8815	0.223	.6795E+0	.1472E+1	.3205E-1	.2816E+2	.1891E+0
8.20	6.80	.734	3.967	0.133	2.833	9.820	12.163	2.9118	0.377	.5139E+0	.1946E+1	.3490E-1	.3542E+2	.3253E+0
8.20	8.20	.738	3.969	0.131	4.231	9.826	14.959	2.9291	0.554	.4131E+0	.2420E+1	.3600E-1	.4382E+2	.4840E+0
8.20	10.00	.740	3.970	0.130	6.030	9.828	18.556	2.9378	0.784	.3373E+0	.2964E+1	.3615E-1	.5451E+2	.6884E+0
8.20	12.00	.742	3.971	0.129	8.029	9.831	22.554	2.9465	1.036	.2845E+0	.3515E+1	.3572E-1	.6645E+2	.9146E+0
8.20	15.00	.743	3.972	0.129	11.029	9.833	28.552	2.9508	1.417	.2342E+0	.4270E+1	.3472E-1	.8425E+2	.1255E+1
8.20	18.00	.744	3.972	0.128	14.028	9.834	34.551	2.9552	1.796	.2013E+0	.4967E+1	.3365E-1	.1021E+3	.1595E+1
8.20	22.00	.745	3.973	0.128	18.028	9.836	42.550	2.9595	2.299	.1715E+0	.5831E+1	.3233E-1	.1259E+3	.2047E+1
8.20	27.00	.745	3.973	0.128	23.028	9.836	52.550	2.9595	2.936	.1462E+0	.6838E+1	.3089E-1	.1555E+3	.2615E+1
8.20	33.00	.746	3.973	0.127	29.027	9.837	64.549	2.9639	3.686	.1256E+0	.7963E+1	.2943E-1	.1913E+3	.3292E+1
10.00	4.70	.283	4.642	0.359	0.058	10.915	8.125	1.3135	0.021	.4347E+1	.2300E+0	.4895E-2	.1067E+2	.3297E-2
10.00	5.60	.704	4.852	0.148	0.748	11.577	9.414	3.4158	0.111	.7990E+0	.1252E+1	.2235E-1	.3215E+2	.1116E+0
10.00	6.80	.728	4.864	0.136	1.936	11.611	11.784	3.5410	0.263	.5146E+0	.1943E+1	.2850E-1	.4173E+2	.2843E+0
10.00	8.20	.734	4.867	0.133	3.333	11.620	14.577	3.5724	0.443	.3909E+0	.2558E+1	.3120E-1	.5207E+2	.4869E+0
10.00	10.00	.738	4.869	0.131	5.131	11.626	18.172	3.5933	0.672	.3093E+0	.3233E+1	.3233E-1	.6530E+2	.7465E+0
10.00	12.00	.739	4.870	0.131	7.131	11.627	22.171	3.5986	0.931	.2563E+0	.3901E+1	.3251E-1	.7978E+2	.1036E+1
10.00	15.00	.741	4.871	0.130	10.130	11.630	28.169	3.6090	1.312	.2080E+0	.4807E+1	.3204E-1	.1017E+3	.1469E+1
10.00	18.00	.742	4.871	0.129	13.129	11.631	34.167	3.6143	1.694	.1774E+0	.5636E+1	.3131E-1	.1235E+3	.1902E+1
10.00	22.00	.742	4.871	0.129	17.129	11.631	42.167	3.6143	2.210	.1501E+0	.6661E+1	.3028E-1	.1524E+3	.2482E+1
10.00	27.00	.743	4.872	0.129	22.129	11.633	52.166	3.6195	2.844	.1274E+0	.7848E+1	.2907E-1	.1888E+3	.3202E+1
10.00	33.00	.743	4.872	0.129	28.129	11.633	64.166	3.6195	3.615	.1090E+0	.9172E+1	.2779E-1	.2323E+3	.4071E+1

CASE # 3   COPYRIGHT 1982 STEVE SMITH   DATA DATE 1/ 3/82   TRANSFORMER DESIGN TABLE

SCALE BY 1.000E-04   FUNCTION 1/f(D) $f(D)=[(U/(FG(DE)^2))^5 * (1/(DEUm))^4]^{(1/13)}$

```
Q/P    0.33  0.39  0.47  0.56  0.68  0.82  1.00  1.20  1.50  1.80  2.20  2.70  3.30  3.90  4.70  5.60  6.80  8.20  10.00

0.33    83    95   108   119   129   135   137   130   108    72    .     .     .     .     .     .     .     .     .
0.39   101   117   135   150   165   174   178   171   144   101    53    .     .     .     .     .     .     .     .
0.47   122   144   170   193   215   231   238   232   201   147    94    .     .     .     .     .     .     .     .
0.56   145   174   208   240   273   298   312   309   275   211   168   149    .     .     .     .     .     .     .
0.68   174   210   256   301   351   392   420   424   389   314   284   353   135    .     .     .     .     .     .
0.82   204   249   307   367   438   501   552   572   544   460   480   566   566   359    .     .     .     .     .
1.00   241   296   368   446   541   635   723   774   769   685   757   801   .     .     .     .     .     .     .
1.20   279   344   431   527   648   775   906  1001  1043   976  1271  1399  1452  1610  1411    .     .     .     .
1.50   331   411   518   638   795   966  1159  1326  1467  1464  1866  1975  2904  .     .    665    .     .     .
1.80   381   473   599   742   930  1141  1390  1625  1874  1975  2357  2693  3717    .     .     .     .     .     .
2.20   442   551   700   870  1098  1358  1674  2005  2371  2693  2904  3632  3663  4698  3887    .     .     .     .
2.70   514   642   817  1020  1292  1608  2000  2405  2934  3342  3663  4792  4792  5972  5561    .  3621    .     .
3.30   595   744   949  1187  1510  1887  2362  2864  3548  4125  4698  5043  6513  6513  7561  9363    .     .     .
3.90   672   840  1073  1344  1714  2148  2700  3292  4115  4841  5629  6288 10459  5972  8498 11099  1776    .     .
4.70   768   961  1229  1542  1970  2476  3123  3825  4820  5724  6764  7770  8619  7561 14938    .  1974    .     .
5.60   870  1089  1395  1752  2242  2823  3570  4387  5560  6647  7939  9280 10459 10459 11129 14036 14765  3073    .
6.80   998  1251  1603  2016  2583  3257  4130  5090  6482  7792  9389 11122 12803 11099 17059 14938 19235  4370    .
8.20  1139  1428  1832  2305  2957  3734  4744  5859  7488  9039 10959 13102 15290 15459 18765 17059 26481  3073   816
10.00 1310  1643  2105  2651  3409  4310  5484  6786  8699 10536 12838 15459 17059 18765 23128 25167 26481 14600 18238
12.00 1488  1868  2398  3021  3881  4912  6257  7753  9962 12094 14789 17898 21245 20587 27526 30509 33251 25618 32738
15.00 1740  2184  2806  3536  4546  5758  7344  9112 11732 14277 17517 21298 25439 24191 30509 33730 34592 34592 47753
18.00 1976  2481  3188  4020  5170  6553  8363 10385 13390 16319 20066 24468 29373 27730 33550 37730 42157 44592 60082
22.00 2273  2854  3669  4627  5954  7549  9641 11982 15469 18878 23255 28429 34199 33782 39100 39100 44330 45734 74563
27.00 2621  3293  4233  5340  6873  8719 11141 13855 17905 21874 26987 33058 39872 39521 45981 44330 50177 57290 98030
33.00 3014  3786  4869  6143  7909 10036 12829 15963 20646 25244 31182 38257 46236 53693 62910 72406 83825 95602    ++
```

CASE # 3   COPYRIGHT 1982 STEVE SMITH   DATA DATE 1/ 3/82   TRANSFORMER DESIGN TABLE

FUNCTION   1/(P*Q*f(D))

$$f(D)=[(U/(FG(DE)^2))^5 * (1/(DEUm))^4]^{(1/13)}$$

SCALE BY 1.000E-04

Q/P	0.33	0.39	0.47	0.56	0.68	0.82	1.00	1.20	1.50	1.80	2.20	2.70	3.30	3.90	4.70	5.60	6.80	8.20	10.00
0.33	763	739	696	643	575	500	414	329	218	122	51								
0.39	781	768	734	687	621	545	455	366	247	144	76	55							
0.47	789	788	769	733	672	598	506	412	285	174	112	109	27						
0.56	786	796	790	767	718	649	558	460	327	209	157	198	95	42					
0.68	774	792	800	791	759	702	617	520	382	256	218	288	200	153	91				
0.82	755	779	796	800	785	745	673	581	442	311	287	397	326	289	212	30			
1.00	730	758	783	797	796	775	723	645	513	380	385	498	440	393	342	166	95		
1.20	704	734	764	784	794	787	755	695	580	452	471	566	506	470	422	299	255	55	
1.50	670	702	735	760	779	785	772	737	652	543	556	597	548	510	467	388	345	217	10
1.80	641	674	708	736	760	773	772	752	694	609	617	612	566	529	492	430	389	312	182
2.20	609	642	677	706	734	753	761	754	719	661	647	614	571	533	488	449	407	352	273
2.70	577	609	644	674	704	726	741	742	724	688	656	606	565	528	476	454	410	372	318
3.30	547	578	612	642	673	697	716	723	717	694	654	592	552	517	462	449	413	376	334
3.90	522	552	585	616	646	672	692	703	703	690	644	573	536	499	445	440	401	373	339
4.70	495	524	557	586	616	642	664	678	662	677	628	552	514	481	425	426	388	365	336
5.60	445	499	530	559	589	615	638	653	635	659	607	536	494	461	406	409	374	353	329
6.80	421	472	502	529	559	584	607	624	609	637	584	514	479	439		392			
8.20	397	447	475	502	530	555	578	595	580	612	560	494	471	417					
10.00	376	421	449	474	501	526	548	565	553	585	531	479	453						
12.00	351	399	425	450	476	499	521	538	521	560	507	453	447						
15.00	333	373	398	426	446	468	490	506	496	529	480	429	425						
18.00	313	353	377	399	422	444	465	481	469	504	454								
22.00	294	333	355	376	398	418	438	454	442	477	430								
27.00	277	313	334	353	374	394	413	428	417	450									
33.00		294	314	332	352	371	389	403		425									

CASE # 3   COPYRIGHT 1982 STEVE SMITH   DATA DATE 1/ 3/82   TRANSFORMER DESIGN TABLE

$$f(D)=\left[\left(U/(FG(DE)^2)\right)^5 * \left(1/(PQf(D))\right)\ \left(1/(DEUm)\right)^4\right]^{(1/13)}$$

P	Q	D	E	F	G	U	Um	DE	FG	f(D)	1/f(D)	1/(PQf(D))	DEUm	FG(DE)^2/U
0.33	0.33	.929	.094	.071	.142	2.158	0.721	0.0873	0.010	.1203E+3	.8312E-2	.7633E-1	.6299E-1	.3564E-4
0.33	0.39	.937	.102	.063	.186	2.177	0.818	0.0956	0.012	.9944E+2	.1006E-1	.7814E-1	.7822E-1	.4917E-4
0.33	0.47	.943	.108	.057	.254	2.192	0.961	0.1018	0.014	.8168E+2	.1224E-1	.7893E-1	.9790E-1	.6852E-4
0.33	0.56	.947	.112	.053	.336	2.201	1.130	0.1061	0.018	.6882E+2	.1453E-1	.7863E-1	.1198E+0	.9101E-4
0.33	0.68	.950	.115	.050	.450	2.209	1.361	0.1092	0.023	.5760E+2	.1736E-1	.7736E-1	.1487E+0	.1217E-3
0.33	0.82	.952	.117	.048	.586	2.213	1.636	0.1114	0.028	.4897E+2	.2042E-1	.7546E-1	.1822E+0	.1577E-3
0.33	1.00	.954	.119	.046	.762	2.218	1.990	0.1135	0.035	.4154E+2	.2407E-1	.7295E-1	.2259E+0	.2037E-3
0.33	1.20	.955	.120	.045	.960	2.221	2.387	0.1146	0.043	.3589E+2	.2786E-1	.7036E-1	.2735E+0	.2555E-3
0.33	1.50	.956	.121	.044	1.258	2.223	2.984	0.1157	0.055	.3017E+2	.3315E-1	.6696E-1	.3452E+0	.3332E-3
0.33	1.80	.956	.121	.044	1.558	2.223	3.584	0.1157	0.069	.2626E+2	.3807E-1	.6410E-1	.4146E+0	.4126E-3
0.33	2.20	.957	.122	.043	1.956	2.226	4.381	0.1168	0.084	.2261E+2	.4424E-1	.6093E-1	.5115E+0	.5152E-3
0.33	2.70	.957	.122	.043	2.456	2.226	5.381	0.1168	0.106	.1944E+2	.5144E-1	.5773E-1	.6283E+0	.6469E-3
0.33	3.30	.958	.123	.042	3.054	2.228	6.578	0.1178	0.128	.1680E+2	.5953E-1	.5466E-1	.7752E+0	.7994E-3
0.33	3.90	.958	.123	.042	3.654	2.228	7.778	0.1178	0.153	.1489E+2	.6715E-1	.5218E-1	.9166E+0	.9564E-3
0.33	4.70	.958	.123	.042	4.454	2.230	9.378	0.1178	0.187	.1303E+2	.7676E-1	.4949E-1	.1105E+1	.1166E-2
0.33	5.60	.958	.123	.042	5.354	2.228	11.178	0.1178	0.225	.1150E+2	.8696E-1	.4706E-1	.1317E+1	.1401E-2
0.33	6.80	.959	.124	.042	6.554	2.228	13.578	0.1189	0.275	.1002E+2	.9979E-1	.4447E-1	.1600E+1	.1715E-2
0.33	8.20	.959	.124	.041	7.952	2.230	16.376	0.1189	0.326	.8780E+1	.1139E+0	.4209E-1	.1947E+1	.2067E-2
0.33	10.00	.959	.124	.041	9.752	2.230	19.976	0.1189	0.400	.7636E+1	.1310E+0	.3969E-1	.2375E+1	.2535E-2
0.33	12.00	.959	.124	.041	11.752	2.230	23.976	0.1189	0.482	.6719E+1	.1488E+0	.3758E-1	.2851E+1	.3055E-2
0.33	15.00	.959	.124	.041	14.752	2.230	29.976	0.1189	0.605	.5748E+1	.1740E+0	.3515E-1	.3565E+1	.3885E-2
0.33	18.00	.959	.124	.041	17.752	2.230	35.976	0.1189	0.728	.5060E+1	.1976E+0	.3327E-1	.4278E+1	.4615E-2
0.33	22.00	.959	.124	.041	21.752	2.230	43.976	0.1189	0.892	.4408E+1	.2273E+0	.3131E-1	.5229E+1	.5654E-2
0.33	27.00	.959	.124	.041	26.752	2.230	53.976	0.1189	1.097	.3815E+1	.2621E+0	.2942E-1	.6419E+1	.6954E-2
0.33	33.00	.959	.124	.041	32.752	2.230	65.976	0.1189	1.343	.3318E+1	.3014E+0	.2768E-1	.7846E+1	.8514E-2
0.39	0.33	.907	.102	.093	.126	2.164	0.758	0.0925	0.012	.1052E+3	.9507E-2	.7387E-1	.7017E-1	.4634E-4
0.39	0.39	.917	.112	.083	.166	2.188	0.850	0.1027	0.014	.8564E+2	.1168E-1	.7677E-1	.8728E-1	.6641E-4
0.39	0.47	.927	.122	.073	.226	2.213	0.981	0.1131	0.016	.6921E+2	.1445E-1	.7883E-1	.1110E+0	.9537E-4
0.39	0.56	.934	.129	.066	.302	2.230	1.141	0.1205	0.020	.5755E+2	.1738E-1	.7956E-1	.1375E+0	.1298E-3
0.39	0.68	.939	.134	.061	.412	2.244	1.367	0.1258	0.025	.4763E+2	.2100E-1	.7917E-1	.1720E+0	.1775E-3
0.39	0.82	.942	.137	.058	.546	2.249	1.638	0.1291	0.032	.4015E+2	.2490E-1	.7778E-1	.2114E+0	.2345E-3
0.39	1.00	.945	.140	.055	.720	2.256	1.990	0.1323	0.040	.3383E+2	.2956E-1	.7579E-1	.2633E+0	.3072E-3
0.39	1.20	.946	.141	.054	.918	2.259	2.387	0.1334	0.050	.2910E+2	.3437E-1	.7343E-1	.3184E+0	.3905E-3
0.39	1.50	.948	.143	.052	1.214	2.264	2.981	0.1356	0.063	.2435E+2	.4106E-1	.7019E-1	.4042E+0	.5125E-3
0.39	1.80	.949	.144	.051	1.512	2.266	3.578	0.1367	0.077	.2114E+2	.4730E-1	.6737E-1	.4890E+0	.6355E-3
0.39	2.20	.949	.144	.051	1.912	2.266	4.378	0.1367	0.098	.1816E+2	.5508E-1	.6420E-1	.5983E+0	.8036E-3

CASE # 3   COPYRIGHT 1982 STEVE SMITH   DATA DATE 1/ 3/82   TRANSFORMER DESIGN TABLE

$$f(D)=[(U/(FG(DE)^2))^5 * (1/(DEUm))^4]^{(1/13)}$$

P	Q	D	E	F	G	U	Um	DE	FG	f(D)	1/f(D)	1/(PQf(D))	DEUm	FG(DE)^2/U
0.39	2.70	.950	0.145	0.050	2.410	2.269	5.376	0.1377	0.121	.1558E+2	.6417E-1	.6094E-1	.7405E+0	.1008E-2
0.39	3.30	.950	0.145	0.050	2.410	2.269	6.576	0.1377	0.151	.1345E+2	.7437E-1	.5778E-1	.9058E+0	.1259E-2
0.39	3.90	.951	0.146	0.049	3.608	2.271	7.773	0.1388	0.177	.1191E+2	.8398E-1	.5521E-1	.1079E+1	.1501E-2
0.39	4.70	.951	0.146	0.049	4.408	2.271	9.373	0.1388	0.216	.1041E+2	.9608E-1	.5242E-1	.1301E+1	.1834E-2
0.39	5.60	.951	0.146	0.049	5.308	2.271	11.173	0.1388	0.260	.9181E+1	.1089E+0	.4987E-1	.1551E+1	.2208E-2
0.39	6.80	.951	0.146	0.049	6.508	2.271	13.573	0.1388	0.319	.7995E+1	.1251E+0	.4716E-1	.1885E+1	.2707E-2
0.39	8.20	.952	0.147	0.048	7.906	2.273	16.370	0.1399	0.379	.7002E+1	.1428E+0	.4466E-1	.2291E+1	.3269E-2
0.39	10.00	.952	0.147	0.048	9.706	2.273	19.970	0.1399	0.466	.6087E+1	.1643E+0	.4213E-1	.2795E+1	.4013E-2
0.39	12.00	.952	0.147	0.048	11.706	2.273	23.970	0.1399	0.562	.5354E+1	.1866E+0	.3991E-1	.3354E+1	.4840E-2
0.39	15.00	.952	0.147	0.048	14.706	2.273	29.970	0.1399	0.706	.4579E+1	.2184E+0	.3733E-1	.4194E+1	.6081E-2
0.39	18.00	.952	0.147	0.048	17.706	2.273	35.970	0.1399	0.850	.4030E+1	.2481E+0	.3534E-1	.5034E+1	.7321E-2
0.39	22.00	.952	0.147	0.048	21.706	2.273	43.970	0.1399	1.042	.3503E+1	.2854E+0	.3327E-1	.6153E+1	.8975E-2
0.39	27.00	.952	0.147	0.048	26.706	2.273	53.970	0.1399	1.282	.3037E+1	.3293E+0	.3127E-1	.7553E+1	.1104E-1
0.39	33.00	.952	0.147	0.048	32.706	2.273	65.970	0.1399	1.570	.2641E+1	.3786E+0	.2942E-1	.9232E+1	.1352E-1
0.47	0.33	.873	0.108	0.127	0.114	2.161	0.821	0.0943	0.014	.9267E+2	.1079E-1	.6958E-1	.7743E-1	.5954E-4
0.47	0.39	.887	0.122	0.113	0.146	2.196	0.901	0.1082	0.016	.7428E+2	.1346E-1	.7345E-1	.9753E-1	.8800E-4
0.47	0.47	.902	0.137	0.098	0.196	2.232	1.018	0.1236	0.019	.5886E+2	.1699E-1	.7691E-1	.1258E+0	.1314E-3
0.47	0.56	.913	0.148	0.087	0.264	2.259	1.167	0.1351	0.023	.4808E+2	.2080E-1	.7903E-1	.1577E+0	.1857E-3
0.47	0.68	.922	0.157	0.078	0.366	2.281	1.381	0.1448	0.029	.3913E+2	.2556E-1	.7997E-1	.1999E+0	.2623E-3
0.47	0.82	.927	0.162	0.073	0.496	2.293	1.647	0.1502	0.036	.3258E+2	.3069E-1	.7964E-1	.2473E+0	.3562E-3
0.47	1.00	.932	0.167	0.068	0.666	2.305	1.993	0.1556	0.045	.2718E+2	.3679E-1	.7787E-1	.3101E+0	.4760E-3
0.47	1.20	.934	0.169	0.066	0.862	2.310	2.387	0.1578	0.057	.2322E+2	.4307E-1	.7636E-1	.3768E+0	.6137E-3
0.47	1.50	.937	0.172	0.063	1.156	2.317	2.978	0.1612	0.073	.1931E+2	.5178E-1	.7345E-1	.4800E+0	.8164E-3
0.47	1.80	.938	0.173	0.062	1.454	2.319	3.575	0.1623	0.090	.1670E+2	.5988E-1	.7078E-1	.5802E+0	.1023E-2
0.47	2.20	.939	0.174	0.061	1.852	2.322	4.373	0.1634	0.113	.1429E+2	.6997E-1	.6767E-1	.7144E+0	.1299E-2
0.47	2.60	.940	0.175	0.060	2.350	2.324	5.370	0.1645	0.141	.1224E+2	.8173E-1	.6440E-1	.8833E+0	.1642E-2
0.47	3.30	.941	0.176	0.059	2.948	2.327	6.567	0.1656	0.174	.1054E+2	.9491E-1	.6119E-1	.1088E+1	.2050E-2
0.47	3.90	.941	0.176	0.059	3.548	2.327	7.767	0.1656	0.209	.9318E+1	.1073E+0	.5855E-1	.1286E+1	.2468E-2
0.47	4.70	.942	0.177	0.058	4.346	2.329	9.364	0.1667	0.252	.8134E+1	.1229E+0	.5565E-1	.1561E+1	.3009E-2
0.47	5.60	.942	0.177	0.058	5.246	2.329	11.164	0.1667	0.304	.7168E+1	.1395E+0	.5301E-1	.1861E+1	.3632E-2
0.47	6.80	.942	0.177	0.058	6.446	2.329	13.564	0.1667	0.374	.6237E+1	.1603E+0	.5017E-1	.2262E+1	.4462E-2
0.47	8.20	.943	0.177	0.058	7.846	2.329	16.364	0.1667	0.455	.5458E+1	.1832E+0	.4754E-1	.2728E+1	.5432E-2
0.47	10.00	.943	0.177	0.057	9.446	2.332	19.961	0.1679	0.550	.4742E+1	.2109E+0	.4487E-1	.3351E+1	.6643E-2
0.47	12.00	.943	0.178	0.057	11.644	2.332	23.961	0.1679	0.664	.4170E+1	.2398E+0	.4252E-1	.4022E+1	.8020E-2
0.47	15.00	.943	0.178	0.057	14.644	2.332	29.961	0.1679	0.835	.3564E+1	.2808E+0	.3980E-1	.5029E+1	.1009E-1
0.47	18.00	.943	0.178	0.057	17.644	2.332	35.961	0.1679	1.006	.3136E+1	.3188E+0	.3769E-1	.6036E+1	.1215E-1

CASE # 3   COPYRIGHT 1982 STEVE SMITH   DATA DATE 1/ 3/82

TRANSFORMER DESIGN TABLE

$$f(D)=[(U/(FG(DE)^2))^5 * (1/(DEUm))^4]^{(1/13)}$$

P	Q	D	E	F	G	U	Um	DE	FG	f(D)	1/f(D)	1/(PQf(D))	DEUm	FG(DE)^2/U
0.47	22.00	.943	0.178	0.057	21.644	2.332	43.961	0.1679	1.234	.2726E+1	.3669E+0	.3548E-1	.7379E+1	.1491E-1
0.47	27.00	.943	0.178	0.057	26.644	2.332	53.961	0.1679	1.519	.2362E+1	.4233E+0	.3336E-1	.9058E+1	.1835E-1
0.47	33.00	.943	0.178	0.057	32.644	2.332	65.961	0.1679	1.861	.2054E+1	.4869E+0	.3139E-1	.1107E+2	.2249E-1
0.56	0.33	.832	0.112	0.168	0.106	2.152	0.900	0.0932	0.018	.8412E+2	.1189E-1	.6433E-1	.8385E-1	.7186E-4
0.56	0.39	.848	0.128	0.152	0.134	2.191	0.974	0.1085	0.020	.6660E+2	.1501E-1	.6875E-1	.1057E+0	.1095E-3
0.56	0.47	.868	0.148	0.132	0.174	2.239	1.077	0.1285	0.023	.5186E+2	.1928E-1	.7326E-1	.1384E+0	.1693E-3
0.56	0.56	.884	0.164	0.116	0.232	2.278	1.211	0.1450	0.027	.4159E+2	.2404E-1	.7667E-1	.1756E+0	.2483E-3
0.56	0.68	.899	0.179	0.101	0.322	2.315	1.408	0.1609	0.033	.3320E+2	.3012E-1	.7911E-1	.2266E+0	.3638E-3
0.56	0.82	.909	0.189	0.091	0.442	2.339	1.660	0.1718	0.040	.2721E+2	.3675E-1	.8002E-1	.2851E+0	.5076E-3
0.56	1.00	.916	0.196	0.084	0.608	2.356	2.000	0.1795	0.051	.2242E+2	.4461E-1	.7965E-1	.3590E+0	.6987E-3
0.56	1.20	.919	0.200	0.080	0.800	2.366	2.388	0.1840	0.064	.1898E+2	.5268E-1	.7839E-1	.4395E+0	.9119E-3
0.56	1.50	.924	0.204	0.076	1.092	2.375	2.977	0.1885	0.083	.1567E+2	.6384E-1	.7599E-1	.5611E+0	.1241E-2
0.56	1.80	.926	0.206	0.074	1.388	2.380	3.571	0.1908	0.103	.1348E+2	.7417E-1	.7358E-1	.6812E+0	.1570E-2
0.56	2.20	.928	0.208	0.072	1.784	2.385	4.365	0.1930	0.128	.1149E+2	.8702E-1	.7063E-1	.8426E+0	.2007E-2
0.56	2.70	.929	0.209	0.071	2.282	2.388	5.363	0.1942	0.162	.9807E+1	.1020E+0	.6744E-1	.1041E+1	.2558E-2
0.56	3.30	.930	0.210	0.070	2.880	2.390	6.560	0.1953	0.202	.8425E+1	.1187E+0	.6423E-1	.1281E+1	.3217E-2
0.56	3.90	.931	0.211	0.069	3.478	2.392	7.757	0.1964	0.240	.7438E+1	.1344E+0	.6156E-1	.1524E+1	.3871E-2
0.56	4.70	.931	0.211	0.069	4.278	2.392	9.357	0.1964	0.295	.6484E+1	.1542E+0	.5860E-1	.1838E+1	.4761E-2
0.56	5.60	.932	0.212	0.068	5.176	2.395	11.154	0.1976	0.352	.5707E+1	.1752E+0	.5587E-1	.2204E+1	.5738E-2
0.56	6.60	.932	0.212	0.068	6.376	2.395	13.554	0.1976	0.434	.4961E+1	.2016E+0	.5294E-1	.2678E+1	.7068E-2
0.56	8.20	.932	0.212	0.068	7.776	2.395	16.354	0.1976	0.529	.4338E+1	.2305E+0	.5020E-1	.3231E+1	.8620E-2
0.56	10.00	.933	0.213	0.067	9.574	2.397	19.951	0.1987	0.641	.3766E+1	.2655E+0	.4741E-1	.3965E+1	.1057E-1
0.56	12.00	.933	0.213	0.067	11.574	2.397	23.951	0.1987	0.775	.3310E+1	.3021E+0	.4496E-1	.4760E+1	.1278E-1
0.56	15.00	.933	0.213	0.067	14.574	2.397	29.951	0.1987	0.976	.2828E+1	.3536E+0	.4210E-1	.5952E+1	.1609E-1
0.56	18.00	.933	0.213	0.067	17.574	2.397	35.951	0.1987	1.177	.2488E+1	.4020E+0	.3988E-1	.7145E+1	.1940E-1
0.56	22.00	.933	0.213	0.067	21.574	2.397	43.951	0.1987	1.445	.2161E+1	.4627E+0	.3756E-1	.8734E+1	.2381E-1
0.56	27.00	.933	0.213	0.067	26.574	2.397	53.951	0.1987	1.780	.1873E+1	.5340E+0	.3532E-1	.1072E+2	.2933E-1
0.56	33.00	.933	0.213	0.067	32.574	2.397	65.951	0.1987	2.182	.1528E+1	.6143E+0	.3324E-1	.1311E+2	.3595E-1
0.68	0.33	.775	0.115	0.225	0.100	2.133	1.011	0.0891	0.023	.7556E+2	.1289E-1	.5746E-1	.9013E-1	.8377E-3
0.68	0.39	.794	0.134	0.206	0.122	2.180	1.077	0.1064	0.025	.6074E+2	.1646E-1	.6208E-1	.1146E+0	.1305E-3
0.68	0.47	.817	0.157	0.183	0.156	2.235	1.171	0.1283	0.029	.4653E+2	.2149E-1	.6724E-1	.1502E+0	.2101E-3
0.68	0.56	.839	0.179	0.161	0.202	2.289	1.288	0.1502	0.033	.3660E+2	.2733E-1	.7176E-1	.1935E+0	.3205E-3
0.68	0.68	.862	0.202	0.138	0.276	2.345	1.463	0.1741	0.038	.2851E+2	.3508E-1	.7587E-1	.2547E+0	.4925E-3
0.68	0.82	.880	0.220	0.120	0.380	2.388	1.691	0.1936	0.046	.2285E+2	.4376E-1	.7847E-1	.3274E+0	.7156E-3
0.68	1.00	.892	0.232	0.108	0.536	2.418	2.017	0.2069	0.058	.1847E+2	.5415E-1	.7963E-1	.4174E+0	.1025E-2
0.68	1.20	.900	0.240	0.100	0.720	2.437	2.394	0.2160	0.072	.1543E+2	.6481E-1	.7942E-1	.5171E+0	.1378E-2

CASE # 3   COPYRIGHT 1982 STEVE SMITH   DATA DATE 1/ 3/82   TRANSFORMER DESIGN TABLE

$$f(D)=[(U/(FG(DE)^2))^5 * (1/(DEUm))^4]^{(1/13)}$$

P	Q	D	E	F	G	U	Um	DE	FG	f(D)	1/f(D)	1/(PQf(D))	DEUm	FG(DE)^2/U
0.68	1.50	.907	0.247	0.093	1.006	2.454	2.974	0.2240	0.094	.1258E+2	.7948E-1	.7792E-1	.6663E+0	.1913E-2
0.68	1.80	.910	0.250	0.090	1.300	2.461	3.565	0.2461	0.117	.1075E+2	.9302E-1	.7599E-1	.8111E+0	.2460E-2
0.68	2.20	.913	0.253	0.087	1.694	2.469	4.357	0.2310	0.147	.9109E+1	.1098E+0	.7338E-1	.1006E+1	.3185E-2
0.68	2.70	.915	0.255	0.085	2.190	2.474	5.351	0.2333	0.186	.7738E+1	.1292E+0	.7039E-1	.1249E+1	.4097E-2
0.68	3.30	.916	0.256	0.084	2.788	2.476	6.548	0.2345	0.234	.6621E+1	.1510E+0	.6728E-1	.1536E+1	.5201E-2
0.68	3.90	.917	0.257	0.083	3.386	2.478	7.745	0.2357	0.281	.5835E+1	.1714E+0	.6462E-1	.1825E+1	.6298E-2
0.68	4.70	.918	0.258	0.082	4.184	2.481	9.343	0.2368	0.343	.5076E+1	.1970E+0	.6164E-1	.2213E+1	.7758E-2
0.68	5.60	.919	0.259	0.081	5.082	2.483	11.140	0.2380	0.412	.4461E+1	.2242E+0	.5887E-1	.2651E+1	.9391E-2
0.68	6.80	.920	0.259	0.081	6.282	2.483	13.540	0.2380	0.509	.3872E+1	.2583E+0	.5585E-1	.3223E+1	.1161E-1
0.68	8.20	.920	0.260	0.080	7.680	2.486	16.337	0.2392	0.614	.3382E+1	.2957E+0	.5302E-1	.3908E+1	.1414E-1
0.68	10.00	.920	0.260	0.080	9.480	2.486	19.937	0.2392	0.758	.2934E+1	.3409E+0	.5013E-1	.4769E+1	.1746E-1
0.68	12.00	.920	0.260	0.080	11.480	2.486	23.937	0.2392	0.918	.2576E+1	.3881E+0	.4757E-1	.5726E+1	.2114E-1
0.68	15.00	.921	0.261	0.079	14.478	2.488	29.934	0.2404	1.144	.2200E+1	.4546E+0	.4457E-1	.7196E+1	.2656E-1
0.68	18.00	.921	0.261	0.079	17.478	2.488	35.934	0.2404	1.381	.1934E+1	.5170E+0	.4224E-1	.8638E+1	.3207E-1
0.68	22.00	.921	0.261	0.079	21.478	2.488	43.934	0.2404	1.697	.1680E+1	.5954E+0	.3980E-1	.1055E+2	.3941E-1
0.68	27.00	.921	0.261	0.079	26.478	2.488	53.934	0.2404	2.092	.1455E+1	.6873E+0	.3744E-1	.1296E+2	.4858E-1
0.68	33.00	.921	0.261	0.079	32.478	2.488	65.934	0.2404	2.566	.1264E+1	.7909E+0	.3552E-1	.1585E+2	.5959E-1
0.82	0.33	.708	0.118	0.292	0.094	2.111	1.143	0.0035	0.027	.7388E+2	.1334E-1	.5002E-1	.9547E-1	.9077E-4
0.82	0.39	.728	0.138	0.272	0.114	2.159	1.206	0.1005	0.031	.5735E+2	.1744E-1	.5452E-1	.1211E+0	.1449E-3
0.82	0.47	.753	0.163	0.247	0.144	2.220	1.294	0.1227	0.036	.4337E+2	.2306E-1	.5983E-1	.1588E+0	.2414E-3
0.82	0.56	.780	0.190	0.220	0.180	2.286	1.397	0.1482	0.040	.3355E+2	.2981E-1	.6491E-1	.2070E+0	.3805E-3
0.82	0.68	.810	0.220	0.190	0.240	2.358	1.551	0.1782	0.046	.2554E+2	.3916E-1	.7023E-1	.2764E+0	.6140E-3
0.82	0.82	.838	0.248	0.162	0.324	2.426	1.751	0.2078	0.052	.1997E+2	.5009E-1	.7449E-1	.3639E+0	.9343E-3
0.82	1.00	.860	0.270	0.140	0.460	2.480	2.048	0.2322	0.064	.1574E+2	.6354E-1	.7749E-1	.4756E+0	.1400E-2
0.82	1.20	.874	0.284	0.126	0.632	2.514	2.408	0.2482	0.080	.1291E+2	.7746E-1	.7872E-1	.5978E+0	.1952E-2
0.82	1.50	.885	0.295	0.115	0.910	2.541	2.977	0.2611	0.105	.1035E+2	.9658E-1	.7852E-1	.7772E+0	.2008E-2
0.82	1.80	.891	0.301	0.109	1.198	2.555	3.560	0.2682	0.131	.8762E+1	.1141E+0	.7732E-1	.9547E+0	.3676E-2
0.82	2.20	.895	0.305	0.105	1.590	2.565	4.348	0.2730	0.167	.7365E+1	.1355E+0	.7526E-1	.1187E+1	.4850E-2
0.82	2.70	.898	0.308	0.102	2.084	2.572	5.340	0.2766	0.213	.6219E+1	.1608E+0	.7263E-1	.1477E+1	.6322E-2
0.82	3.30	.901	0.311	0.099	2.678	2.580	6.531	0.2802	0.265	.5300E+1	.1887E+0	.6972E-1	.1830E+1	.8070E-2
0.82	3.90	.902	0.312	0.098	3.276	2.582	7.728	0.2814	0.321	.4655E+1	.2148E+0	.6717E-1	.2175E+1	.9848E-2
0.82	4.70	.903	0.313	0.097	4.074	2.584	9.325	0.2826	0.395	.4039E+1	.2476E+0	.6424E-1	.2636E+1	.1222E-1
0.82	5.60	.904	0.314	0.096	4.972	2.587	11.122	0.2839	0.477	.3543E+1	.2823E+0	.6147E-1	.3157E+1	.1487E-1
0.82	6.80	.905	0.315	0.095	6.170	2.589	13.520	0.2851	0.586	.3070E+1	.3257E+0	.5842E-1	.3854E+1	.1840E-1
0.82	8.20	.906	0.316	0.094	7.568	2.592	16.317	0.2863	0.711	.2678E+1	.3734E+0	.5554E-1	.4671E+1	.2250E-1
0.82	10.00	.906	0.316	0.094	9.368	2.592	19.917	0.2863	0.881	.2320E+1	.4310E+0	.5256E-1	.5702E+1	.2785E-1

CASE # 3   COPYRIGHT 1982 STEVE SMITH   DATA DATE 1/ 3/82

TRANSFORMER DESIGN TABLE

$$f(D)=[(U/(FG(DE)^2))^5 * (1/(DEUm)^4]^{(1/13)}$$

P	Q	D	E	F	G	U	Um	DE	FG	f(D)	1/f(D)	1/(PQf(D))	DEUm	FG(DE)^2/U
0.82	12.00	.907	0.317	0.093	11.366	2.594	23.914	0.2875	1.057	.2036E+1	.4912E+0	.4992E-1	.6876E+1	.3369E-1
0.82	15.00	.907	0.317	0.093	14.366	2.594	29.914	0.2875	1.336	.1737E+1	.5758E+0	.4682E-1	.8601E+1	.4258E-1
0.82	18.00	.907	0.317	0.093	17.366	2.594	35.914	0.2875	1.615	.1526E+1	.6553E+0	.4439E-1	.1033E+2	.5147E-1
0.82	22.00	.908	0.318	0.092	21.364	2.597	43.911	0.2887	1.965	.1325E+1	.7549E+0	.4185E-1	.1268E+2	.6311E-1
0.82	27.00	.908	0.318	0.092	26.364	2.597	53.911	0.2887	2.425	.1147E+1	.8719E+0	.3938E-1	.1557E+2	.7788E-1
0.82	33.00	.908	0.318	0.092	32.364	2.597	65.911	0.2887	2.977	.9964E+0	.1004E+1	.3709E-1	.1903E+2	.9561E-1
1.00	0.33	.620	0.120	0.380	0.090	2.077	1.317	0.0744	0.034	.7317E+2	.1367E-1	.4141E-1	.9798E-1	.9115E-4
1.00	0.39	.641	0.141	0.359	0.108	2.128	1.377	0.0904	0.039	.5630E+2	.1776E-1	.4555E-1	.1245E+0	.1488E-3
1.00	0.47	.669	0.169	0.331	0.132	2.196	1.457	0.1131	0.044	.4203E+2	.2379E-1	.5063E-1	.1647E+0	.2543E-3
1.00	0.56	.698	0.198	0.302	0.164	2.266	1.554	0.1382	0.050	.3201E+2	.3124E-1	.5578E-1	.2148E+0	.4174E-3
1.00	0.68	.735	0.235	0.265	0.210	2.356	1.688	0.1727	0.062	.2382E+2	.4198E-1	.6173E-1	.2916E+0	.7046E-3
1.00	0.82	.773	0.273	0.227	0.274	2.449	1.860	0.2110	0.062	.1812E+2	.5518E-1	.6729E-1	.3924E+0	.1131E-2
1.00	1.00	.809	0.309	0.191	0.382	2.536	2.117	0.2500	0.073	.1383E+2	.7229E-1	.7229E-1	.5291E+0	.1799E-2
1.00	1.20	.835	0.335	0.165	0.530	2.599	2.442	0.2797	0.087	.1104E+2	.9056E-1	.7547E-1	.6832E+0	.2633E-2
1.00	1.50	.856	0.356	0.144	0.788	2.650	2.982	0.3047	0.113	.8631E+1	.1159E+0	.7724E-1	.9088E+0	.3976E-2
1.00	1.80	.866	0.366	0.134	1.068	2.674	3.554	0.3170	0.143	.7195E+1	.1390E+0	.7722E-1	.1126E+1	.5376E-2
1.00	2.20	.873	0.373	0.127	1.454	2.691	4.334	0.3256	0.185	.5975E+1	.1674E+0	.7607E-1	.1411E+1	.7275E-2
1.00	2.70	.878	0.378	0.122	1.944	2.704	5.320	0.3319	0.237	.5000E+1	.2000E+0	.7407E-1	.1765E+1	.9662E-2
1.00	3.30	.882	0.382	0.118	2.536	2.713	6.508	0.3369	0.299	.4234E+1	.2362E+0	.7157E-1	.2193E+1	.1252E-1
1.00	3.90	.884	0.384	0.116	3.132	2.718	7.702	0.3395	0.363	.3704E+1	.2700E+0	.6923E-1	.2615E+1	.1540E-1
1.00	4.70	.886	0.386	0.114	3.928	2.723	9.297	0.3420	0.448	.3202E+1	.3123E+0	.6645E-1	.3179E+1	.1923E-1
1.00	5.60	.887	0.387	0.113	4.826	2.726	11.094	0.3433	0.545	.2801E+1	.3570E+0	.6376E-1	.3808E+1	.2358E-1
1.00	6.80	.888	0.388	0.112	6.024	2.728	13.491	0.3445	0.675	.2421E+1	.4130E+0	.6074E-1	.4648E+1	.2916E-1
1.00	8.20	.889	0.389	0.111	7.422	2.730	16.288	0.3458	0.824	.2108E+1	.4744E+0	.5785E-1	.5633E+1	.3609E-1
1.00	10.00	.890	0.390	0.110	9.220	2.733	19.885	0.3471	1.014	.1824E+1	.5484E+0	.5484E-1	.6902E+1	.4471E-1
1.00	12.00	.891	0.391	0.109	11.218	2.735	23.882	0.3484	1.223	.1598E+1	.6257E+0	.5214E-1	.8320E+1	.5426E-1
1.00	15.00	.891	0.391	0.109	14.218	2.735	29.882	0.3484	1.550	.1362E+1	.7344E+0	.4896E-1	.1041E+2	.6877E-1
1.00	18.00	.892	0.392	0.108	17.216	2.738	35.880	0.3497	1.859	.1196E+1	.8363E+0	.4646E-1	.1255E+2	.8304E-1
1.00	22.00	.892	0.392	0.108	21.216	2.738	43.880	0.3497	2.291	.1037E+1	.9641E+0	.4338E-1	.1534E+2	.1023E+0
1.00	27.00	.892	0.392	0.108	26.216	2.738	53.880	0.3497	2.831	.8976E+0	.1114E+1	.4126E-1	.1884E+2	.1264E+0
1.00	33.00	.892	0.392	0.108	32.216	2.738	65.880	0.3497	3.479	.7795E+0	.1283E+1	.3888E-1	.2304E+2	.1554E+0
1.20	0.33	.522	0.122	0.478	0.086	2.039	1.511	0.0637	0.041	.7672E+2	.1304E-1	.3292E-1	.9624E-1	.8177E-4
1.20	0.39	.544	0.144	0.456	0.102	2.092	1.568	0.0783	0.047	.5845E+2	.1711E-1	.3656E-1	.1229E+0	.1364E-3
1.20	0.47	.573	0.173	0.427	0.124	2.163	1.645	0.0991	0.053	.4307E+2	.2322E-1	.4117E-1	.1631E+0	.2406E-3
1.20	0.56	.605	0.205	0.395	0.150	2.240	1.734	0.1240	0.059	.3232E+2	.3094E-1	.4604E-1	.2151E+0	.4068E-3
1.20	0.68	.646	0.246	0.354	0.188	2.340	1.857	0.1589	0.067	.2356E+2	.4244E-1	.5201E-1	.2951E+0	.7182E-3

CASE # 3   COPYRIGHT 1982 STEVE SMITH   DATA DATE 1/ 3/82

TRANSFORMER DESIGN TABLE

$$f(D) = \left[\left(\frac{U}{FG(DE)^2}\right)^5 \cdot \left(\frac{1}{DEU_m}\right)^4\right]^{1/13}$$

P	Q	D	E	F	G	U	U_m	DE	FG	$f(D)$	$1/f(D)$	$1/(PQf(D))$	DEU_m	$FG(DE)^2/U$
1.20	0.82	.690	0.290	0.310	0.240	2.447	2.011	0.2001	0.074	.1748E+2	.5722E-1	.5813E-1	.4024E+0	.1217E-2
1.20	1.00	.739	0.339	0.261	0.322	2.566	2.231	0.2505	0.084	.1292E+2	.7741E-1	.6450E-1	.5589E+0	.2056E-2
1.20	1.20	.781	0.381	0.219	0.438	2.668	2.511	0.2976	0.096	.9986E+1	.1001E+0	.6954E-1	.7472E+0	.3183E-2
1.20	1.50	.818	0.418	0.182	0.664	2.758	3.005	0.3419	0.121	.7539E+1	.1326E+0	.7369E-1	.1028E+1	.5123E-2
1.20	1.80	.836	0.436	0.164	0.928	2.802	3.554	0.3645	0.152	.6154E+1	.1625E+0	.7524E-1	.1295E+1	.7217E-2
1.20	2.20	.849	0.449	0.151	1.302	2.833	4.317	0.3812	0.197	.5027E+1	.1989E+0	.7535E-1	.1645E+1	.1008E-1
1.20	2.70	.857	0.457	0.143	1.786	2.853	5.294	0.3916	0.255	.4157E+1	.2405E+0	.7424E-1	.2073E+1	.1373E-1
1.20	3.30	.862	0.462	0.138	2.376	2.865	6.479	0.3982	0.328	.3491E+1	.2864E+0	.7233E-1	.2580E+1	.1815E-1
1.20	4.00	.865	0.465	0.135	2.970	2.872	7.671	0.4022	0.401	.3038E+1	.3292E+0	.7033E-1	.3085E+1	.2259E-1
1.20	4.70	.868	0.468	0.132	3.764	2.879	9.262	0.4062	0.497	.2614E+1	.3825E+0	.6782E-1	.3763E+1	.2847E-1
1.20	5.60	.870	0.470	0.130	4.660	2.884	11.057	0.4089	0.606	.2279E+1	.4387E+0	.6529E-1	.4521E+1	.3512E-1
1.20	6.80	.872	0.472	0.128	5.856	2.889	13.451	0.4116	0.750	.1965E+1	.5090E+0	.6238E-1	.5536E+1	.4395E-1
1.20	8.20	.873	0.473	0.127	7.254	2.891	16.248	0.4129	0.921	.1707E+1	.5859E+0	.5954E-1	.6709E+1	.5433E-1
1.20	10.00	.874	0.474	0.126	9.052	2.894	19.845	0.4143	1.141	.1474E+1	.6786E+0	.5655E-1	.8221E+1	.6764E-1
1.20	12.00	.875	0.475	0.125	11.050	2.896	23.842	0.4156	1.381	.1291E+1	.7753E+0	.5384E-1	.9909E+1	.8238E-1
1.20	15.00	.875	0.475	0.125	14.050	2.896	29.842	0.4156	1.756	.1097E+1	.9112E+0	.5062E-1	.1240E+2	.1047E+0
1.20	18.00	.876	0.476	0.124	17.048	2.899	35.839	0.4170	2.114	.9629E+0	.1039E+1	.4808E-1	.1494E+2	.1268E+0
1.20	22.00	.876	0.476	0.124	21.048	2.899	43.839	0.4170	2.610	.8346E+0	.1198E+1	.4539E-1	.1828E+2	.1565E+0
1.20	27.00	.877	0.477	0.123	26.046	2.901	53.837	0.4183	3.204	.7218E+0	.1386E+1	.4276E-1	.2252E+2	.1932E+0
1.20	33.00	.877	0.477	0.123	32.046	2.901	65.837	0.4183	3.942	.6264E+0	.1596E+1	.4031E-1	.2754E+2	.2378E+0
1.50	0.33	.376	0.126	0.624	0.078	1.984	1.800	0.0474	0.049	.9271E+2	.1079E-1	.2179E-1	.8527E-1	.5506E-4
1.50	0.39	.399	0.149	0.601	0.092	2.040	1.854	0.0595	0.055	.6923E+2	.1444E-1	.2469E-1	.1102E+0	.9579E-4
1.50	0.47	.429	0.179	0.571	0.112	2.113	1.928	0.0768	0.064	.4997E+2	.2009E-1	.2858E-1	.1481E+0	.1785E-3
1.50	0.56	.463	0.213	0.537	0.134	2.196	2.011	0.0986	0.072	.3639E+2	.2748E-1	.3271E-1	.1983E+0	.3108E-3
1.50	0.68	.508	0.258	0.492	0.164	2.305	2.123	0.1311	0.081	.2569E+2	.3893E-1	.3816E-1	.2782E+0	.6014E-3
1.50	0.82	.558	0.308	0.442	0.204	2.426	2.260	0.1719	0.090	.1839E+2	.5437E-1	.4420E-1	.3883E+0	.1096E-2
1.50	1.00	.619	0.369	0.381	0.262	2.574	2.445	0.2284	0.100	.1300E+2	.7691E-1	.5128E-1	.5585E+0	.2023E-2
1.50	1.20	.678	0.428	0.322	0.344	2.718	2.677	0.2902	0.111	.9586E+1	.1043E+0	.5796E-1	.7767E+0	.3432E-2
1.50	1.50	.745	0.495	0.255	0.510	2.881	3.085	0.3688	0.130	.6815E+1	.1467E+0	.6522E-1	.1138E+1	.6140E-2
1.50	1.80	.784	0.534	0.216	0.732	2.975	3.574	0.4187	0.158	.5337E+1	.1874E+0	.6940E-1	.1496E+1	.9314E-2
1.50	2.20	.810	0.560	0.190	1.080	3.038	4.299	0.4536	0.205	.4217E+1	.2371E+0	.7186E-1	.1950E+1	.1390E-1
1.50	2.70	.826	0.576	0.174	1.548	3.077	5.254	0.4758	0.269	.3409E+1	.2934E+0	.7243E-1	.2500E+1	.1981E-1
1.50	3.30	.835	0.585	0.165	2.130	3.099	6.428	0.4885	0.351	.2819E+1	.3548E+0	.7167E-1	.3140E+1	.2706E-1
1.50	3.90	.840	0.590	0.160	2.720	3.111	7.614	0.4956	0.435	.2430E+1	.4115E+0	.7034E-1	.3773E+1	.3436E-1
1.50	4.70	.844	0.594	0.156	3.512	3.121	9.202	0.5013	0.548	.2075E+1	.4820E+0	.6836E-1	.4613E+1	.4412E-1
1.50	5.60	.847	0.597	0.153	4.406	3.128	10.994	0.5057	0.674	.1799E+1	.5560E+0	.6619E-1	.5559E+1	.5510E-1

CASE # 3   COPYRIGHT 1982 STEVE SMITH   DATA DATE 1/3/82

TRANSFORMER DESIGN TABLE

$$f(D)=[(U/(FG(DE)^2))^5 * (1/(DEUm))^4]^{(1/13)}$$

P	Q	D	E	F	G	U	Um	DE	FG	f(D)	1/f(D)	1/(PQf(D))	DEUm	FG(DE)^2/U
1.50	6.80	.850	0.600	0.150	5.600	3.136	13.385	0.5100	0.840	.1543E+1	.6482E+0	.6355E-1	.6826E+1	.6968E-1
1.50	8.20	.852	0.602	0.148	6.996	3.140	16.179	0.5129	1.035	.1335E+1	.7488E+0	.6888E-1	.8298E+1	.8673E-1
1.50	10.00	.853	0.603	0.147	8.794	3.143	19.776	0.5144	1.293	.1150E+1	.8699E+0	.5799E-1	.1017E+2	.1088E+0
1.50	12.00	.854	0.604	0.146	10.792	3.145	23.774	0.5158	1.576	.1004E+1	.9962E+0	.5534E-1	.1226E+2	.1333E+0
1.50	15.00	.855	0.605	0.145	13.790	3.148	29.771	0.5173	2.000	.8524E+0	.1173E+1	.5214E-1	.1540E+2	.1700E+0
1.50	18.00	.856	0.606	0.144	16.788	3.150	35.768	0.5187	2.417	.7468E+0	.1339E+1	.4959E-1	.1855E+2	.2065E+0
1.50	22.00	.856	0.606	0.144	20.788	3.150	43.768	0.5187	2.993	.6464E+0	.1547E+1	.4688E-1	.2270E+2	.2557E+0
1.50	27.00	.857	0.607	0.143	25.786	3.153	53.765	0.5202	3.687	.5585E+0	.1790E+1	.4421E-1	.2797E+2	.3165E+0
1.50	33.00	.857	0.607	0.143	31.786	3.153	65.765	0.5202	4.545	.4844E+0	.2065E+1	.4171E-1	.3421E+2	.3902E+0
1.80	0.33	.231	0.131	0.769	0.068	1.932	2.086	0.0303	0.052	.1382E+3	.7233E-2	.1218E-1	.6311E-1	.2479E-4
1.80	0.39	.255	0.155	0.745	0.080	1.990	2.137	0.0395	0.060	.9899E+2	.1010E-1	.1439E-1	.8446E-1	.4677E-4
1.80	0.47	.288	0.188	0.712	0.094	2.070	2.203	0.0541	0.067	.6786E+2	.1474E-1	.1742E-1	.1193E+0	.9477E-4
1.80	0.56	.323	0.223	0.677	0.114	2.155	2.283	0.0720	0.077	.4745E+2	.2107E-1	.2091E-1	.1644E+0	.1858E-3
1.80	0.68	.370	0.270	0.630	0.140	2.270	2.388	0.0999	0.088	.3188E+2	.3137E-1	.2563E-1	.2386E+0	.3876E-3
1.80	0.82	.424	0.324	0.576	0.172	2.401	2.514	0.1374	0.099	.2176E+2	.4595E-1	.3113E-1	.3453E+0	.7788E-3
1.80	1.00	.490	0.390	0.510	0.220	2.561	2.685	0.1911	0.112	.1461E+2	.6847E-1	.3804E-1	.5131E+0	.1600E-2
1.80	1.20	.560	0.460	0.440	0.280	2.731	2.885	0.2576	0.123	.1024E+2	.9764E-1	.4520E-1	.7432E+0	.2993E-2
1.80	1.50	.650	0.550	0.350	0.400	2.950	3.228	0.3575	0.140	.6817E+1	.1467E+0	.5433E-1	.1154E+1	.6066E-2
1.80	1.80	.716	0.616	0.284	0.568	3.110	3.639	0.4411	0.161	.5064E+1	.1975E+0	.6095E-1	.1605E+1	.1009E-1
1.80	2.20	.766	0.666	0.234	0.868	3.232	4.296	0.5102	0.203	.3821E+1	.2617E+0	.6609E-1	.2192E+1	.1636E-1
1.80	2.70	.794	0.694	0.206	1.312	3.300	5.216	0.5510	0.270	.2992E+1	.3342E+0	.6877E-1	.2874E+1	.2487E-1
1.80	3.30	.818	0.718	0.190	1.880	3.338	6.371	0.5751	0.357	.2424E+1	.4125E+0	.6944E-1	.3664E+1	.3539E-1
1.80	3.90	.821	0.721	0.182	2.464	3.358	7.548	0.5873	0.448	.2066E+1	.4841E+0	.6896E-1	.4433E+1	.4607E-1
1.80	4.70	.824	0.724	0.176	3.252	3.372	9.131	0.5966	0.572	.1747E+1	.5724E+0	.6766E-1	.5447E+1	.6040E-1
1.80	5.60	.828	0.728	0.172	4.144	3.382	10.919	0.6028	0.713	.1505E+1	.6647E+0	.6594E-1	.6582E+1	.7657E-1
1.80	6.80	.831	0.731	0.169	5.338	3.389	13.311	0.6075	0.902	.1283E+1	.7792E+0	.6366E-1	.8008E+1	.9821E-1
1.80	8.20	.834	0.734	0.166	6.732	3.397	16.102	0.6124	1.118	.1106E+1	.9039E+0	.6124E-1	.9857E+1	.1233E+0
1.80	10.00	.836	0.736	0.164	8.528	3.402	19.696	0.6153	1.399	.9491E+0	.1054E+1	.5853E-1	.1212E+2	.1557E+0
1.80	12.00	.837	0.737	0.163	10.526	3.404	23.693	0.6169	1.716	.8268E+0	.1209E+1	.5599E-1	.1462E+2	.1918E+0
1.80	15.00	.839	0.739	0.161	13.522	3.409	29.688	0.6200	2.177	.7004E+0	.1428E+1	.5288E-1	.1841E+2	.2455E+0
1.80	18.00	.839	0.739	0.161	16.522	3.409	35.688	0.6200	2.660	.6128E+0	.1632E+1	.5037E-1	.2211E+2	.3000E+0
1.80	22.00	.840	0.740	0.160	20.520	3.411	43.685	0.6216	3.283	.5297E+0	.1888E+1	.4767E-1	.2715E+2	.3719E+0
1.80	27.00	.841	0.741	0.159	25.518	3.414	53.682	0.6232	4.057	.4572E+0	.2187E+1	.4501E-1	.3345E+2	.4616E+0
1.80	33.00	.841	0.741	0.159	31.518	3.414	65.682	0.6232	5.011	.3961E+0	.2524E+1	.4250E-1	.4093E+2	.5701E+0
2.20	0.47	.107	0.207	0.893	0.056	2.031	2.548	0.0221	0.050	.1886E+3	.5301E-2	.5127E-2	.5644E-1	.1208E-4
2.20	0.56	.144	0.244	0.856	0.072	2.121	2.623	0.0351	0.062	.1067E+3	.9369E-2	.7605E-2	.9215E-1	.3588E-4

CASE # 3   COPYRIGHT 1982 STEVE SMITH   DATA DATE 1/ 3/82   TRANSFORMER DESIGN TABLE

$$f(D)=[(U/(FG(DE)^2))^5 * (1/(DEUm))^4]^{(1/13)}$$

P	Q	D	E	F	G	U	Um	DE	FG	f(D)	1/f(D)	1/(PQf(D))	DEUm	FG(DE)^2/U
2.20	0.68	.192	0.292	0.808	0.096	2.237	2.725	0.0561	0.078	.5959E+2	.1678E-1	.1122E-1	.1528E+0	.1090E-3
2.20	0.82	.248	0.348	0.752	0.124	2.373	2.845	0.0863	0.093	.3522E+2	.2840E-1	.1574E-1	.2456E+0	.2927E-3
2.20	1.00	.319	0.419	0.681	0.162	2.546	3.002	0.1337	0.110	.2083E+2	.4802E-1	.2183E-1	.4013E+0	.7742E-3
2.20	1.20	.395	0.495	0.605	0.210	2.730	3.185	0.1955	0.127	.1321E+2	.7570E-1	.2867E-1	.6228E+0	.1779E-2
2.20	1.50	.502	0.602	0.498	0.296	2.990	3.479	0.3022	0.147	.7867E+1	.1271E+0	.3852E-1	.1051E+1	.4502E-2
2.20	1.80	.596	0.696	0.404	0.408	3.219	3.811	0.4148	0.165	.5360E+1	.1866E+0	.4711E-1	.1581E+1	.8812E-2
2.20	2.20	.687	0.787	0.313	0.626	3.440	4.350	0.5407	0.196	.3713E+1	.2693E+0	.5564E-1	.2352E+1	.1665E-1
2.20	2.70	.747	0.847	0.253	1.006	3.585	5.179	0.6327	0.255	.2730E+1	.3663E+0	.6166E-1	.3277E+1	.2842E-1
2.20	3.30	.777	0.877	0.223	1.546	3.658	6.293	0.6814	0.345	.2129E+1	.4698E+0	.6471E-1	.4288E+1	.4376E-1
2.20	3.90	.791	0.891	0.209	2.118	3.692	7.453	0.7048	0.443	.1776E+1	.5629E+0	.6561E-1	.5253E+1	.5955E-1
2.20	4.70	.800	0.900	0.200	2.900	3.714	9.207	0.7200	0.580	.1478E+1	.6764E+0	.6542E-1	.6500E+1	.8095E-1
2.20	5.60	.807	0.907	0.193	3.786	3.731	10.807	0.7319	0.731	.1260E+1	.7939E+0	.6444E-1	.7910E+1	.1049E+0
2.20	6.80	.811	0.911	0.189	4.978	3.741	13.196	0.7388	0.941	.1065E+1	.9389E+0	.6276E-1	.9749E+1	.1373E+0
2.20	8.20	.815	0.915	0.185	6.370	3.751	15.985	0.7457	1.178	.9125E+0	.1096E+1	.6075E-1	.1192E+2	.1747E+0
2.20	10.00	.817	0.917	0.183	8.166	3.755	19.579	0.7492	1.494	.7789E+0	.1284E+1	.5835E-1	.1467E+2	.2233E+0
2.20	12.00	.819	0.919	0.181	10.162	3.760	23.573	0.7527	1.839	.6762E+0	.1479E+1	.5602E-1	.1774E+2	.2771E+0
2.20	15.00	.821	0.921	0.179	13.158	3.765	29.567	0.7561	2.355	.5799E+0	.1752E+1	.5388E-1	.2236E+2	.3577E+0
2.20	18.00	.822	0.922	0.178	16.156	3.768	35.565	0.7579	2.876	.4984E+0	.2007E+1	.5067E-1	.2695E+2	.4384E+0
2.20	22.00	.823	0.923	0.177	20.154	3.770	43.562	0.7596	3.567	.4300E+0	.2326E+1	.4805E-1	.3308E+2	.5460E+0
2.20	27.00	.823	0.923	0.177	25.154	3.770	53.562	0.7596	4.452	.3705E+0	.2699E+1	.4543E-1	.4069E+2	.6615E+0
2.20	33.00	.824	0.924	0.176	31.152	3.772	65.559	0.7614	5.483	.3207E+0	.3118E+1	.4295E-1	.4991E+2	.8425E+0
2.70	1.00	.114	0.464	0.886	0.072	2.548	3.374	0.0529	0.064	.6733E+2	.1485E-1	.5501E-2	.1785E+0	.7006E-4
2.70	1.20	.191	0.541	0.809	0.118	2.835	3.639	0.1433	0.095	.2835E+2	.3527E-1	.1089E-1	.3672E+0	.3727E-3
2.70	1.50	.306	0.656	0.694	0.188	3.014	3.825	0.2007	0.130	.1248E+2	.8013E-1	.1978E-1	.7678E+0	.1744E-2
2.70	1.80	.415	0.765	0.585	0.270	3.279	4.113	0.3175	0.158	.7149E+1	.1399E+0	.2878E-1	.1306E+1	.4855E-2
2.70	2.20	.545	0.895	0.455	0.410	3.595	4.542	0.4878	0.187	.4243E+1	.2357E+0	.3967E-1	.2215E+1	.1235E-1
2.70	2.70	.663	1.013	0.337	0.674	3.881	5.204	0.6716	0.227	.2753E+1	.3632E+0	.4982E-1	.3495E+1	.2640E-1
2.70	3.30	.730	1.080	0.270	1.140	4.044	6.213	0.7884	0.308	.1983E+1	.5043E+0	.5660E-1	.4898E+1	.4731E-1
2.70	3.90	.758	1.108	0.242	1.684	4.112	7.333	0.8399	0.408	.1590E+1	.6288E+0	.5971E-1	.6159E+1	.6999E-1
2.70	4.70	.775	1.125	0.225	2.450	4.153	8.884	0.8719	0.551	.1287E+1	.7770E+0	.6123E-1	.7746E+1	.1009E+0
2.70	5.60	.784	1.134	0.216	3.332	4.175	10.659	0.8891	0.720	.1078E+1	.9280E+0	.6138E-1	.9476E+1	.1362E+0
2.70	6.80	.791	1.141	0.209	4.518	4.192	13.039	0.9025	0.944	.8991E+0	.1112E+1	.6058E-1	.1177E+2	.1835E+0
2.70	8.20	.796	1.146	0.204	5.908	4.204	15.824	0.9122	1.205	.7632E+0	.1310E+1	.5918E-1	.1444E+2	.2383E+0
2.70	10.00	.799	1.149	0.201	7.702	4.212	19.416	0.9181	1.548	.6469E+0	.1546E+1	.5726E-1	.1782E+2	.3098E+0
2.70	12.00	.802	1.152	0.198	9.696	4.219	23.647	0.9239	1.920	.5587E+0	.1790E+1	.5524E-1	.2163E+2	.3884E+0
2.70	15.00	.804	1.154	0.196	12.692	4.224	29.401	0.9278	2.488	.4695E+0	.2130E+1	.5259E-1	.2728E+2	.5070E+0

CASE # 3   COPYRIGHT 1982 STEVE SMITH   DATA DATE 1/ 3/82

TRANSFORMER DESIGN TABLE

$$f(D)=[(U/(FG(DE)^2))^5 * (1/(DEUm))^4]^{(1/13)}$$

P	Q	D	E	F	G	U	Um	DE	FG	f(D)	1/f(D)	1/(PQf(D))	DEUm	FG(DE)^2/U
2.70	18.00	.805	1.155	0.195	15.690	4.226	35.399	0.9298	3.060	.4087E+0	.2447E+1	.5035E-1	.3291E+2	.6258E+0
2.70	22.00	.806	1.156	0.194	19.688	4.229	43.396	0.9317	3.819	.3518E+0	.2843E+1	.4786E-1	.4043E+2	.7841E+0
2.70	27.00	.807	1.157	0.193	24.686	4.231	53.393	0.9337	4.764	.3026E+0	.3306E+1	.4535E-1	.4985E+2	.9817E+0
3.30	33.00	.808	1.158	0.192	30.684	4.234	65.390	0.9357	5.891	.2614E+0	.3826E+1	.4294E-1	.6118E+2	.1218E+1
3.30	1.50	.074	0.724	0.926	0.052	3.051	4.231	0.0536	0.048	.7397E+2	.1352E-1	.2731E-1	.2267E+0	.4531E-4
3.30	1.80	.187	0.837	0.813	0.126	3.325	4.508	0.1565	0.102	.1768E+2	.5656E-1	.9523E-2	.7055E+0	.7547E-3
3.30	2.20	.334	0.984	0.666	0.232	3.682	4.887	0.3287	0.155	.6888E+1	.1452E+0	.2000E-1	.1606E+1	.4533E-2
3.30	2.70	.500	1.150	0.500	0.400	4.085	5.413	0.5750	0.200	.3444E+1	.2904E+0	.3259E-1	.3112E+1	.1619E-1
3.30	3.30	.643	1.293	0.357	0.714	4.433	6.204	0.8314	0.255	.2087E+1	.4792E+0	.4400E-1	.5158E+1	.3975E-1
3.30	3.90	.711	1.361	0.289	1.178	4.598	7.210	0.9677	0.340	.1535E+1	.6513E+0	.5061E-1	.6977E+1	.6933E-1
3.30	4.70	.746	1.396	0.254	1.908	4.683	8.710	1.0414	0.485	.1177E+1	.8498E+0	.5479E-1	.9070E+1	.1122E+0
3.30	5.60	.762	1.412	0.238	2.776	4.722	10.464	1.0759	0.661	.9561E+0	.1046E+1	.5660E-1	.1126E+2	.1620E+0
3.30	6.80	.772	1.422	0.228	3.956	4.746	12.835	1.0978	0.902	.7811E+0	.1280E+1	.5706E-1	.1409E+2	.2290E+0
3.30	8.20	.779	1.429	0.221	5.342	4.763	15.615	1.1132	1.181	.6548E+0	.1529E+1	.5655E-1	.1738E+2	.3071E+0
3.30	10.00	.783	1.433	0.217	7.134	4.773	19.204	1.1220	1.548	.5487E+0	.1823E+1	.5523E-1	.2155E+2	.4083E+0
3.30	12.00	.786	1.436	0.214	9.128	4.780	23.195	1.1287	1.953	.4707E+0	.2125E+1	.5360E-1	.2618E+2	.5206E+0
3.30	15.00	.789	1.439	0.211	12.122	4.787	29.187	1.1354	2.558	.3931E+0	.2544E+1	.5139E-1	.3314E+2	.6887E+0
3.30	18.00	.791	1.441	0.209	15.118	4.792	35.181	1.1398	3.160	.3409E+0	.2934E+1	.4939E-1	.4010E+2	.8566E+0
3.30	22.00	.792	1.442	0.208	19.116	4.795	43.178	1.1421	3.976	.2924E+0	.3420E+1	.4711E-1	.4931E+2	.1082E+1
3.30	27.00	.793	1.443	0.207	24.114	4.797	53.175	1.1443	4.992	.2508E+0	.3987E+1	.4475E-1	.6085E+2	.1362E+1
3.30	33.00	.794	1.444	0.206	30.112	4.800	65.172	1.1465	6.203	.2163E+0	.4624E+1	.4246E-1	.7472E+2	.1699E+1
3.90	2.20	.111	1.061	0.889	0.078	3.740	5.267	0.1178	0.069	.2783E+2	.3593E-1	.4188E-2	.6203E+0	.2571E-3
3.90	2.70	.294	1.244	0.706	0.212	4.185	5.744	0.3657	0.150	.6211E+1	.1610E+0	.1529E-1	.2101E+1	.4784E-2
3.90	3.30	.491	1.441	0.509	0.418	4.664	6.381	0.7075	0.213	.2691E+1	.3717E+0	.2888E-1	.4515E+1	.2284E-1
3.90	3.90	.630	1.580	0.370	0.740	5.001	7.184	0.9954	0.274	.1675E+1	.5972E+0	.3926E-1	.7151E+1	.5424E-1
3.90	4.70	.710	1.660	0.290	1.380	5.196	8.555	1.1786	0.400	.1160E+1	.8619E+0	.4702E-1	.1008E+2	.1070E+0
3.90	5.60	.740	1.690	0.260	2.220	5.268	10.269	1.2506	0.577	.8985E+0	.1111E+1	.5096E-1	.1284E+2	.1713E+0
3.90	6.80	.757	1.707	0.243	3.386	5.310	12.621	1.2922	0.823	.7125E+0	.1404E+1	.5293E-1	.1631E+2	.2588E+0
3.90	8.20	.766	1.716	0.234	4.768	5.332	15.395	1.3145	1.116	.5862E+0	.1706E+1	.5334E-1	.2024E+2	.3616E+0
3.90	10.00	.771	1.721	0.229	6.558	5.344	18.981	1.3269	1.502	.4857E+0	.2059E+1	.5279E-1	.2519E+2	.4948E+0
3.90	12.00	.775	1.725	0.225	8.550	5.353	22.969	1.3369	1.924	.4134E+0	.2419E+1	.5169E-1	.3071E+2	.6422E+0
3.90	15.00	.778	1.728	0.222	11.544	5.361	28.961	1.3444	2.563	.3428E+0	.2917E+1	.4986E-1	.3893E+2	.8640E+0
3.90	18.00	.780	1.730	0.220	14.540	5.366	34.955	1.3494	3.199	.2960E+0	.3378E+1	.4812E-1	.4717E+2	.1086E+1
3.90	22.00	.782	1.732	0.218	18.536	5.370	42.949	1.3544	4.041	.2530E+0	.3952E+1	.4606E-1	.5817E+2	.1380E+1
3.90	27.00	.783	1.733	0.217	23.534	5.373	52.946	1.3569	5.107	.2164E+0	.4620E+1	.4388E-1	.7185E+2	.1758E+1
3.90	33.00	.784	1.734	0.216	29.532	5.375	64.944	1.3595	6.379	.1862E+0	.5369E+1	.4172E-1	.8829E+2	.2193E+1

CASE # 3   DATA DATE 1/ 3/82

TRANSFORMER DESIGN TABLE

$$f(D) = [(U/(FG(DE)^2))^5 * (1/(DEUm))^4]^{1/13}$$

P	Q	D	E	F	G	U	Um	DE	FG	f(D)	1/f(D)	1/(PQf(D))	DEUm	FG(DE)^2/U
4.70	3.30	.219	1.569	0.781	0.162	4.803	6.815	0.3436	0.127	.7089E+1	.1411E+0	.9095E-2	.2342E+1	.3110E-2
4.70	3.90	.424	1.774	0.576	0.352	5.301	7.429	0.7522	0.203	.2573E+1	.3887E+0	.2121E-1	.5588E+1	.2164E-1
4.70	4.70	.619	1.969	0.381	0.762	5.774	8.472	1.2188	0.290	.1333E+1	.7561E+0	.3423E-1	.1033E+2	.7469E-1
4.70	5.60	.703	2.053	0.297	1.494	5.979	10.032	1.4433	0.444	.9010E+0	.1110E+1	.4217E-1	.1448E+2	.1546E+0
4.70	6.80	.737	2.087	0.263	2.626	6.061	12.335	1.5381	0.691	.6694E+0	.1494E+1	.4674E-1	.1897E+2	.2696E+0
4.70	8.20	.751	2.101	0.249	3.998	6.095	15.495	1.5779	0.996	.5329E+0	.1876E+1	.4869E-1	.2382E+2	.4066E+0
4.70	10.00	.759	2.109	0.241	5.782	6.115	18.672	1.6007	1.393	.4324E+0	.2313E+1	.4921E-1	.2999E+2	.5839E+0
4.70	12.00	.764	2.114	0.236	7.772	6.127	22.657	1.6151	1.834	.3633E+0	.2753E+1	.4881E-1	.3659E+2	.7809E+0
4.70	15.00	.768	2.118	0.232	10.764	6.136	28.646	1.6266	2.497	.2981E+0	.3355E+1	.4759E-1	.4660E+2	.1077E+1
4.70	18.00	.770	2.120	0.230	13.760	6.141	34.640	1.6324	3.165	.2558E+0	.3910E+1	.4622E-1	.5655E+2	.1373E+1
4.70	22.00	.772	2.122	0.228	17.756	6.146	42.634	1.6382	4.048	.2175E+0	.4598E+1	.4447E-1	.6984E+2	.1768E+1
4.70	27.00	.773	2.123	0.227	22.754	6.149	52.632	1.6411	5.165	.1853E+0	.5397E+1	.4253E-1	.8637E+2	.2262E+1
4.70	33.00	.774	2.124	0.226	28.752	6.151	64.629	1.6440	6.498	.1590E+0	.6291E+1	.4056E-1	.1062E+3	.2855E+1
5.60	3.90	.110	1.910	0.890	0.080	7.940	7.940	0.2101	0.071	.1503E+2	.6652E-1	.3046E-2	.1668E+1	.5780E-3
5.60	4.70	.389	2.189	0.611	0.322	8.743	8.743	0.8515	0.197	.2288E+1	.4370E+0	.1660E-1	.7445E+1	.2333E-1
5.60	5.60	.611	2.411	0.389	0.778	6.655	9.908	1.4731	0.303	.1068E+1	.9363E+0	.2986E-1	.1460E+2	.9869E-1
5.60	6.80	.707	2.507	0.293	1.786	6.888	12.034	1.7724	0.523	.6767E+0	.1478E+1	.3881E-1	.2133E+2	.2387E+0
5.60	8.20	.735	2.535	0.265	3.130	6.956	14.754	1.8632	0.829	.5064E+0	.1975E+1	.4300E-1	.2749E+2	.4139E+0
5.60	10.00	.748	2.548	0.252	4.904	6.988	18.317	1.9059	1.236	.3973E+0	.2517E+1	.4494E-1	.3491E+2	.6424E+0
5.60	12.00	.754	2.554	0.246	6.892	7.002	22.300	1.9257	1.695	.3278E+0	.3051E+1	.4548E-1	.4294E+2	.8979E+0
5.60	15.00	.759	2.559	0.241	9.882	7.015	28.285	1.9423	2.382	.2658E+0	.3773E+1	.4492E-1	.5494E+2	.1281E+1
5.60	18.00	.762	2.562	0.238	12.876	7.022	34.277	1.9522	3.064	.2256E+0	.4433E+1	.4398E-1	.6692E+2	.1663E+1
5.60	22.00	.764	2.564	0.236	16.872	7.027	42.271	1.9589	3.982	.1906E+0	.5247E+1	.4259E-1	.8280E+2	.2174E+1
5.60	27.00	.766	2.566	0.234	21.868	7.032	52.265	1.9656	5.117	.1616E+0	.6189E+1	.4094E-1	.1027E+3	.2812E+1
5.60	33.00	.767	2.567	0.233	27.866	7.034	64.262	1.9689	6.493	.1381E+0	.7241E+1	.3918E-1	.1265E+3	.3578E+1
6.80	5.60	.287	2.687	0.713	0.226	7.068	10.319	0.7712	0.161	.2762E+1	.3621E+0	.9508E-2	.7958E+1	.1356E+0
6.80	6.80	.605	3.005	0.395	0.790	7.840	11.811	1.8180	0.312	.8492E+0	.1178E+1	.2547E-1	.2147E+2	.1315E+0
6.80	8.20	.706	3.106	0.294	1.988	8.086	14.322	2.1928	0.584	.5199E+0	.1924E+1	.3450E-1	.3141E+2	.3476E+0
6.80	10.00	.733	3.133	0.267	3.734	8.151	17.845	2.2965	0.997	.3776E+0	.2648E+1	.3894E-1	.4098E+2	.6458E+0
6.80	12.00	.744	3.144	0.256	5.712	8.178	21.813	2.3391	1.462	.3007E+0	.3325E+1	.4075E-1	.5102E+2	.9783E+0
6.80	15.00	.751	3.151	0.249	8.698	8.195	27.793	2.3664	2.166	.2372E+0	.4216E+1	.4133E-1	.6577E+2	.1480E+1
6.80	18.00	.754	3.154	0.246	11.692	8.202	33.785	2.3781	2.876	.1993E+0	.5018E+1	.4099E-1	.8034E+2	.1983E+1
6.80	22.00	.757	3.157	0.243	15.686	8.212	41.776	2.3898	3.812	.1667E+0	.5999E+1	.4014E-1	.9984E+2	.2652E+1
6.80	27.00	.758	3.158	0.242	20.684	8.212	51.773	2.3938	5.006	.1403E+0	.7128E+1	.3882E-1	.1239E+3	.3493E+1
6.80	33.00	.760	3.160	0.240	26.680	8.217	63.767	2.4016	6.403	.1193E+0	.8382E+1	.3735E-1	.1531E+3	.4495E+1
8.20	6.80	.217	3.317	0.783	0.166	8.298	12.319	0.7198	0.130	.3255E+1	.3073E+0	.5510E-2	.8867E+1	.8115E-2

CASE # 3   COPYRIGHT 1982 STEVE SMITH   DATA DATE 1/ 3/82   TRANSFORMER DESIGN TABLE

$$f(D)=[(U/(FG(DE)^2))^5 * (1/(DEUm))^4]^{(1/13)}$$

P	Q	D	E	F	G	U	Um	DE	FG	f(D)	1/f(D)	1/(PQf(D))	DEUm	FG(DE)^2/U
8.20	8.20	.600	3.700	0.400	0.800	9.228	14.024	2.2200	0.320	.6849E+0	.1460E+1	.2171E-1	.3113E+2	.1709E+0
8.20	10.00	.710	3.810	0.290	2.380	9.496	17.389	2.7051	0.690	.3904E+0	.2562E+1	.3124E-1	.4682E+2	.5319E+0
8.20	12.00	.732	3.832	0.268	4.336	9.549	21.247	2.8050	1.162	.2891E+0	.3459E+1	.3515E-1	.5960E+2	.9575E+0
8.20	15.00	.743	3.843	0.257	7.314	9.576	27.215	2.8553	1.880	.2187E+0	.4573E+1	.3718E-1	.7771E+2	.1600E+1
8.20	18.00	.747	3.847	0.253	10.306	9.585	33.204	2.8737	2.607	.1802E+0	.5550E+1	.3760E-1	.9542E+2	.2246E+1
8.20	22.00	.751	3.851	0.249	14.298	9.595	41.192	2.8921	3.560	.1486E+0	.6729E+1	.3730E-1	.1191E+3	.3103E+1
8.20	27.00	.753	3.853	0.247	19.294	9.600	51.187	2.9013	4.766	.1239E+0	.8074E+1	.3647E-1	.1485E+3	.4179E+1
8.20	33.00	.754	3.854	0.246	25.292	9.602	63.184	2.9059	6.222	.1046E+0	.9560E+1	.3533E-1	.1836E+3	.5471E+1
10.00	8.20	.073	4.073	0.927	0.054	9.748	14.758	2.2973	0.050	.1225E+2	.8163E-1	.9955E-3	.4338E+1	.4540E-3
10.00	10.00	.596	4.596	0.404	0.808	11.019	16.863	2.7392	0.326	.5483E+0	.1824E+1	.1824E-1	.4619E+2	.2223E+0
10.00	12.00	.709	4.709	0.291	2.582	11.293	20.540	3.3387	0.751	.3055E+0	.3274E+1	.2728E-1	.6858E+2	.7416E+0
10.00	15.00	.733	4.733	0.267	5.534	11.351	26.471	3.4693	1.478	.2094E+0	.4775E+1	.3184E-1	.9184E+2	.1567E+1
10.00	18.00	.740	4.740	0.260	8.520	11.368	32.451	3.5076	2.215	.1664E+0	.6008E+1	.3338E-1	.1138E+3	.2397E+1
10.00	22.00	.745	4.745	0.255	12.510	11.381	40.437	3.5350	3.190	.1341E+0	.7456E+1	.3389E-1	.1429E+3	.3503E+1
10.00	27.00	.748	4.748	0.252	17.504	11.388	50.428	3.5515	4.411	.1101E+0	.9083E+1	.3364E-1	.1791E+3	.4886E+1
10.00	33.00	.749	4.749	0.251	23.502	11.390	62.425	3.5570	5.899	.9205E-1	.1086E+2	.3292E-1	.2220E+3	.6553E+1

CASE # 4   COPYRIGHT 1982 STEVE SMITH   DATA DATE 1/ 3/82   TRANSFORMER DESIGN TABLE

SCALE BY 1.00E-04   FUNCTION   1/F(D) $f(D) = [(U/(FG(DE)^2)]^5 * (1/(DEUm))^4]^{(1/13)}$

Q/P	0.33	0.39	0.47	0.56	0.68	0.82	1.00	1.20	1.50	1.80	2.20	2.70	3.30	3.90	4.70	5.60	6.80	8.20	10.00
0.33	64	77	93	108	124	139	153	161	165	158	135	88	.	.	.	.	.	.	.
0.39	75	91	111	131	154	175	194	207	213	206	179	123	64	.	.	.	.	.	.
0.47	89	108	134	161	192	223	251	272	284	278	248	178	114	95	.	.	.	.	.
0.56	103	127	158	192	234	275	317	348	371	369	336	254	160	120	90	.	.	.	.
0.68	120	149	188	230	285	342	404	454	496	504	472	377	203	180	150	120	.	.	.
0.82	140	173	220	272	340	415	500	575	648	675	653	549	341	260	210	180	130	.	.
1.00	163	203	258	321	405	501	614	723	844	908	914	814	574	440	360	290	240	190	.
1.20	187	233	298	372	473	589	732	876	1052	1169	1226	1154	900	760	620	500	400	310	240
1.50	220	276	353	443	566	711	893	1085	1339	1539	1701	1721	1501	1280	1080	900	740	600	480
1.80	251	316	405	510	653	824	1042	1276	1601	1878	2150	2304	2190	1950	1700	1460	1230	1020	830
2.20	291	365	470	592	762	964	1226	1513	1921	2290	2697	3037	3141	2900	2630	2340	2040	1740	1450
2.70	337	424	546	689	888	1128	1440	1786	2290	2760	3315	3860	4248	4050	3780	3450	3080	2680	2280
3.30	389	489	631	798	1030	1311	1679	2091	2699	3279	3989	4746	5424	5400	5180	4820	4380	3870	3340
3.90	438	551	711	900	1164	1483	1904	2377	3081	3762	4612	5556	6479	6650	6550	6250	5770	5180	4540
4.70	499	629	812	1029	1332	1700	2187	2736	3560	4365	5387	6554	7763	8100	8180	7940	7430	6760	6000
5.60	565	712	920	1166	1511	1930	2486	3116	4066	5001	6201	7598	9092	9700	9850	9500	8500	7000	5132
6.80	647	816	1055	1338	1735	2219	2862	3593	4699	5796	7216	8893	10731	12200	13600	14500	14700	14600	14346
8.20	738	931	1203	1527	1982	2536	3274	4115	5393	6665	8324	10302	12506	14500	16800	19000	20900	21900	22483
10.00	848	1070	1383	1756	2280	2919	3773	4747	6230	7714	9657	11995	14631	17065	19989	22844	25973	28640	30462
12.00	963	1215	1572	1996	2592	3321	4294	5407	7105	8808	11048	13758	16838	19712	23221	26740	30759	34525	37948
15.00	1125	1420	1837	2334	3032	3886	5028	6335	8335	10346	13000	16227	19923	23406	27716	32123	37117	43381	47813
18.00	1277	1612	2086	2651	3445	4416	5717	7207	9489	11787	14828	18538	22807	26853	31898	37324	42491	49779	56705
22.00	1468	1853	2399	3049	3963	5082	6581	8300	10936	13595	17121	21434	26416	31162	37118	43335	50896	58776	67592
27.00	1692	2137	2767	3517	4572	5864	7596	9584	12634	15716	19808	24825	30640	36200	43213	50583	59633	69198	80130
33.00	1945	2457	3181	4043	5257	6745	8739	11029	14547	18103	22832	28641	35389	41860	50053	58707	69407	80828	94067

CASE # 4   COPYRIGHT 1982 STEVE SMITH   DATA DATE 1/ 3/82   TRANSFORMER DESIGN TABLE

SCALE BY 1.00E-04   FUNCTION   1/[P*Q*f(D)]   f(D)=[(U/(FG(DE)^2))^5 * (1/(DEUm))^4]^(1/13)

Q/P	0.33	0.39	0.47	0.56	0.68	0.82	1.00	1.20	1.50	1.80	2.20	2.70	3.30	3.90	4.70	5.60	6.80	8.20	10.00
0.33	591	599	597	583	554	514	462	407	332	265	186	99							
0.39	584	599	606	601	580	546	497	442	364	294	209	116							
0.47	572	592	607	611	602	577	534	481	403	329	240	141	42						
0.56	557	580	600	612	614	600	566	518	442	366	273	168	61						
0.68	537	562	587	605	616	614	594	556	486	412	316	205	90						
0.82	516	542	570	592	610	617	609	585	527	458	362	248	126	30					
1.00	493	520	549	573	596	611	614	603	563	505	415	301	174	66					
1.20	471	498	528	554	579	598	610	608	584	541	464	356	227	112					
1.50	445	471	501	528	555	578	595	603	595	570	515	425	303	185	52				
1.80	423	449	479	506	534	558	579	591	593	579	543	474	369	255	113				
2.20	401	426	454	481	509	534	557	573	582	578	557	511	433	336	199	63			
2.70	378	402	430	456	484	509	533	551	565	568	558	529	477	407	293	156			
3.30	357	380	407	432	459	484	508	528	545	552	549	533	498	451	371	259	101		
3.90	340	362	388	412	439	464	485	508	527	536	538	528	503	470	412	331	196	13	
4.70	322	343	368	391	417	441	464	485	505	516	521	517	501	477	436	381	288	101	
5.60	306	326	349	372	397	420	444	464	484	496	503	502	492	475	445	405	341	196	41
6.80	288	308	330	351	375	398	421	440	461	473	482	484	478	466	444	415	370	288	151
8.00	273	291	312	333	355	377	399	418	438	452	461	465	462	454	437	408	381	309	248
10.00	257	274	294	314	335	356	377	396	415	429	439	444	443	438	425	412	382	349	305
12.00	243	260	279	297	318	337	358	375	394	408	419	425	425	421	412	398	377	351	316
15.00	227	243	261	278	297	316	335	352	370	383	394	401	402	401	393	382	366	345	319
18.00	215	230	247	263	281	299	318	334	351	364	374	381	384	383	377	368	354	337	315
22.00	202	216	232	247	265	282	299	314	331	343	354	361	364	363	359	352	340	326	307
27.00	191	203	218	233	249	265	281	296	312	325	333	341	344	344	344	335	325	313	297
33.00	179	191	205	219	234	249	265	279	294	305	314	321	325	325	323	318	309	299	285

CASE # 4   COPYRIGHT 1982 STEVE SMITH   DATA DATE 1/ 3/82   TRANSFORMER DESIGN TABLE

$$f(D)=[(U/(FG(DE)^2))^5 * (1/(DEUm))^4]^{(1/13)}$$

P	Q	D	E	F	G	U	Um	DE	FG	f(D)	1/f(D)	1/(PQf(D))	DEUm	FG(DE)^2/U
0.33	0.33	.962	0.072	0.038	0.186	2.128	1.080	0.0693	0.007	.1555E+3	.6431E-2	.5906E-1	.7482E-1	.1594E-4
0.33	0.39	.964	0.074	0.036	0.242	2.133	1.250	0.0713	0.009	.1330E+3	.7520E-2	.5843E-1	.8920E-1	.2079E-4
0.33	0.47	.966	0.076	0.034	0.318	2.137	1.481	0.0734	0.011	.1127E+3	.8870E-2	.5719E-1	.1087E+0	.2726E-4
0.33	0.56	.967	0.077	0.033	0.406	2.140	1.746	0.0745	0.013	.9724E+2	.1028E-1	.5565E-1	.1300E+0	.3471E-4
0.33	0.68	.968	0.078	0.032	0.524	2.142	2.101	0.0755	0.017	.8304E+2	.1204E-1	.5367E-1	.1586E+0	.4462E-4
0.33	0.82	.969	0.079	0.031	0.662	2.145	2.516	0.0766	0.021	.7165E+2	.1396E-1	.5158E-1	.1926E+0	.5507E-4
0.33	1.00	.970	0.080	0.030	0.840	2.147	3.051	0.0776	0.025	.6151E+2	.1626E-1	.4927E-1	.2368E+0	.7068E-4
0.33	1.20	.970	0.080	0.030	1.040	2.147	3.651	0.0776	0.031	.5361E+2	.1865E-1	.4710E-1	.2833E+0	.8750E-4
0.33	1.50	.971	0.081	0.029	1.338	2.150	4.546	0.0787	0.039	.4544E+2	.2201E-1	.4446E-1	.3576E+0	.1117E-3
0.33	1.80	.971	0.081	0.029	1.638	2.150	5.446	0.0787	0.048	.3977E+2	.2515E-1	.4233E-1	.4284E+0	.1367E-3
0.33	2.20	.971	0.081	0.029	2.038	2.150	6.646	0.0787	0.059	.3439E+2	.2908E-1	.4005E-1	.5228E+0	.1701E-3
0.33	2.70	.971	0.081	0.029	2.538	2.150	8.146	0.0787	0.074	.2969E+2	.3368E-1	.3782E-1	.6407E+0	.2118E-3
0.33	3.30	.971	0.081	0.029	3.138	2.150	9.946	0.0787	0.091	.2573E+2	.3886E-1	.3569E-1	.7823E+0	.2619E-3
0.33	3.90	.972	0.082	0.028	3.736	2.152	11.742	0.0797	0.105	.2285E+2	.4375E-1	.3400E-1	.9359E+0	.3088E-3
0.33	4.70	.972	0.082	0.028	4.536	2.152	14.142	0.0797	0.127	.2003E+2	.4992E-1	.3219E-1	.1127E+1	.3749E-3
0.33	5.60	.972	0.082	0.028	5.436	2.152	16.842	0.0797	0.152	.1771E+2	.5648E-1	.3056E-1	.1342E+1	.4493E-3
0.33	6.80	.972	0.082	0.028	6.636	2.152	20.442	0.0797	0.186	.1545E+2	.6472E-1	.2884E-1	.1629E+1	.5485E-3
0.33	8.20	.972	0.082	0.028	8.036	2.152	24.642	0.0797	0.225	.1355E+2	.7379E-1	.2727E-1	.1964E+1	.6642E-3
0.33	10.00	.972	0.082	0.028	9.836	2.152	30.042	0.0797	0.275	.1180E+2	.8477E-1	.2569E-1	.2394E+1	.8130E-3
0.33	12.00	.972	0.082	0.028	11.836	2.152	36.042	0.0797	0.331	.1039E+2	.9627E-1	.2431E-1	.2873E+1	.9783E-3
0.33	15.00	.972	0.082	0.028	14.836	2.152	45.042	0.0797	0.415	.8891E+1	.1125E+0	.2272E-1	.3590E+1	.1226E-2
0.33	18.00	.972	0.082	0.028	17.836	2.152	54.042	0.0797	0.499	.7832E+1	.1277E+0	.2150E-1	.4307E+1	.1474E-2
0.33	22.00	.972	0.082	0.028	21.836	2.152	66.042	0.0797	0.611	.6812E+1	.1468E+0	.2022E-1	.5264E+1	.1805E-2
0.33	27.00	.972	0.082	0.028	26.836	2.152	81.042	0.0797	0.751	.5909E+1	.1692E+0	.1900E-1	.6459E+1	.2218E-2
0.33	33.00	.972	0.082	0.028	32.836	2.152	99.042	0.0797	0.919	.5140E+1	.1945E+0	.1786E-1	.7894E+1	.2714E-2
0.39	0.33	.952	0.082	0.048	0.166	2.143	1.112	0.0781	0.008	.1298E+3	.7707E-2	.5988E-1	.8678E-1	.2265E-4
0.39	0.39	.955	0.085	0.045	0.220	2.151	1.277	0.0812	0.010	.1098E+3	.9107E-2	.5988E-1	.1037E+0	.3033E-4
0.39	0.47	.958	0.088	0.042	0.294	2.158	1.502	0.0843	0.012	.9223E+2	.1084E-1	.5915E-1	.1267E+0	.4067E-4
0.39	0.56	.960	0.090	0.040	0.380	2.163	1.763	0.0864	0.015	.7901E+2	.1266E-1	.5795E-1	.1523E+0	.5246E-4
0.39	0.68	.962	0.092	0.038	0.496	2.168	2.113	0.0885	0.019	.6709E+2	.1491E-1	.5620E-1	.1870E+0	.6811E-4
0.39	0.82	.963	0.093	0.037	0.634	2.170	2.528	0.0896	0.023	.5765E+2	.1735E-1	.5424E-1	.2264E+0	.8670E-4
0.39	1.00	.964	0.094	0.036	0.812	2.173	3.063	0.0906	0.029	.4933E+2	.2027E-1	.5198E-1	.2776E+0	.1105E-3
0.39	1.20	.965	0.095	0.035	1.010	2.175	3.658	0.0917	0.035	.4289E+2	.2332E-1	.4982E-1	.3354E+0	.1366E-3
0.39	1.50	.965	0.095	0.035	1.310	2.175	4.558	0.0917	0.046	.3627E+2	.2757E-1	.4713E-1	.4179E+0	.1772E-3
0.39	1.80	.966	0.096	0.034	1.608	2.177	5.454	0.0927	0.055	.3169E+2	.3155E-1	.4495E-1	.5057E+0	.2159E-3

CASE # 4   COPYRIGHT 1982 STEVE SMITH   DATA DATE 1/ 3/82   TRANSFORMER DESIGN TABLE

$$f(D)=[(U/(FG(DE)^2))^5 * (1/(DEUm))^4]^{(1/13)}$$

P	Q	D	E	F	G	U	Um	DE	FG	f(D)	1/f(D)	1/(PQf(D))	DEUm	FG(DE)^2/U
0.39	2.20	.966	.096	.034	2.008	2.177	6.654	0.0927	0.068	.2737E+2	.3653E-1	.4258E-1	.6170E+0	.2696E-3
0.39	2.70	.966	.096	.034	2.508	2.177	8.154	0.0927	0.085	.2360E+2	.4237E-1	.4023E-1	.7561E+0	.3368E-3
0.39	3.30	.966	.096	.034	3.108	2.177	9.954	0.0927	0.106	.2044E+2	.4892E-1	.3801E-1	.9231E+0	.4174E-3
0.39	3.90	.967	.097	.033	3.706	2.180	11.749	0.0938	0.122	.1815E+2	.5511E-1	.3623E-1	.1102E+1	.4936E-3
0.39	4.70	.967	.097	.033	4.506	2.180	14.149	0.0938	0.149	.1590E+2	.6291E-1	.3432E-1	.1327E+1	.6002E-3
0.39	5.60	.967	.097	.033	5.406	2.180	16.849	0.0938	0.178	.1405E+2	.7120E-1	.3260E-1	.1580E+1	.7201E-3
0.39	6.80	.967	.097	.033	6.606	2.180	20.449	0.0938	0.218	.1225E+2	.8162E-1	.3078E-1	.1918E+1	.8799E-3
0.39	8.20	.967	.097	.033	8.006	2.180	24.649	0.0938	0.264	.1074E+2	.9309E-1	.2911E-1	.2312E+1	.1066E-2
0.39	10.00	.967	.097	.033	9.806	2.180	30.049	0.0938	0.324	.9349E+1	.1070E+0	.2743E-1	.2819E+1	.1306E-2
0.39	12.00	.967	.097	.033	11.806	2.180	36.049	0.0938	0.390	.8231E+1	.1215E+0	.2596E-1	.3381E+1	.1572E-2
0.39	15.00	.967	.097	.033	14.806	2.180	45.049	0.0938	0.489	.7044E+1	.1420E+0	.2427E-1	.4226E+1	.1972E-2
0.39	18.00	.967	.097	.033	17.806	2.180	54.049	0.0938	0.588	.6204E+1	.1612E+0	.2296E-1	.5070E+1	.2372E-2
0.39	22.00	.967	.097	.033	21.806	2.180	66.049	0.0938	0.720	.5396E+1	.1853E+0	.2160E-1	.6195E+1	.2904E-2
0.39	27.00	.967	.097	.033	26.806	2.180	81.049	0.0938	0.885	.4680E+1	.2137E+0	.2029E-1	.7602E+1	.3557E-2
0.39	33.00	.967	.097	.033	32.806	2.180	99.049	0.0938	1.083	.4071E+1	.2457E+0	.1909E-1	.9291E+1	.4370E-2
0.47	0.33	.935	.092	.065	.147	2.155	1.171	0.0857	0.010	.1080E+3	.9259E-2	.5970E-1	.1004E+0	.3249E-4
0.47	0.39	.947	.099	.058	.193	2.172	1.317	0.0929	0.011	.9007E+2	.1110E-1	.6057E-1	.1224E+0	.4444E-4
0.47	0.47	.947	.104	.053	.263	2.185	1.533	0.0982	0.014	.7464E+2	.1340E-1	.6065E-1	.1505E+0	.6142E-4
0.47	0.56	.950	.107	.050	.347	2.192	1.788	0.1013	0.017	.6331E+2	.1580E-1	.6001E-1	.1812E+0	.8120E-4
0.47	0.68	.953	.110	.047	.461	2.199	2.134	0.1045	0.022	.5331E+2	.1876E-1	.5869E-1	.2230E+0	.1075E-3
0.47	0.82	.955	.112	.045	.597	2.204	2.544	0.1066	0.027	.4553E+2	.2196E-1	.5699E-1	.2713E+0	.1385E-3
0.47	1.00	.956	.113	.044	.775	2.206	3.079	0.1077	0.034	.3877E+2	.2579E-1	.5488E-1	.3317E+0	.1792E-3
0.47	1.20	.957	.114	.042	.973	2.209	3.674	0.1088	0.042	.3359E+2	.2977E-1	.5278E-1	.3997E+0	.2241E-3
0.47	1.50	.958	.115	.042	1.271	2.211	4.570	0.1099	0.053	.2831E+2	.3552E-1	.5010E-1	.5020E+0	.2912E-3
0.47	1.80	.959	.116	.041	1.569	2.214	5.465	0.1109	0.064	.2469E+2	.4050E-1	.4787E-1	.6062E+0	.3575E-3
0.47	2.20	.959	.116	.041	1.969	2.214	6.665	0.1109	0.081	.2123E+2	.4698E-1	.4543E-1	.7393E+0	.4486E-3
0.47	2.70	.960	.117	.040	2.467	2.216	8.160	0.1120	0.099	.1833E+2	.5455E-1	.4299E-1	.9139E+0	.5585E-3
0.47	3.30	.960	.117	.040	3.067	2.216	9.960	0.1120	0.123	.1586E+2	.6307E-1	.4066E-1	.1116E+1	.6943E-3
0.47	3.90	.960	.117	.040	3.667	2.216	11.760	0.1120	0.147	.1406E+2	.7110E-1	.3879E-1	.1317E+1	.8302E-3
0.47	4.70	.960	.117	.040	4.467	2.216	14.160	0.1120	0.179	.1231E+2	.8122E-1	.3677E-1	.1586E+1	.1011E-2
0.47	5.60	.960	.117	.040	5.367	2.216	16.860	0.1120	0.215	.1087E+2	.9197E-1	.3494E-1	.1888E+1	.1215E-2
0.47	6.80	.960	.117	.040	6.567	2.216	20.460	0.1120	0.263	.9480E+1	.1055E+0	.3301E-1	.2292E+1	.1487E-2
0.47	8.20	.960	.117	.040	7.967	2.216	24.660	0.1120	0.319	.8309E+1	.1203E+0	.3123E-1	.2762E+1	.1804E-2
0.47	10.00	.961	.118	.039	9.765	2.219	30.055	0.1131	0.381	.7229E+1	.1383E+0	.2943E-1	.3399E+1	.2195E-2
0.47	12.00	.961	.118	.039	11.765	2.219	36.055	0.1131	0.459	.6362E+1	.1572E+0	.2787E-1	.4077E+1	.2644E-2
0.47	15.00	.961	.118	.039	14.765	2.219	45.055	0.1131	0.576	.5444E+1	.1837E+0	.2606E-1	.5095E+1	.3319E-2

CASE # 4   DATA DATE 1/ 3/82   TRANSFORMER DESIGN TABLE

$$f(D) = [(U/(FG(DE)^2))^5 \times (1/(DEUm))^4]^{1/13}$$

P	Q	D	E	F	G	U	Um	DE	FG	f(D)	1/f(D)	1/(PQf(D))	DEUm	FG(DE)^2/U
0.47	18.00	.961	0.118	0.039	17.765	2.219	54.055	0.1131	0.693	.4794E+1	.2086E+0	.2466E-1	.6112E+1	.3993E-2
0.47	22.00	.961	0.118	0.039	21.765	2.219	66.055	0.1131	0.849	.4168E+1	.2399E+0	.2330E-1	.7469E+1	.4892E-2
0.47	27.00	.961	0.118	0.039	26.765	2.219	81.055	0.1131	1.044	.3615E+1	.2767E+0	.2180E-1	.9166E+1	.6016E-2
0.47	33.00	.961	0.118	0.039	32.765	2.219	99.055	0.1131	1.278	.3144E+1	.3181E+0	.2051E-1	.1120E+2	.7365E-2
0.56	0.33	.913	0.100	0.087	0.131	2.162	1.252	0.0910	0.011	.9281E+2	.1078E-1	.5831E-1	.1140E+0	.4354E-4
0.56	0.39	.923	0.110	0.077	0.171	2.186	1.384	0.1012	0.013	.7622E+2	.1312E-1	.6007E-1	.1401E+0	.6159E-4
0.56	0.47	.932	0.119	0.068	0.233	2.208	1.580	0.1106	0.016	.6217E+2	.1608E-1	.6111E-1	.1748E+0	.8764E-4
0.56	0.56	.937	0.124	0.063	0.313	2.220	1.826	0.1159	0.020	.5209E+2	.1920E-1	.6121E-1	.2116E+0	.1191E-3
0.56	0.68	.942	0.129	0.058	0.423	2.232	2.162	0.1212	0.025	.4341E+2	.2304E-1	.6050E-1	.2620E+0	.1613E-3
0.56	0.82	.945	0.132	0.055	0.557	2.240	2.567	0.1244	0.031	.3679E+2	.2718E-1	.5919E-1	.3194E+0	.2116E-3
0.56	1.00	.947	0.134	0.053	0.733	2.245	3.097	0.1266	0.039	.3114E+2	.3212E-1	.5735E-1	.3921E+0	.2772E-3
0.56	1.20	.948	0.135	0.052	0.931	2.247	3.692	0.1277	0.048	.2687E+2	.3722E-1	.5539E-1	.4714E+0	.3510E-3
0.56	1.50	.950	0.137	0.050	1.227	2.254	4.583	0.1298	0.061	.2256E+2	.4433E-1	.5278E-1	.5950E+0	.4591E-3
0.56	1.80	.951	0.138	0.049	1.525	2.254	5.478	0.1309	0.075	.1962E+2	.5096E-1	.5056E-1	.7172E+0	.5680E-3
0.56	2.20	.951	0.138	0.049	1.925	2.254	6.678	0.1309	0.094	.1688E+2	.5924E-1	.4809E-1	.8743E+0	.7171E-3
0.56	2.70	.952	0.139	0.048	2.423	2.257	8.173	0.1320	0.116	.1451E+2	.6892E-1	.4558E-1	.1079E+1	.8980E-3
0.56	3.30	.952	0.139	0.048	3.023	2.257	9.973	0.1320	0.145	.1253E+2	.7978E-1	.4317E-1	.1317E+1	.1120E-2
0.56	3.90	.953	0.140	0.047	3.621	2.259	11.768	0.1331	0.170	.1111E+2	.9002E-1	.4122E-1	.1566E+1	.1334E-2
0.56	4.70	.953	0.140	0.047	4.421	2.259	14.168	0.1331	0.208	.9717E+1	.1029E+0	.3910E-1	.1886E+1	.1629E-2
0.56	5.60	.953	0.140	0.047	5.321	2.259	16.868	0.1331	0.250	.8578E+1	.1166E+0	.3718E-1	.2245E+1	.1961E-2
0.56	6.80	.953	0.140	0.047	6.521	2.259	20.468	0.1331	0.306	.7472E+1	.1338E+0	.3514E-1	.2724E+1	.2403E-2
0.56	8.20	.953	0.140	0.047	7.921	2.259	24.668	0.1331	0.372	.6547E+1	.1527E+0	.3326E-1	.3283E+1	.2919E-2
0.56	10.00	.954	0.141	0.046	9.719	2.262	30.063	0.1342	0.447	.5693E+1	.1756E+0	.3137E-1	.4034E+1	.3560E-2
0.56	12.00	.954	0.141	0.046	11.719	2.262	36.063	0.1342	0.539	.5009E+1	.1996E+0	.2971E-1	.4840E+1	.4292E-2
0.56	15.00	.954	0.141	0.046	14.719	2.262	45.063	0.1342	0.677	.4285E+1	.2334E+0	.2778E-1	.6047E+1	.5391E-2
0.56	18.00	.954	0.141	0.046	17.719	2.262	54.063	0.1342	0.815	.3772E+1	.2651E+0	.2630E-1	.7255E+1	.6490E-2
0.56	22.00	.954	0.141	0.046	21.719	2.262	66.063	0.1342	0.999	.3280E+1	.3049E+0	.2475E-1	.8865E+1	.7955E-2
0.56	27.00	.954	0.141	0.046	26.719	2.262	81.063	0.1342	1.229	.2844E+1	.3517E+0	.2326E-1	.1088E+2	.9787E-2
0.56	33.00	.954	0.141	0.046	32.719	2.262	99.063	0.1342	1.505	.2473E+1	.4043E+0	.2188E-1	.1329E+2	.1198E-1
0.68	0.33	.880	0.107	0.120	0.117	2.162	1.378	0.0939	0.014	.8043E+2	.1243E-1	.5540E-1	.1294E+0	.5706E-4
0.68	0.39	.894	0.121	0.106	0.149	2.196	1.490	0.1079	0.016	.6498E+2	.1539E-1	.5803E-1	.1608E+0	.8352E-4
0.68	0.47	.907	0.134	0.093	0.203	2.227	1.667	0.1212	0.019	.5196E+2	.1925E-1	.6022E-1	.2021E+0	.1244E-3
0.68	0.56	.917	0.144	0.083	0.273	2.252	1.889	0.1317	0.023	.4280E+2	.2337E-1	.6136E-1	.2488E+0	.1744E-3
0.68	0.68	.925	0.152	0.075	0.377	2.271	2.210	0.1403	0.028	.3511E+2	.2848E-1	.6160E-1	.3100E+0	.2448E-3
0.68	0.82	.930	0.157	0.070	0.507	2.283	2.606	0.1457	0.035	.2942E+2	.3399E-1	.6097E-1	.3796E+0	.3297E-3
0.68	1.00	.934	0.161	0.066	0.679	2.293	3.126	0.1501	0.045	.2467E+2	.4054E-1	.5961E-1	.4691E+0	.4399E-3

CASE # 4   COPYRIGHT 1982 STEVE SMITH   DATA DATE 1/ 3/82   TRANSFORMER DESIGN TABLE

$$f(D)=[(U/(FG(DE)^2))^5 * (1/(DEUm))^4]^{(1/13)}$$

P	Q	D	E	F	G	U	Um	DE	FG	f(D)	1/f(D)	1/(PQf(D))	DEUm	FG(DE)^2/U
0.68	1.20	.937	.164	0.063	0.873	2.300	3.712	0.1534	0.055	.2115E+2	.4727E-1	.5793E-1	.5692E+0	.5621E-3
0.68	1.50	.939	.166	0.061	1.169	2.305	4.602	0.1556	0.071	.1766E+2	.5663E-1	.5552E-1	.7159E+0	.7484E-3
0.68	1.80	.940	.167	0.060	1.467	2.308	5.497	0.1567	0.088	.1531E+2	.6533E-1	.5338E-1	.8612E+0	.9360E-3
0.68	2.20	.941	.168	0.059	1.865	2.310	6.692	0.1578	0.110	.1313E+2	.7618E-1	.5092E-1	.1056E+1	.1186E-2
0.68	2.70	.942	.169	0.058	2.363	2.312	8.187	0.1589	0.137	.1126E+2	.8883E-1	.4838E-1	.1301E+1	.1496E-2
0.68	3.30	.943	.170	0.057	2.961	2.315	9.982	0.1600	0.169	.9707E+1	.1030E+0	.4591E-1	.1597E+1	.1866E-2
0.68	3.90	.943	.170	0.057	3.561	2.315	11.782	0.1600	0.203	.8592E+1	.1164E+0	.4389E-1	.1885E+1	.2244E-2
0.68	4.70	.943	.170	0.057	4.361	2.315	14.182	0.1600	0.249	.7507E+1	.1332E+0	.4168E-1	.2269E+1	.2749E-2
0.68	5.60	.944	.171	0.056	5.259	2.317	16.878	0.1611	0.294	.6619E+1	.1511E+0	.3967E-1	.2719E+1	.3299E-2
0.68	6.80	.944	.171	0.056	6.459	2.317	20.478	0.1611	0.362	.5763E+1	.1735E+0	.3753E-1	.3299E+1	.4051E-2
0.68	8.20	.944	.171	0.056	7.859	2.317	24.678	0.1611	0.440	.5046E+1	.1982E+0	.3554E-1	.3976E+1	.4929E-2
0.68	10.00	.944	.171	0.056	9.659	2.317	30.077	0.1611	0.541	.4386E+1	.2280E+0	.3353E-1	.4846E+1	.6058E-2
0.68	12.00	.944	.171	0.056	11.659	2.317	36.078	0.1611	0.653	.3858E+1	.2592E+0	.3177E-1	.5812E+1	.7313E-2
0.68	15.00	.945	.172	0.055	14.657	2.320	45.073	0.1622	0.806	.3298E+1	.3032E+0	.2972E-1	.7312E+1	.9145E-2
0.68	18.00	.945	.172	0.055	17.657	2.320	54.073	0.1622	0.971	.2903E+1	.3445E+0	.2814E-1	.8772E+1	.1102E-1
0.68	22.00	.945	.172	0.055	21.657	2.320	66.073	0.1622	1.191	.2523E+1	.3963E+0	.2649E-1	.1072E+2	.1351E-1
0.68	27.00	.945	.172	0.055	26.657	2.320	81.073	0.1622	1.466	.2187E+1	.4572E+0	.2490E-1	.1315E+2	.1663E-1
0.68	33.00	.945	.172	0.055	32.657	2.320	99.073	0.1622	1.796	.1902E+1	.5257E+0	.2343E-1	.1607E+2	.2038E-1
0.82	0.33	.838	.111	0.162	0.107	2.153	1.542	0.0933	0.017	.7184E+2	.1392E-1	.5144E-1	.1439E+0	.7029E-4
0.82	0.39	.854	.127	0.146	0.135	2.192	1.645	0.1087	0.020	.5725E+2	.1747E-1	.5462E-1	.1788E+0	.1066E-3
0.82	0.47	.873	.146	0.127	0.177	2.238	1.792	0.1277	0.023	.4494E+2	.2225E-1	.5774E-1	.2298E+0	.1642E-3
0.82	0.56	.888	.161	0.112	0.237	2.275	1.990	0.1433	0.027	.3631E+2	.2754E-1	.5997E-1	.2850E+0	.2399E-3
0.82	0.68	.902	.175	0.098	0.329	2.309	2.281	0.1582	0.032	.2921E+2	.3423E-1	.6139E-1	.3608E+0	.3497E-3
0.82	0.82	.911	.184	0.089	0.451	2.330	2.658	0.1679	0.040	.2411E+2	.4148E-1	.6169E-1	.4463E+0	.4861E-3
0.82	1.00	.918	.191	0.082	0.617	2.347	3.164	0.1756	0.051	.1997E+2	.5007E-1	.6106E-1	.5557E+0	.6653E-3
0.82	1.20	.922	.195	0.078	0.809	2.357	3.744	0.1801	0.063	.1698E+2	.5888E-1	.5988E-1	.6743E+0	.8686E-3
0.82	1.50	.926	.199	0.074	1.101	2.367	4.625	0.1846	0.081	.1407E+2	.7107E-1	.5778E-1	.8537E+0	.1173E-2
0.82	1.80	.928	.201	0.072	1.397	2.372	5.515	0.1868	0.101	.1214E+2	.8236E-1	.5580E-1	.1030E+1	.1481E-2
0.82	2.20	.929	.202	0.071	1.795	2.374	6.710	0.1880	0.127	.1037E+2	.9641E-1	.5344E-1	.1261E+1	.1897E-2
0.82	2.70	.930	.204	0.070	2.293	2.377	8.205	0.1891	0.161	.8868E+1	.1128E+0	.5093E-1	.1552E+1	.2415E-2
0.82	3.30	.931	.204	0.069	2.891	2.379	10.001	0.1902	0.200	.7629E+1	.1311E+0	.4844E-1	.1902E+1	.3035E-2
0.82	3.90	.932	.205	0.068	3.489	2.381	11.796	0.1914	0.237	.6743E+1	.1483E+0	.4638E-1	.2257E+1	.3649E-2
0.82	4.70	.933	.206	0.067	4.287	2.384	14.191	0.1925	0.287	.5883E+1	.1700E+0	.4411E-1	.2732E+1	.4466E-2
0.82	5.60	.933	.206	0.067	5.187	2.384	16.891	0.1925	0.348	.5182E+1	.1930E+0	.4202E-1	.3252E+1	.5403E-2
0.82	6.80	.933	.206	0.067	6.387	2.384	20.441	0.1925	0.428	.4507E+1	.2219E+0	.3979E-1	.3945E+1	.6653E-2
0.82	8.20	.934	.207	0.066	7.785	2.386	24.686	0.1936	0.514	.3944E+1	.2536E+0	.3771E-1	.4780E+1	.8075E-2

CASE # 4   COPYRIGHT 1982 STEVE SMITH   DATA DATE 1/3/82   TRANSFORMER DESIGN TABLE

$$f(D)=\left[\left(\frac{U}{FG(DE)^2}\right)^5 * \left(\frac{1}{DEUm}\right)^4\right]^{(1/13)}$$

P	Q	D	E	F	G	U	Um	DE	FG	f(D)	1/f(D)	1/(PQf(D))	DEUm	FG(DE)^2/U
0.82	10.00	.934	0.207	0.066	9.585	2.386	30.086	0.1936	0.633	.3425E+1	.2919E+0	.3560E-1	.5826E+1	.9941E-2
0.82	12.00	.934	0.207	0.066	11.585	2.386	36.086	0.1936	0.765	.3011E+1	.3321E+0	.3375E-1	.6988E+1	.1202E-1
0.82	15.00	.934	0.207	0.066	14.585	2.386	45.086	0.1936	0.963	.2574E+1	.3886E+0	.3159E-1	.8731E+1	.1513E-1
0.82	18.00	.935	0.208	0.065	17.583	2.389	54.081	0.1948	1.143	.2264E+1	.4416E+0	.2992E-1	.1053E+2	.1815E-1
0.82	22.00	.935	0.208	0.065	21.583	2.389	66.081	0.1948	1.403	.1968E+1	.5088E+0	.2817E-1	.1287E+2	.2228E-1
0.82	27.00	.935	0.208	0.065	26.583	2.389	81.081	0.1948	1.728	.1705E+1	.5864E+0	.2644E-1	.1579E+2	.2745E-1
0.82	33.00	.935	0.208	0.065	32.583	2.389	99.081	0.1948	2.118	.1483E+1	.6745E+0	.2493E-1	.1930E+2	.3364E-1
1.00	0.33	.782	0.115	0.218	0.099	2.137	1.763	0.0902	0.022	.6553E+2	.1526E-1	.4624E-1	.1590E+0	.8242E-4
1.00	0.39	.800	0.133	0.200	0.123	2.181	1.856	0.1067	0.025	.5162E+2	.1937E-1	.4968E-1	.1979E+0	.1287E-3
1.00	0.47	.823	0.156	0.156	0.157	2.237	1.984	0.1287	0.028	.3982E+2	.2511E-1	.5343E-1	.2552E+0	.2061E-3
1.00	0.56	.844	0.177	0.156	0.205	2.288	2.152	0.1497	0.032	.3154E+2	.3170E-1	.5661E-1	.3221E+0	.3137E-3
1.00	0.68	.866	0.199	0.134	0.281	2.341	2.405	0.1726	0.038	.2477E+2	.4037E-1	.5936E-1	.4151E+0	.4798E-3
1.00	0.82	.883	0.216	0.117	0.387	2.382	2.742	0.1910	0.045	.2001E+2	.4998E-1	.6095E-1	.5238E+0	.6941E-3
1.00	1.00	.895	0.228	0.105	0.543	2.412	3.224	0.2044	0.057	.1628E+2	.6143E-1	.6143E-1	.6589E+0	.9884E-3
1.00	1.20	.902	0.235	0.098	0.729	2.429	3.790	0.2123	0.071	.1367E+2	.7315E-1	.6096E-1	.8045E+0	.1326E-2
1.00	1.50	.908	0.241	0.092	1.017	2.443	4.661	0.2191	0.094	.1120E+2	.8929E-1	.5952E-1	.1021E+1	.1840E-2
1.00	1.80	.911	0.244	0.089	1.311	2.450	5.546	0.2226	0.117	.9600E+1	.1042E+0	.5777E-1	.1235E+1	.2360E-2
1.00	2.20	.914	0.247	0.086	1.705	2.458	6.732	0.2261	0.147	.8156E+1	.1226E+0	.5573E-1	.1522E+1	.3040E-2
1.00	2.70	.916	0.249	0.084	2.201	2.463	8.222	0.2284	0.185	.6944E+1	.1440E+0	.5334E-1	.1878E+1	.3917E-2
1.00	3.30	.918	0.251	0.082	2.797	2.467	10.012	0.2307	0.229	.5955E+1	.1679E+0	.5089E-1	.2310E+1	.4949E-2
1.00	3.90	.918	0.251	0.082	3.397	2.467	11.812	0.2307	0.279	.5252E+1	.1904E+0	.4882E-1	.2725E+1	.6010E-2
1.00	4.70	.919	0.252	0.081	4.195	2.470	14.207	0.2319	0.340	.4573E+1	.2187E+0	.4652E-1	.3295E+1	.7399E-2
1.00	5.60	.920	0.253	0.080	5.093	2.472	16.903	0.2331	0.407	.4023E+1	.2486E+0	.4439E-1	.3939E+1	.8953E-2
1.00	6.80	.921	0.254	0.079	6.291	2.475	20.498	0.2342	0.497	.3494E+1	.2862E+0	.4209E-1	.4801E+1	.1102E-1
1.00	8.20	.921	0.254	0.079	7.691	2.475	24.698	0.2342	0.608	.3054E+1	.3274E+0	.3993E-1	.5785E+1	.1347E-1
1.00	10.00	.921	0.254	0.079	9.491	2.475	30.098	0.2342	0.750	.2651E+1	.3773E+0	.3773E-1	.7050E+1	.1662E-1
1.00	12.00	.922	0.255	0.078	11.489	2.477	36.093	0.2354	0.896	.2329E+1	.4294E+0	.3579E-1	.8497E+1	.2005E-1
1.00	15.00	.922	0.255	0.079	14.489	2.477	45.093	0.2354	1.130	.1989E+1	.5028E+0	.3352E-1	.1062E+2	.2528E-1
1.00	18.00	.922	0.255	0.078	17.489	2.477	54.093	0.2354	1.364	.1749E+1	.5717E+0	.3176E-1	.1273E+2	.3052E-1
1.00	22.00	.922	0.255	0.078	21.489	2.477	66.093	0.2354	1.676	.1519E+1	.6581E+0	.2992E-1	.1556E+2	.3750E-1
1.00	27.00	.922	0.255	0.078	26.489	2.477	81.093	0.2354	2.066	.1316E+1	.7596E+0	.2813E-1	.1909E+2	.4623E-1
1.00	33.00	.922	0.255	0.078	32.489	2.477	99.093	0.2354	2.534	.1144E+1	.8739E+0	.2648E-1	.2333E+2	.5670E-1
1.20	0.33	.718	0.118	0.282	0.094	2.115	2.017	0.0847	0.027	.6198E+2	.1614E-1	.4075E-1	.1709E+0	.8997E-4
1.20	0.39	.737	0.137	0.263	0.116	2.161	2.104	0.1010	0.031	.4838E+2	.2067E-1	.4417E-1	.2125E+0	.1439E-3
1.20	0.47	.763	0.163	0.237	0.144	2.224	2.218	0.1244	0.034	.3683E+2	.2715E-1	.4814E-1	.2759E+0	.2373E-3
1.20	0.56	.789	0.189	0.211	0.182	2.287	2.362	0.1491	0.038	.2877E+2	.3484E-1	.5185E-1	.3522E+0	.3733E-3

CASE # 4   COPYRIGHT 1982 STEVE SMITH   DATA DATE 1/ 3/82   TRANSFORMER DESIGN TABLE

$$f(D) = [(U/(FG(DE)^2))^5 * (1/(DEUm))^4]^{1/13}$$

P	Q	D	E	F	G	U	Um	DE	FG	f(D)	1/f(D)	1/(PQf(D))	DEUm	$FG(DE)^2/U$
1.20	0.68	.818	0.218	0.182	0.244	2.358	2.581	0.1783	0.044	.2204E+2	.4538E-1	.5561E-1	.4602E+0	.5989E-3
1.20	0.82	.844	0.244	0.156	0.332	2.421	2.875	0.2059	0.052	.1738E+2	.5752E-1	.5846E-1	.5920E+0	.9072E-3
1.20	1.00	.865	0.265	0.135	0.470	2.472	3.313	0.2292	0.063	.1383E+2	.7233E-1	.6028E-1	.7593E+0	.1349E-2
1.20	1.20	.878	0.278	0.122	0.644	2.504	3.849	0.2441	0.079	.1142E+2	.8757E-1	.6081E-1	.9396E+0	.1870E-2
1.20	1.50	.888	0.288	0.112	0.924	2.528	4.701	0.2557	0.103	.9220E+1	.1085E+0	.6026E-1	.1202E+1	.2678E-2
1.20	1.80	.893	0.293	0.107	1.214	2.540	5.576	0.2616	0.130	.7835E+1	.1276E+0	.5909E-1	.1459E+1	.3501E-2
1.20	2.20	.897	0.297	0.103	1.606	2.550	6.757	0.2664	0.165	.6610E+1	.1513E+0	.5730E-1	.1800E+1	.4604E-2
1.20	2.70	.900	0.300	0.100	2.100	2.557	8.242	0.2700	0.210	.5998E+1	.1786E+0	.5514E-1	.2225E+1	.5987E-2
1.20	3.30	.903	0.303	0.097	2.694	2.564	10.028	0.2736	0.261	.4782E+1	.2091E+0	.5201E-1	.2744E+1	.7629E-2
1.20	3.90	.904	0.304	0.096	3.292	2.567	11.823	0.2748	0.316	.4206E+1	.2377E+0	.5080E-1	.3249E+1	.9299E-2
1.20	4.70	.905	0.305	0.095	4.090	2.569	14.218	0.2760	0.389	.3655E+1	.2736E+0	.4851E-1	.3925E+1	.1152E-1
1.20	5.60	.906	0.306	0.094	4.988	2.572	16.913	0.2772	0.469	.3209E+1	.3116E+0	.4637E-1	.4689E+1	.1401E-1
1.20	6.80	.907	0.307	0.093	6.186	2.574	20.508	0.2784	0.575	.2783E+1	.3593E+0	.4443E-1	.5711E+1	.1733E-1
1.20	8.20	.908	0.308	0.092	7.584	2.577	24.704	0.2797	0.698	.2430E+1	.4115E+0	.4182E-1	.6999E+1	.2118E-1
1.20	10.00	.908	0.308	0.092	9.384	2.577	30.104	0.2797	0.863	.2107E+1	.4747E+0	.3956E-1	.8419E+1	.2621E-1
1.20	12.00	.909	0.309	0.091	11.382	2.579	36.099	0.2809	1.036	.1849E+1	.5407E+0	.3755E-1	.1014E+2	.3169E-1
1.20	15.00	.909	0.309	0.091	14.382	2.579	45.099	0.2809	1.309	.1578E+1	.6335E+0	.3502E-1	.1267E+2	.4004E-1
1.20	18.00	.909	0.309	0.091	17.382	2.579	54.099	0.2809	1.582	.1388E+1	.7207E+0	.3336E-1	.1520E+2	.4833E-1
1.20	22.00	.909	0.309	0.091	21.382	2.579	66.099	0.2809	1.946	.1205E+1	.8300E+0	.3144E-1	.1857E+2	.5952E-1
1.20	27.00	.910	0.310	0.090	26.382	2.579	81.094	0.2821	2.374	.1043E+1	.9584E+0	.2958E-1	.2288E+2	.7319E-1
1.20	33.00	.910	0.310	0.090	32.380	2.581	99.094	0.2821	2.914	.9067E+0	.1103E+1	.2785E-1	.2795E+2	.8984E-1
1.50	0.33	.620	0.120	0.380	0.090	2.077	2.407	0.0744	0.034	.6078E+2	.1645E-1	.3324E-1	.1791E+0	.9115E-4
1.50	0.39	.641	0.141	0.359	0.108	2.128	2.485	0.0904	0.039	.4695E+2	.2130E-1	.3641E-1	.2246E+0	.1488E-3
1.50	0.47	.669	0.169	0.331	0.132	2.196	2.589	0.1131	0.044	.3521E+2	.2840E-1	.4028E-1	.2927E+0	.2543E-3
1.50	0.56	.699	0.199	0.301	0.162	2.269	2.713	0.1391	0.049	.2695E+2	.3710E-1	.4417E-1	.3774E+0	.4159E-3
1.50	0.68	.735	0.235	0.265	0.210	2.356	2.898	0.1727	0.056	.2017E+2	.4957E-1	.4860E-1	.5006E+0	.7047E-3
1.50	0.82	.772	0.272	0.228	0.276	2.446	3.139	0.2100	0.063	.1544E+2	.6479E-1	.5267E-1	.6590E+0	.1134E-2
1.50	1.00	.809	0.309	0.191	0.382	2.536	3.499	0.2500	0.073	.1185E+2	.8438E-1	.5625E-1	.8746E+0	.1798E-2
1.50	1.20	.834	0.334	0.166	0.532	2.597	3.977	0.2786	0.088	.9506E+1	.1052E+0	.5843E-1	.1108E+1	.2639E-2
1.50	1.50	.855	0.355	0.145	0.790	2.648	4.775	0.3035	0.115	.7469E+1	.1339E+0	.5950E-1	.1449E+1	.3986E-2
1.50	1.80	.865	0.365	0.135	1.070	2.672	5.627	0.3157	0.144	.6248E+1	.1601E+0	.5928E-1	.1776E+1	.5389E-2
1.50	2.20	.873	0.373	0.127	1.454	2.691	6.788	0.3256	0.185	.5205E+1	.1921E+0	.5822E-1	.2210E+1	.7775E-2
1.50	2.70	.878	0.378	0.122	1.944	2.704	8.264	0.3319	0.237	.4367E+1	.2290E+0	.5654E-1	.2743E+1	.9662E-2
1.50	3.30	.882	0.382	0.118	2.536	2.713	10.044	0.3369	0.299	.3705E+1	.2699E+0	.5453E-1	.3388E+1	.1252E-1
1.50	3.90	.884	0.384	0.116	3.112	2.718	11.834	0.3395	0.363	.3245E+1	.3081E+0	.5267E-1	.4017E+1	.1540E-1
1.50	4.70	.886	0.386	0.114	3.928	2.723	14.225	0.3420	0.448	.2809E+1	.3560E+0	.5049E-1	.4865E+1	.1923E-1

CASE # 4 COPYRIGHT 1982 STEVE SMITH DATA DATE 1/ 3/82 TRANSFORMER DESIGN TABLE

$$f(D)=[(U/(FG(DE)^2))^5 * (1/(DEUm)^4)]^{(1/13)}$$

P	Q	D	E	F	G	U	Um	DE	FG	f(D)	1/f(D)	1/(PQf(D))	DEUm	FG(DE)^2/U
1.50	5.60	.887	0.387	0.113	4.826	2.726	16.920	0.3433	0.545	.2460E+1	.4066E+0	.4840E-1	.5808E+1	.2358E-1
1.50	6.80	.888	0.388	0.112	6.024	2.728	20.515	0.3445	0.675	.2128E+1	.4699E+0	.4607E-1	.7068E+1	.2936E-1
1.50	8.20	.889	0.389	0.111	7.422	2.730	24.710	0.3458	0.824	.1854E+1	.5393E+0	.4384E-1	.8545E+1	.3693E-1
1.50	10.00	.890	0.390	0.110	9.220	2.733	30.105	0.3471	1.014	.1605E+1	.6230E+0	.4153E-1	.1045E+2	.4471E-1
1.50	12.00	.891	0.391	0.109	11.218	2.735	36.100	0.3484	1.223	.1407E+1	.7105E+0	.3947E-1	.1258E+2	.5426E-1
1.50	15.00	.891	0.391	0.109	14.218	2.735	45.100	0.3484	1.550	.1200E+1	.8335E+0	.3704E-1	.1571E+2	.6877E-1
1.50	18.00	.891	0.391	0.109	17.218	2.735	54.100	0.3484	1.877	.1054E+1	.9489E+0	.3514E-1	.1885E+2	.8328E-1
1.50	22.00	.892	0.392	0.108	21.216	2.738	66.096	0.3497	2.291	.9144E+0	.1094E+1	.3314E-1	.2311E+2	.1023E+0
1.50	27.00	.892	0.392	0.108	26.216	2.738	81.096	0.3497	2.831	.7915E+0	.1263E+1	.3120E-1	.2836E+2	.1264E+0
1.50	33.00	.892	0.392	0.108	32.216	2.738	99.096	0.3497	3.479	.6874E+0	.1455E+1	.2939E-1	.3465E+2	.1554E+0
1.80	0.33	.522	0.122	0.478	0.086	2.039	2.797	0.0637	0.041	.6348E+2	.1575E-1	.2652E-1	.1781E+0	.8177E-4
1.80	0.39	.545	0.145	0.455	0.100	2.095	2.866	0.0790	0.045	.4853E+2	.2061E-1	.2935E-1	.2264E+0	.1356E-3
1.80	0.47	.573	0.173	0.427	0.124	2.163	2.969	0.0991	0.053	.3591E+2	.2785E-1	.3291E-1	.2944E+0	.2406E-3
1.80	0.56	.605	0.205	0.395	0.150	2.240	3.084	0.1240	0.059	.2707E+2	.3694E-1	.3665E-1	.3825E+0	.4068E-3
1.80	0.68	.646	0.246	0.354	0.188	2.340	3.245	0.1589	0.067	.1984E+2	.5039E-1	.4117E-1	.5157E+0	.7182E-3
1.80	0.82	.690	0.290	0.310	0.240	2.447	3.451	0.2001	0.074	.1481E+2	.6754E-1	.4576E-1	.6906E+0	.1217E-2
1.80	1.00	.739	0.339	0.261	0.322	2.566	3.753	0.2505	0.084	.1101E+2	.9084E-1	.5047E-1	.9402E+0	.2056E-2
1.80	1.20	.780	0.380	0.220	0.440	2.666	4.154	0.2964	0.097	.8556E+1	.1169E+0	.5411E-1	.1231E+1	.3190E-2
1.80	1.50	.817	0.417	0.183	0.666	2.755	4.874	0.3407	0.122	.6493E+1	.1539E+0	.5699E-1	.1661E+1	.5134E-2
1.80	1.80	.835	0.435	0.165	0.930	2.687	5.687	0.3632	0.153	.5326E+1	.1878E+0	.5795E-1	.2066E+1	.7232E-2
1.80	2.20	.848	0.448	0.152	1.304	2.831	6.823	0.3799	0.198	.4367E+1	.2290E+0	.5782E-1	.2592E+1	.1011E-1
1.80	2.70	.856	0.456	0.144	1.788	2.850	8.285	0.3903	0.257	.3623E+1	.2760E+0	.5680E-1	.3234E+1	.1376E-1
1.80	3.30	.862	0.462	0.138	2.376	2.865	10.055	0.3982	0.328	.3050E+1	.3279E+0	.5522E-1	.4005E+1	.1815E-1
1.80	3.90	.865	0.465	0.135	2.970	2.872	11.841	0.4022	0.401	.2658E+1	.3762E+0	.5359E-1	.4763E+1	.2259E-1
1.80	4.70	.868	0.468	0.132	3.764	2.879	14.226	0.4062	0.497	.2291E+1	.4365E+0	.5160E-1	.5779E+1	.2847E-1
1.80	5.60	.870	0.470	0.130	4.660	2.884	16.917	0.4089	0.606	.2000E+1	.5001E+0	.4961E-1	.6917E+1	.3512E-1
1.80	6.80	.871	0.471	0.129	5.858	2.887	20.512	0.4102	0.756	.1725E+1	.5796E+0	.4735E-1	.8415E+1	.4406E-1
1.80	8.20	.873	0.473	0.127	7.254	2.891	24.702	0.4129	0.921	.1500E+1	.6665E+0	.4516E-1	.1020E+2	.5433E-1
1.80	10.00	.874	0.474	0.126	9.052	2.894	30.097	0.4143	1.141	.1296E+1	.7714E+0	.4285E-1	.1247E+2	.6764E-1
1.80	12.00	.875	0.475	0.125	11.050	2.896	36.092	0.4156	1.381	.1135E+1	.8808E+0	.4078E-1	.1508E+2	.8238E-1
1.80	15.00	.875	0.475	0.125	14.050	2.896	45.092	0.4156	1.756	.9666E+0	.1035E+1	.3832E-1	.1874E+2	.1047E+0
1.80	18.00	.876	0.476	0.124	17.048	2.899	54.087	0.4170	2.114	.8484E+0	.1179E+1	.3638E-1	.2255E+2	.1268E+0
1.80	22.00	.876	0.476	0.124	21.048	2.899	66.087	0.4170	2.610	.7355E+0	.1360E+1	.3433E-1	.2756E+2	.1565E+0
1.80	27.00	.877	0.477	0.123	26.046	2.901	81.083	0.4183	3.204	.6363E+0	.1572E+1	.3234E-1	.3392E+2	.1932E+0
1.80	33.00	.877	0.477	0.123	32.046	2.901	99.083	0.4183	3.942	.5524E+0	.1810E+1	.3048E-1	.4145E+2	.2378E+0
2.20	0.33	.392	0.125	0.608	0.079	1.990	3.314	0.0491	0.048	.7422E+2	.1347E-1	.1856E-1	.1628E+0	.5852E-4

CASE # 4   COPYRIGHT 1982 STEVE SMITH   DATA DATE 1/ 3/82   TRANSFORMER DESIGN TABLE

$$f(D) = \left[\left(\frac{U}{FG(DE)^2}\right)^5 * \left(\frac{1}{DEUm}\right)^4 \right]^{1/13}$$

P	Q	D	E	F	G	U	Um	DE	FG	f(D)	1/f(D)	1/(PQf(D))	DEUm	FG(DE)^2/U
2.20	0.39	.415	0.148	0.585	0.093	2.046	3.383	0.0616	0.055	.5575E+2	.1794E-1	.2091E-1	.2082E+0	.1011E-3
2.20	0.47	.446	0.179	0.554	0.114	2.121	3.472	0.0800	0.062	.4036E+2	.2478E-1	.2396E-1	.2777E+0	.1660E-3
2.20	0.56	.479	0.212	0.521	0.135	2.201	3.582	0.1017	0.071	.2974E+2	.3363E-1	.2730E-1	.3643E+0	.3314E-3
2.20	0.68	.524	0.257	0.476	0.165	2.310	3.723	0.1348	0.079	.2118E+2	.4721E-1	.3156E-1	.5020E+0	.6194E-3
2.20	0.82	.573	0.306	0.427	0.207	2.429	3.905	0.1755	0.089	.1531E+2	.6532E-1	.3621E-1	.6854E+0	.1123E-2
2.20	1.00	.633	0.366	0.367	0.267	2.575	4.154	0.2319	0.098	.1094E+2	.9144E-1	.4154E-1	.9632E+0	.2049E-2
2.20	1.20	.690	0.423	0.310	0.353	2.714	4.477	0.2921	0.110	.8155E+1	.1226E+0	.4645E-1	.1308E+1	.3444E-2
2.20	1.50	.753	0.486	0.247	0.527	2.867	5.071	0.3662	0.130	.5879E+1	.1701E+0	.5155E-1	.1857E+1	.6093E-2
2.20	1.80	.790	0.523	0.210	0.753	2.957	5.791	0.4134	0.158	.4650E+1	.2150E+0	.5430E-1	.2394E+1	.9146E-2
2.20	2.20	.814	0.547	0.186	1.105	3.015	6.874	0.4455	0.206	.3708E+1	.2697E+0	.5573E-1	.3063E+1	.1354E-1
2.20	2.70	.828	0.561	0.172	1.577	3.049	8.306	0.4648	0.271	.3017E+1	.3315E+0	.5580E-1	.3861E+1	.1922E-1
2.20	3.30	.837	0.570	0.163	2.159	3.071	10.062	0.4774	0.352	.2507E+1	.3989E+0	.5494E-1	.4803E+1	.2612E-1
2.20	3.90	.842	0.575	0.158	2.749	3.083	11.838	0.4844	0.434	.2168E+1	.4612E+0	.5376E-1	.5735E+1	.3307E-1
2.20	4.70	.846	0.579	0.154	3.541	3.093	14.219	0.4901	0.545	.1856E+1	.5387E+0	.5210E-1	.6969E+1	.4236E-1
2.20	5.60	.849	0.582	0.151	4.435	3.100	16.904	0.4944	0.670	.1613E+1	.6201E+0	.5033E-1	.8357E+1	.5281E-1
2.20	6.80	.852	0.585	0.148	5.629	3.107	20.490	0.4987	0.833	.1386E+1	.7216E+0	.4823E-1	.1022E+2	.6669E-1
2.20	8.00	.854	0.587	0.146	7.025	3.112	24.680	0.5016	1.026	.1201E+1	.8324E+0	.4614E-1	.1238E+2	.8292E-1
2.20	10.00	.855	0.588	0.145	8.823	3.114	30.075	0.5030	1.279	.1035E+1	.9657E+0	.4390E-1	.1513E+2	.1039E+0
2.20	12.00	.856	0.589	0.144	10.821	3.117	36.070	0.5045	1.558	.9051E+0	.1105E+1	.4185E-1	.1820E+2	.1272E+0
2.20	15.00	.857	0.590	0.143	13.819	3.119	45.065	0.5059	1.976	.7692E+0	.1300E+1	.3939E-1	.2280E+2	.1622E+0
2.20	18.00	.858	0.591	0.142	16.817	3.122	54.060	0.5074	2.388	.6744E+0	.1483E+1	.3745E-1	.2743E+2	.1969E+0
2.20	22.00	.858	0.591	0.142	20.817	3.122	66.060	0.5074	2.956	.5841E+0	.1712E+1	.3537E-1	.3352E+2	.2438E+0
2.20	27.00	.859	0.592	0.141	25.815	3.124	81.056	0.5088	3.640	.5049E+0	.1981E+1	.3335E-1	.4124E+2	.3016E+0
2.20	33.00	.859	0.592	0.141	31.815	3.124	99.056	0.5088	4.486	.4380E+0	.2283E+1	.3145E-1	.5040E+2	.3717E+0
2.70	0.33	.231	0.131	0.769	0.068	1.932	3.954	0.0303	0.052	.1136E+3	.8807E-2	.9884E-2	.1196E+0	.2479E-4
2.70	0.39	.256	0.156	0.744	0.078	1.993	4.012	0.0399	0.058	.8152E+2	.1227E-1	.1165E-1	.1602E+0	.4124E-4
2.70	0.47	.288	0.188	0.712	0.094	2.070	4.097	0.0541	0.067	.5606E+2	.1784E-1	.1406E-1	.2218E+0	.9477E-4
2.70	0.56	.323	0.223	0.677	0.114	2.155	4.197	0.0720	0.077	.3934E+2	.2542E-1	.1681E-1	.3023E+0	.1858E-3
2.70	0.68	.370	0.270	0.630	0.140	2.270	4.328	0.0999	0.088	.2655E+2	.3766E-1	.2051E-1	.4324E+0	.3878E-3
2.70	0.82	.424	0.324	0.576	0.172	2.401	4.486	0.1374	0.099	.1821E+2	.5492E-1	.2480E-1	.6163E+0	.7788E-3
2.70	1.00	.491	0.391	0.509	0.218	2.564	4.700	0.1920	0.111	.1229E+2	.8137E-1	.3014E-1	.9024E+0	.1595E-2
2.70	1.20	.560	0.460	0.440	0.280	2.731	4.965	0.2576	0.123	.8666E+1	.1154E+0	.3561E-1	.1279E+1	.2999E-2
2.70	1.50	.650	0.550	0.350	0.400	2.950	5.428	0.3575	0.140	.5810E+1	.1721E+0	.4250E-1	.1940E+1	.6066E-2
2.70	1.80	.716	0.616	0.284	0.568	3.110	6.007	0.4411	0.161	.4340E+1	.2304E+0	.4741E-1	.2650E+1	.1009E-1
2.70	2.20	.765	0.665	0.235	0.870	3.229	6.969	0.5087	0.204	.3293E+1	.3037E+0	.5112E-1	.3545E+1	.1639E-1
2.70	2.70	.793	0.693	0.207	1.314	3.297	8.333	0.5495	0.272	.2591E+1	.3860E+0	.5295E-1	.4579E+1	.2491E-1

CASE # 4   COPYRIGHT 1982 STEVE SMITH   DATA DATE 1/ 3/82   TRANSFORMER DESIGN TABLE

$$f(D)=[(U/(FG(DE)^2))^5 * (1/(DEUm))^4]^{(1/13)}$$

P	Q	D	E	F	G	U	Um	DE	FG	f(D)	1/f(D)	1/(PQf(D))	DEUm	FG(DE)^2/U
2.70	3.30	.809	0.709	0.191	1.882	3.336	10.055	0.5736	0.359	.2107E+1	.4746E+0	.5326E-1	.5768E+1	.3545E-1
2.70	3.90	.817	0.717	0.183	2.466	3.355	11.817	0.5858	0.451	.1800E+1	.5556E+0	.5276E-1	.6922E+1	.4615E-1
2.70	4.70	.823	0.723	0.177	3.254	3.370	14.187	0.5950	0.576	.1526E+1	.6554E+0	.5165E-1	.8442E+1	.6051E-1
2.70	5.60	.828	0.728	0.172	4.144	3.382	16.863	0.6028	0.713	.1316E+1	.7598E+0	.5025E-1	.1016E+2	.7657E-1
2.70	6.80	.831	0.731	0.169	5.338	3.389	20.449	0.6075	0.902	.1124E+1	.8893E+0	.4884E-1	.1242E+2	.9821E-1
2.70	8.20	.834	0.734	0.166	6.732	3.397	24.634	0.6122	1.118	.9707E+0	.1033E+1	.4653E-1	.1508E+2	.1233E+0
2.70	10.00	.836	0.736	0.164	8.528	3.402	30.024	0.6153	1.399	.8337E+0	.1200E+1	.4443E-1	.1847E+2	.1557E+0
2.70	12.00	.837	0.737	0.163	10.526	3.404	36.019	0.6169	1.716	.7269E+0	.1376E+1	.4246E-1	.2222E+2	.1918E+0
2.70	15.00	.838	0.738	0.162	13.524	3.406	45.014	0.6184	2.191	.6162E+0	.1623E+1	.4007E-1	.2784E+2	.2460E+0
2.70	18.00	.839	0.739	0.161	16.522	3.409	54.010	0.6200	2.660	.5394E+0	.1854E+1	.3814E-1	.3349E+2	.3000E+0
2.70	22.00	.840	0.740	0.160	20.520	3.411	66.005	0.6216	3.283	.4655E+0	.2143E+1	.3608E-1	.4103E+2	.3719E+0
2.70	27.00	.841	0.741	0.159	25.518	3.414	81.000	0.6232	4.057	.4028E+0	.2483E+1	.3405E-1	.5048E+2	.4616E+0
2.70	33.00	.841	0.741	0.159	31.518	3.414	99.000	0.6232	5.011	.3492E+0	.2864E+1	.3214E-1	.6169E+2	.5701E+0
3.30	0.47	.107	0.207	0.893	0.056	2.121	4.800	0.0221	0.050	.1552E+3	.6443E-2	.4154E-2	.1064E+0	.1208E-4
3.30	0.56	.144	0.244	0.856	0.072	2.237	4.895	0.0351	0.062	.8809E+2	.1135E-1	.6143E-2	.1720E+0	.3588E-4
3.30	0.68	.192	0.292	0.808	0.096	2.373	5.021	0.0561	0.078	.4937E+2	.2025E-1	.9026E-2	.2815E+0	.1090E-3
3.30	0.82	.248	0.348	0.752	0.124	2.546	5.169	0.0863	0.093	.2931E+2	.3412E-1	.1261E-1	.4461E+0	.2927E-3
3.30	1.00	.319	0.419	0.681	0.162	2.730	5.364	0.1337	0.110	.1742E+2	.5741E-1	.1740E-1	.7170E+0	.7742E-3
3.30	1.20	.395	0.495	0.605	0.210	2.990	5.595	0.1955	0.127	.1111E+2	.9003E-1	.2273E-1	.1094E+1	.1779E-2
3.30	1.50	.502	0.602	0.498	0.296	3.219	5.975	0.3022	0.147	.6661E+1	.1501E+0	.3033E-1	.1806E+1	.4502E-2
3.30	1.80	.596	0.696	0.404	0.408	3.440	6.419	0.4148	0.165	.4566E+1	.2190E+0	.3687E-1	.2663E+1	.8812E-2
3.30	2.20	.687	0.787	0.313	0.626	3.583	7.176	0.5407	0.196	.3183E+1	.3141E+0	.4327E-1	.3880E+1	.1665E-1
3.30	2.70	.746	0.846	0.254	1.008	3.656	8.390	0.6311	0.256	.2354E+1	.4248E+0	.4768E-1	.5295E+1	.2846E-1
3.30	3.30	.776	0.876	0.224	1.548	3.690	10.044	0.6798	0.347	.1844E+1	.5424E+0	.4988E-1	.6828E+1	.4383E-1
3.30	3.90	.790	0.890	0.210	2.120	3.714	11.776	0.7031	0.445	.1543E+1	.6479E+0	.5035E-1	.8280E+1	.5965E-1
3.30	4.70	.800	0.900	0.200	2.900	3.729	14.127	0.7200	0.580	.1288E+1	.7763E+0	.5005E-1	.1017E+2	.8095E-1
3.30	5.60	.806	0.906	0.194	3.788	3.741	16.798	0.7302	0.735	.1100E+1	.9092E+0	.4920E-1	.1227E+2	.1051E+0
3.30	6.80	.811	0.911	0.189	4.978	3.748	20.374	0.7388	0.941	.9318E+0	.1073E+1	.4782E-1	.1505E+2	.1373E+0
3.30	8.20	.814	0.914	0.186	6.372	3.755	24.559	0.7440	1.185	.7996E+0	.1251E+1	.4622E-1	.1827E+2	.1750E+0
3.30	10.00	.817	0.917	0.183	8.166	3.760	29.945	0.7492	1.494	.6835E+0	.1463E+1	.4434E-1	.2243E+2	.2233E+0
3.30	12.00	.819	0.919	0.181	10.162	3.765	35.935	0.7527	1.839	.5939E+0	.1684E+1	.4252E-1	.2705E+2	.2771E+0
3.30	15.00	.821	0.921	0.179	13.158	3.768	44.925	0.7561	2.355	.5019E+0	.1991E+1	.4025E-1	.3397E+2	.3577E+0
3.30	18.00	.822	0.922	0.178	16.156	3.770	53.921	0.7579	2.876	.4385E+0	.2281E+1	.3840E-1	.4087E+2	.4384E+0
3.30	22.00	.823	0.923	0.177	20.154	3.770	65.916	0.7596	3.567	.3786E+0	.2642E+1	.3639E-1	.5007E+2	.5460E+0
3.30	27.00	.823	0.923	0.177	25.154	3.770	80.916	0.7596	4.452	.3264E+0	.3064E+1	.3439E-1	.6147E+2	.6815E+0
3.30	33.00	.824	0.924	0.176	31.152	3.772	98.911	0.7614	5.483	.2826E+0	.3539E+1	.3250E-1	.7531E+2	.8425E+0

CASE # 4   COPYRIGHT 1982 STEVE SMITH   DATA DATE 1/3/82   TRANSFORMER DESIGN TABLE

$$f(D) = \left[\left(\frac{U}{FG(DE)^2}\right)^5 \cdot \left(\frac{1}{DEUm}\right)^4\right]^{1/13}$$

P	Q	D	E	F	G	U	Um	DE	FG	f(D)	1/f(D)	1/(PQf(D))	DEUm	$FG(DE)^2/U$
3.90	0.82	.084	.384	.916	.052	2.375	5.794	0.0323	0.048	.1058E+3	.9455E-2	.2957E-2	.1869E+0	.2087E-4
3.90	1.00	.154	.454	.846	.092	2.545	5.994	0.0699	0.078	.3866E+2	.2585E-1	.6624E-2	.4191E+0	.1495E-3
3.90	1.20	.231	.531	.769	.138	2.732	6.220	0.1227	0.106	.1904E+2	.5252E-1	.1122E-1	.7630E+0	.5844E-3
3.90	1.50	.345	.645	.655	.210	3.009	6.566	0.2225	0.138	.9262E+1	.1080E+0	.1846E-1	.1461E+1	.2264E-2
3.90	1.80	.452	.752	.548	.296	3.269	6.946	0.3399	0.162	.5599E+1	.1789E+0	.2549E-1	.2361E+1	.5733E-2
3.90	2.20	.577	.877	.423	.446	3.572	7.539	0.5060	0.189	.3467E+1	.2885E+0	.3362E-1	.3815E+1	.1352E-1
3.90	2.70	.682	.982	.318	.736	3.828	8.529	0.6697	0.234	.2333E+1	.4287E+0	.4071E-1	.5712E+1	.2743E-1
3.90	3.30	.740	1.040	.260	1.220	3.968	10.047	0.7696	0.317	.1723E+1	.5805E+0	.4511E-1	.7732E+1	.4734E-1
3.90	3.90	.764	1.064	.236	1.772	4.027	11.731	0.8129	0.418	.1400E+1	.7142E+0	.4696E-1	.9536E+1	.6863E-1
3.90	4.70	.779	1.079	.221	2.542	4.063	14.058	0.8405	0.562	.1144E+1	.8739E+0	.4766E-1	.1182E+2	.9768E-1
3.90	5.60	.788	1.088	.212	3.424	4.085	16.714	0.8573	0.726	.9644E+0	.1037E+1	.4748E-1	.1433E+2	.1306E+0
3.90	6.80	.794	1.094	.206	4.612	4.100	20.285	0.8686	0.950	.8089E+0	.1236E+1	.4662E-1	.1762E+2	.1749E+0
3.90	8.20	.799	1.099	.201	6.002	4.112	24.461	0.8781	1.206	.6893E+0	.1451E+1	.4537E-1	.2148E+2	.2262E+0
3.90	10.00	.802	1.102	.198	7.796	4.119	29.846	0.8838	1.544	.5860E+0	.1706E+1	.4376E-1	.2638E+2	.2927E+0
3.90	12.00	.804	1.104	.196	9.792	4.124	35.836	0.8876	1.919	.5073E+0	.1971E+1	.4212E-1	.3181E+2	.3667E+0
3.90	15.00	.807	1.107	.193	12.786	4.131	44.822	0.8933	2.468	.4272E+0	.2341E+1	.4001E-1	.4004E+2	.4767E+0
3.90	18.00	.808	1.108	.192	15.784	4.134	53.817	0.8953	3.031	.3724E+0	.2685E+1	.3825E-1	.4818E+2	.5876E+0
3.90	22.00	.809	1.109	.191	19.782	4.136	65.812	0.8972	3.778	.3299E+0	.3116E+1	.3633E-1	.5905E+2	.7353E+0
3.90	27.00	.810	1.110	.190	24.780	4.138	80.807	0.8991	4.708	.2762E+0	.3620E+1	.3438E-1	.7265E+2	.9197E+0
3.90	33.00	.811	1.111	.190	30.778	4.141	98.802	0.9010	5.817	.2389E+0	.4186E+1	.3253E-1	.8902E+2	.1146E+1
4.70	1.50	.138	.705	.862	.091	3.039	7.343	0.0972	0.078	.2723E+2	.3672E-1	.5209E-2	.7141E+0	.2432E-3
4.70	1.80	.250	.817	.750	.167	3.311	7.699	0.2042	0.125	.1042E+2	.9601E-1	.1135E-1	.1572E+1	.1573E-2
4.70	2.20	.395	.962	.605	.277	3.664	8.194	0.3799	0.167	.4865E+1	.2055E+0	.1988E-1	.3113E+1	.6592E-2
4.70	2.70	.552	1.119	.448	.463	4.045	8.932	0.6175	0.207	.2686E+1	.3722E+0	.2933E-1	.5515E+1	.1954E-1
4.70	3.30	.672	1.239	.328	.823	4.337	10.149	0.8324	0.270	.1738E+1	.5754E+0	.3710E-1	.8448E+1	.4311E-1
4.70	3.90	.725	1.292	.275	1.317	4.465	11.691	0.9365	0.362	.1324E+1	.7555E+0	.4122E-1	.1095E+2	.7111E-1
4.70	4.70	.753	1.320	.247	2.061	4.533	13.955	0.9937	0.509	.1038E+1	.9638E+0	.4363E-1	.1387E+2	.1109E+0
4.70	5.60	.767	1.334	.233	2.933	4.567	16.587	1.0229	0.683	.8539E+0	.1171E+1	.4449E-1	.1697E+2	.1565E+0
4.70	6.80	.777	1.344	.223	4.113	4.592	20.139	1.0440	0.917	.7041E+0	.1420E+1	.4444E-1	.2103E+2	.2177E+0
4.70	8.20	.783	1.350	.217	5.501	4.606	24.309	1.0568	1.194	.5934E+0	.1685E+1	.4373E-1	.2569E+2	.2894E+0
4.70	10.00	.787	1.354	.213	7.293	4.616	29.690	1.0653	1.553	.5003E+0	.1999E+1	.4253E-1	.3163E+2	.3819E+0
4.70	12.00	.790	1.357	.208	9.287	4.623	35.675	1.0718	2.054	.4366E+0	.2322E+1	.4117E-1	.3824E+2	.4845E+0
4.70	15.00	.792	1.359	.206	12.283	4.628	44.666	1.0761	2.555	.3608E+0	.2772E+1	.3931E-1	.4806E+2	.6392E+0
4.70	18.00	.794	1.361	.206	15.279	4.633	53.656	1.0804	3.147	.3135E+0	.3190E+1	.3770E-1	.5797E+2	.7929E+0
4.70	22.00	.795	1.362	.203	19.277	4.635	65.651	1.0825	3.952	.2694E+0	.3712E+1	.3590E-1	.7107E+2	.9999E+0
4.70	27.00	.797	1.364	.203	24.273	4.640	80.641	1.0868	4.927	.2314E+0	.4321E+1	.3405E-1	.8764E+2	.1254E+1

CASE # 4   COPYRIGHT 1982 STEVE SMITH   DATA DATE 1/ 3/82

TRANSFORMER DESIGN TABLE

$$f(D) = \left[\left(\frac{U}{FG(DE)^2}\right)^5 \cdot \left(\frac{1}{DEUm}\right)^4\right]^{1/13}$$

P	Q	D	E	F	G	U	Um	DE	FG	f(D)	1/f(D)	1/(PQf(D))	DEUm	FG(DE)^2/U
4.70	33.00	.797	1.364	0.203	30.273	4.640	98.641	1.0868	6.145	.1998E+0	.5005E+1	.3227E-1	.1072E+3	.1564E+1
5.60	2.20	.173	1.040	0.827	0.121	3.724	9.016	0.1799	0.100	.1298E+2	.7707E-1	.6255E-2	.1622E+1	.8668E-3
5.60	2.70	.353	1.220	0.647	0.261	4.162	9.641	0.4305	0.169	.4235E+1	.2361E+0	.1562E-1	.4151E+1	.7512E-2
5.60	3.30	.540	1.407	0.460	0.487	4.616	10.533	0.7596	0.224	.2087E+1	.4792E+0	.2593E-1	.8000E+1	.2798E-1
5.60	3.90	.658	1.525	0.342	0.851	4.903	11.759	1.0032	0.291	.1383E+1	.7229E+0	.3310E-1	.1180E+2	.5973E-1
5.60	4.70	.721	1.588	0.279	1.525	5.056	13.853	1.1447	0.425	.9977E+0	.1002E+1	.3808E-1	.1586E+2	.1103E+0
5.60	5.60	.746	1.613	0.254	2.375	5.116	16.432	1.2030	0.603	.7881E+0	.1269E+1	.4046E-1	.1977E+2	.1706E+0
5.60	6.80	.760	1.627	0.240	3.547	5.150	19.964	1.2363	0.851	.6333E+0	.1580E+1	.4148E-1	.2468E+2	.2526E+0
5.60	8.20	.768	1.635	0.232	4.931	5.170	24.125	1.2554	1.144	.5250E+0	.1905E+1	.4148E-1	.3029E+2	.3487E+0
5.60	10.00	.774	1.641	0.226	6.719	5.184	29.496	1.2699	1.518	.4377E+0	.2285E+1	.4080E-1	.3746E+2	.4723E+0
5.60	12.00	.778	1.645	0.222	8.711	5.194	35.476	1.2796	1.934	.3740E+0	.2674E+1	.3979E-1	.4539E+2	.6096E+0
5.60	15.00	.781	1.648	0.219	11.705	5.201	44.462	1.2868	2.563	.3113E+0	.3212E+1	.3824E-1	.5721E+2	.8161E+0
5.60	18.00	.783	1.650	0.217	14.701	5.206	53.452	1.2917	3.190	.2694E+0	.3712E+1	.3682E-1	.6904E+2	.1022E+1
5.60	22.00	.784	1.651	0.216	18.699	5.209	65.447	1.2941	4.039	.2308E+0	.4334E+1	.3517E-1	.8470E+2	.1299E+1
5.60	27.00	.785	1.652	0.215	23.697	5.211	80.442	1.2966	5.095	.1977E+0	.5058E+1	.3345E-1	.1043E+3	.1644E+1
5.60	33.00	.786	1.653	0.214	29.695	5.213	98.437	1.2990	6.355	.1703E+0	.5871E+1	.3177E-1	.1279E+3	.2057E+1
6.80	2.70	.061	1.328	0.939	0.045	4.782	10.716	0.0810	0.042	.4267E+2	.2343E-1	.1276E-2	.8679E+0	.6469E-4
6.80	3.30	.279	1.546	0.721	0.209	5.258	11.457	0.4312	0.150	.4419E+1	.2263E+0	.1009E-1	.4941E+1	.5851E-2
6.80	3.90	.475	1.742	0.525	0.417	5.671	12.305	0.8273	0.219	.1925E+1	.5196E+0	.1959E-1	.1018E+2	.2847E-1
6.80	4.70	.645	1.912	0.355	0.877	5.834	13.879	1.2330	0.311	.1085E+1	.9218E+0	.2884E-1	.1711E+2	.8344E-1
6.80	5.60	.712	1.979	0.288	1.643	5.902	16.253	1.4088	0.473	.7704E+0	.1298E+1	.3409E-1	.2290E+2	.1610E+0
6.80	6.80	.740	2.007	0.260	2.787	5.933	19.717	1.4849	0.725	.5848E+0	.1710E+1	.3699E-1	.2928E+2	.2707E+0
6.80	8.20	.753	2.020	0.247	4.161	5.953	23.854	1.5208	1.028	.4709E+0	.2124E+1	.3809E-1	.3628E+2	.4006E+0
6.80	10.00	.761	2.028	0.239	5.945	5.962	29.215	1.5431	1.421	.3851E+0	.2597E+1	.3820E-1	.4508E+2	.5683E+0
6.80	12.00	.765	2.032	0.235	7.937	5.972	35.196	1.5542	1.865	.3251E+0	.3076E+1	.3769E-1	.5470E+2	.7556E+0
6.80	15.00	.769	2.036	0.231	10.929	5.979	44.177	1.5654	2.525	.2679E+0	.3732E+1	.3659E-1	.6916E+2	.1036E+1
6.80	18.00	.772	2.039	0.228	13.923	5.982	53.162	1.5739	3.174	.2305E+0	.4338E+1	.3544E-1	.8367E+2	.1315E+1
6.80	22.00	.773	2.040	0.227	17.921	5.987	65.157	1.5767	4.068	.1965E+0	.5090E+1	.3402E-1	.1027E+3	.1691E+1
6.80	27.00	.775	2.042	0.225	22.917	5.989	78.147	1.5823	5.156	.1694E+0	.5903E+1	.3248E-1	.1268E+3	.2156E+1
6.80	33.00	.776	2.043	0.224	28.915	5.990	98.143	1.5851	6.477	.1441E+0	.6941E+1	.3093E-1	.1556E+3	.2717E+1
8.20	3.90	.158	1.891	0.842	0.117	5.421	13.444	0.2988	0.099	.7703E+1	.1298E+0	.4059E-2	.4018E+1	.1627E-2
8.20	4.70	.431	2.164	0.569	0.371	6.084	14.518	0.9328	0.211	.1723E+1	.5804E+0	.1506E-1	.1354E+2	.3022E-1
8.20	5.60	.631	2.364	0.369	0.871	6.570	16.246	1.4919	0.322	.8797E+0	.1137E+1	.2475E-1	.2424E+2	.1089E+0
8.20	6.80	.712	2.445	0.288	1.989	6.767	19.453	1.7411	0.550	.5799E+0	.1725E+1	.3093E-1	.3387E+2	.2463E+0
8.20	8.20	.737	2.470	0.263	3.259	6.828	23.531	1.8206	0.857	.4409E+0	.2268E+1	.3377E-1	.4284E+2	.4162E+0
8.20	10.00	.749	2.482	0.251	5.035	6.857	28.873	1.8593	1.264	.3492E+0	.2864E+1	.3493E-1	.5368E+2	.6372E+0

CASE # 4   COPYRIGHT 1982 STEVE SMITH   DATA DATE 1/ 3/82

TRANSFORMER DESIGN TABLE

$$f(D)=[(U/(FG(DE)^2))^5 * (1/(DEUm))^4]^{(1/13)}$$

P	Q	D	E	F	G	U	Um	DE	FG	f(D)	1/f(D)	1/(PQf(D))	DEUm	FG(DE)^2/U
8.20	12.00	.755	2.488	0.245	7.023	6.872	34.844	1.8787	1.721	.2896E+0	.3453E+1	.3509E-1	.6546E+2	.8838E+0
8.20	15.00	.760	2.493	0.240	10.013	6.884	43.820	1.8949	2.403	.2353E+0	.4249E+1	.3455E-1	.8304E+2	.1254E+1
8.20	18.00	.763	2.496	0.237	13.007	6.891	52.805	1.9047	3.083	.2009E+0	.4975E+1	.3373E-1	.1006E+3	.1621E+1
8.20	22.00	.765	2.498	0.235	17.003	6.896	64.795	1.9112	3.996	.1701E+0	.5878E+1	.3258E-1	.1238E+3	.2117E+1
8.20	27.00	.766	2.499	0.234	22.001	6.898	79.791	1.9145	5.148	.1445E+0	.6920E+1	.3125E-1	.1528E+3	.2735E+1
8.20	33.00	.768	2.501	0.232	27.997	6.903	97.781	1.9210	6.495	.1237E+0	.8083E+1	.2987E-1	.1878E+3	.3472E+1
10.00	5.60	.332	2.665	0.668	0.269	7.044	17.184	0.8849	0.180	.1949E+1	.5132E+0	.9163E-2	.1521E+2	.2000E-1
10.00	6.80	.624	2.957	0.376	0.885	7.753	19.365	1.8454	0.333	.6971E+0	.1435E+1	.2110E-1	.3574E+2	.1462E+0
10.00	8.20	.710	3.043	0.290	2.113	7.962	23.148	2.1608	0.613	.4448E+0	.2248E+1	.2742E-1	.5002E+2	.3594E+0
10.00	10.00	.735	3.068	0.265	3.863	8.023	28.426	2.2552	1.024	.3283E+0	.3046E+1	.3046E-1	.6411E+2	.6490E+0
10.00	12.00	.744	3.077	0.256	5.845	8.045	34.382	2.2895	1.496	.2635E+0	.3795E+1	.3162E-1	.7872E+2	.9751E+0
10.00	15.00	.751	3.084	0.249	8.831	8.062	43.348	2.3163	2.199	.2091E+0	.4781E+1	.3188E-1	.1004E+3	.1464E+1
10.00	18.00	.754	3.087	0.246	11.825	8.069	52.334	2.3278	2.909	.1763E+0	.5671E+1	.3150E-1	.1218E+3	.1954E+1
10.00	22.00	.757	3.090	0.243	15.819	8.076	64.319	2.3394	3.844	.1479E+0	.6759E+1	.3072E-1	.1505E+3	.2605E+1
10.00	27.00	.759	3.092	0.241	20.815	8.081	79.310	2.3471	5.016	.1248E+0	.8013E+1	.2968E-1	.1861E+3	.3420E+1
10.00	33.00	.760	3.093	0.240	26.813	8.084	97.305	2.3509	6.435	.1063E+0	.9407E+1	.2851E-1	.2288E+3	.4400E+1

APPENDIX C
RECTIFIER CIRCUIT DESIGN

Capacitor input rectifier circuits can be designed accurately and simply with the aid of the following graphs, known as Schade's curves.

Depending on whether your circuit is a half wave rectifier, a full wave (center tap or bridge), or a full wave doubler, refer to Fig. C-1, C-2 or C-3, respectively.

For some ratios of source to load resistance, select the appropriate curve. Form the product ωCR_L, and move upward to where the vertical line intersects the chosen curve. Read horizontally on the left the fraction of peak AC voltage which the load will see (the voltage the filter capacitor charges up to).

If one operates on the horizontal plateau to the right of the knee of the chosen curve, the output voltage will be relatively insensitive to the value of the capacitor, a great convenience when one is using electrolytic capacitors which have rather wide tolerances.

Next, go to Fig. C-4. For the particular circuit, the chosen value of ωCR_L, read the ratio of peak to average rectifier current.

Given the average DC output current, one can obtain the RMS (effective heating value) current from the expression $I_{RMS} = \sqrt{I_{pk}I_{avg}}$.

One can now properly specify the rectifiers. Knowing the RMS current in the transformer, one can compute its temperature rise.

Now go to Fig. C-5. For the chosen value of ωCR_L, the circuit, and the R_S/R_L ratio one can read the RMS ripple voltage on the capacitor as a ratio with the DC output voltage. If some values do not work out to an acceptable situation, one can choose a different value of R_S/R_L or a different capacitance value and recalculate.

One can also work backwards from a desired output voltage and a given capacitance value to the value of source resistance required, and then design the transformer to meet that resistance requirement.

A large fraction of the engineers I have known were not familiar with Schade's curves or their use. I have included this information to make it more available. I have also observed a very common error among circuit design engineers, that of ignoring the actual RMS ripple current in the filter capacitor. Fig. C-4 gives the RMS current in the rectifier, from which one can calculate the RMS ripple current in the capacitor and ensure that it is being operated within the manufacturer's ratings.

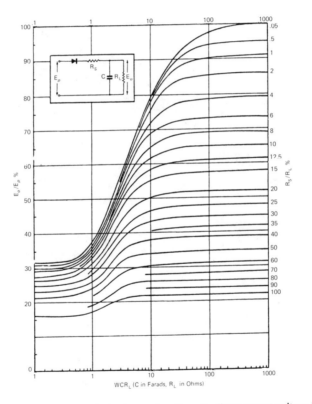

Fig. C-1. Relation of applied alternating peak voltage to direct output voltage in half-wave capacitor-input circuits. (By permission from O. H. Schade, *Proc. I.R.E.*, July 1943, p. 343; © 1943 IRE [now IEEE].)

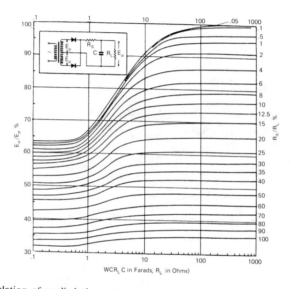

Fig. C-2. Relation of applied alternating peak voltage to direct output voltage in full-wave capacitor-input circuits. (By permission from O. H. Schade, *Proc. I.R.E.*, July 1943, p. 344; © 1943 IRE [now IEEE].)

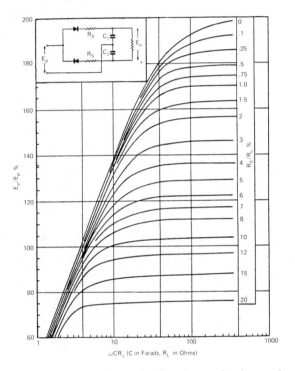

Fig. C-3. Relation of applied alternating peak voltage in capacitor-input voltage doubler circuits. (By permission from O. H. Schade, *Proc. I.R.E.*, July 1943, p. 345; © 1943 IRE [now IEEE].)

338

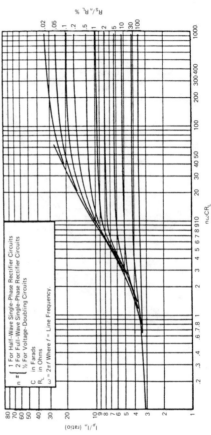

Fig. C–4. Relation of peak current to average current per rectifier in capacitor-input circuits. (By permission from O. H. Schade, *Proc. I.R.E.*, July 1943, p. 346; © 1943 IRE [now IEEE].)

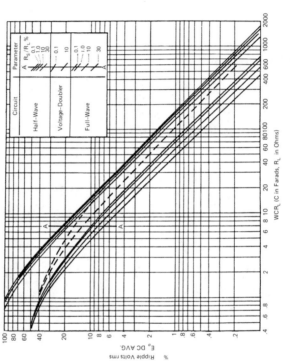

Fig. C-5. Root-mean-square ripple voltage for capacitor-input circuits. (By permission from O. H. Schade, *Proc. I.R.E.*, July 1943, p. 347; © 1943 IRE [now IEEE].)

INDEX

INDEX